ECHINOSTOMES AS EXPERIMENTAL MODELS FOR BIOLOGICAL RESEARCH

ECHINOSTOMES AS EXPERIMENTAL MODELS FOR BIOLOGICAL RESEARCH

Edited by

BERNARD FRIED

Lafayette College,
Easton,
PA, U.S.A.

and

THADDEUS K. GRACZYK

Johns Hopkins University,
Baltimore,
MD, U.S.A.

KLUWER ACADEMIC PUBLISHERS

DORDRECHT / BOSTON / LONDON

Library of Congress Cataloging-in-Publication Data

Echinostomes as experimental models for biological research / edited by Bernard Fried
and Thaddeus K. Graczyk.
p. cm.
ISBN 0-7923-6156-3 (alk. paper)
1. Echinostomatidae. 2. Animal models in research. I. Fried, Bernard, 1933- II.
Graczyk, Thaddeus K.

QL391.P7 E28 2000
592'.48--dc21

99-089293

ISBN 0-7923-6156-3

Published by Kluwer Academic Publishers,
P.O. Box 17, 3300 AA Dordrecht, The Netherlands.

Sold and distributed in North, Central and South America
by Kluwer Academic Publishers,
101 Philip Drive, Norwell, MA 02061, U.S.A.

In all other countries, sold and distributed
by Kluwer Academic Publishers,
P.O. Box 322, 3300 AH Dordrecht, The Netherlands.

Printed on acid-free paper

Cover design:
Life cycle schematic of a stylized echinostome in a human, original drawing by M. A. Haseeb
(Chapter 4. Fig. 2).

Printed in the Netherlands.

To Grace and Sophia

Table of Contents

Preface

Echinostomes are medically- and veterinary-important intestinal parasitic flatworms that invade man, domestic animals and wildlife, and also parasitize in their larval stages numerous invertebrate and cold-blooded vertebrate hosts. They are ubiquitous parasites easy to maintain in the laboratory as adult and larval worm stages. For these reasons, echinostomes have served for decades as excellent experimental model animals at all levels of organization, from molecular to organismic. They have been used in experiments on excystation and in vitro cultivation, larval parasite-host relationships, adult parasite-host relationships, reproductive behavior, and various aspects of host-parasite recognition and interactions. There is no single area of experimental parasitology where echinostomes are not represented. Echinostomes have contributed significantly to developments and discoveries in systematics, ultrastructure, biochemistry, immunology, neurobiology, physiology, and molecular biology. Until now, there was no book that approached the subject of echinostomes comprehensively. The information has been scattered world-wide in numerous scientific journals. The main goal of out book is to acquaint and update readers in developments in various aspects of parasitology and experimental biology which were accomplished based on studies with echinostomes. Thus, our book will have appeal to both molecular and organismic researchers from various disciplines of science.

Specific problems for our targeted audience are: 1) the tremendous amount of information, discoveries, and developments in parasitology and experimental biology, and 2) lack of a book that covered specifically echinostomes and echinostomiasis. In our book, the presentation of information relies on the fact that much of the material is in tables in order to: 1) facilitate instant access to specific information, 2) reduce text space, and 3) increase the amount of available information. Our comprehensive overview also is suitable for developing a lecture program. The most important features of our book making it attractive to the scientific community are: 1) comprehensive review of medically and veterinary important echinostomes, 2) overview of experimental laboratory models with echinostomes, and 3) indication of applicability of echinostomes for further experimental models. Potential book buyers are government and private institutions which provide medical and veterinary services and/or are committed to teaching students or personnel. Further, we believe that the market for our book also includes the same one that has been important in the Fried/Graczyk *"Advances in Trematode Biology"* book, CRC, Boca Raton, Florida, 1997. That market also includes advanced undergraduates in biology, graduate biology students, medical doctors, veterinarians, wildlife biologists, and experimental biologists interested in new models (particularly invertebrates) for their research. The book also should be attractive to graduate students in Parasitology, Immunology, Molecular Biology, Microbiology, Physiology, Invertebrate Biology, and Malacology; researchers concerned with trematodes; faculty committed to teaching parasitology courses; medical and veterinary personnel; managers of captive animal centers; managers of wildlife; and veterinarians.

We hope this book encourages research workers from a variety of disciplines to initiate research with these fascinating organisms.

Bernard Fried
Thaddeus K. Graczyk

AN OVERVIEW OF THE BIOLOGY OF ECHINOSTOMES

IVAN KANEV[1*]

[1] *Institute of Experimental Pathology and Parasitology, Bulgarian Academy of Sciences, Bl. 25, Sofia 1113, Bulgaria.*

MAURITZ STERNER[2],

[2] *Harold W. Manter Laboratory of Parasitology, University of Nebraska - Lincoln, Lincoln, NE. 68588-0514, USA.*

VALENTINE RADEV[1]

[1] *Institute of Experimental Pathology and Parasitology, Bulgarian Academy of Sciences, Bl. 25, Sofia 1113, Bulgaria.*

BERNARD FRIED[3]

[3] *Department of Biology, Lafayette College, Easton PA 18042, USA.*

CONTENTS

---

[*] Present address: Harold W. Manter Laboratory of Parasitology, University of Nebraska - Lincoln, Lincoln, NE. 68588-0514

B. Fried and T.K. Graczyk (eds.),
Echinostomes as Experimental Models for Biological Research, 1–29.
© 2000 *Kluwer Academic Publishers. Printed in the Netherlands.*

1. Introduction, Literature and History

Echinostomes are digenean parasites in the class Trematoda. The adult worms are parasites of fishes, reptiles, birds and mammals, including humans. Their characteristic morphological feature is a crown of spines and a collar-like tegumentary flap surrounding the oral sucker. Larval stages develop in freshwater and marine snails, mussels, insects, amphibians and reptiles. Adult and larval echinostomes are used widely as models for laboratory work for the following reasons:

1). They are wide spred and common and some species are cosmopolitan. Other species are found in several geographical regions. Relatively few species are restricted locally.

2). They are found in many wild and domestic birds and mammals, including humans. Many echinostomes are hematophagous, ingesting lymph and blood from their hosts. Therefore, they are parasites with significant health and economic importance. Some species such as Echinostoma lindoense Sandground & Bonne, 1940 (known after Kanev, 1985 as E. echinatum Zeder, 1803) and E. ilocanum (Garrison, 1908) cause severe health problems in humans in Southeast Asia.

3). Echinostomes infect most organs and sites in the abdominal cavity of birds and mammals, including the alimentary, excretory and reproductive systems.

4). Larval and adult echinostomes provide good material for biological, physiological, biochemical, immunological and other studies in modern biology.

5). Echinostomes can be maintained easily and inexpensively in the laboratory.

Extensive literature is available on echinostomes including texts, monographs, treatises and primary publications. This literature mentions about 2000 names of species, genera, subfamilies, families and superfamilies. Some genera, i.e., Echinostoma, Euparyphium (known also as Isthmiophora) are important disease-producing organisms. Others, such as Echinoparyphium and Hypoderaeum have attracted the attention of helminthologists for use in laboratory research. Table 1.1. Present literature citations from 1970 to 1998 on some of the most important echinostomes used in medical and experimental studies.

Table 1.1. Number of citations listed in CAB International (Helminthological Abstracts) from 1970 to 1998 for medically and experimentally important echinostomes.

Genera	Number of citations
Echinostoma	9689
Echinoparyphium	3752
Euparyphium (Isthmiophora)	2243

Table 1.2. Significant contributions to the biology of echinostomes from the eighteenth to the twentieth century

Date	Author	Contribution
1773	Müller	Described and illustrated as Cercaria lemna the first echinostome cercaria
1776 - 1790	Schrank	Described and illustrated as Fasciola melis, Festucaria anatis, and Festucaria boshadis, the first adult echinostomes
1782	Bloch	Described and illustrated as Cuculanus conoideus adult echinostomes
1782 - 1787	Goetze	Published the first detailed figures of the collar and spines of adult worms named in the

		original as <u>Planaria</u> <u>teres</u> <u>siplici</u> <u>poro</u>
1786	Berman	Adults and larval stages described as "<u>cercous</u> <u>infusories</u>"
1789, 1802	Froelich	Adults from birds in Germany named as <u>Fasciola</u> <u>appendiculata</u>, <u>Fasciola</u> <u>crenata</u> and <u>Fasciola</u> <u>revoluta</u>
1790	Gmelin	Revised listing in Linnaeu's <u>Systema</u> <u>helmintum</u>
1795	Viborg	Desribed adults and hosts from East Prussia
1797, 1819	Abildgaard	Adults and hosts from Denmark
1800 - 1803	Zeder	Established a new name, <u>Distomum</u> <u>echinatum</u>, and published detailed descriptions of the head collar and collar-spines
1809 - 1819	Rudolphi	Erected the genus <u>Echinostoma</u> and published two new systems: one in <u>Entozoorum</u> <u>sive</u> <u>vermium</u> and the second in <u>Entozoorum</u> <u>synopsis</u>
1818	Bojanus	Described and illustrated as "Koningsgelber würmer" the first echinostome redia found in naturally infected freshwater snails
1824	Bremser	Published <u>Icones</u> <u>helminthum</u> with illustrations of <u>Distoma</u> <u>echinatum</u> and other adult worms described in Rudolphi's <u>Entozoorum</u>
1831 - 1839	Mehlis	Described adults and hosts in Germany
1837 - 1842	Siebold	Discovered for the first time life the life cycle pattern of <u>Echinostoma</u> and Trematoda in general. This became possible after his original studies with echinostome cercariae named in the original as <u>Cercaria</u> <u>echinata</u> and eggs, miracidia and sporocysts of <u>Cyclocoelum</u> <u>sp.</u>, named in the original by Siebold as <u>Monostoma</u> <u>mutabile</u>; plus comparative examinations of Bojanus' illustrations of echinostome rediae from naturally infected freshwater snails
1839 - 1864	Creplin	Data on parasitic worms from Central Europe from his collection in Greifswald
1842	Steenstrup	Published information on larvae, adults and life cycles
1844	Belingham	Adults and hosts from Ireland
1845	Dujardin	Revision based on materials from Austria, Germany and France
1850 - 1858	Diesing	Revision based on materials from Vienna and Berlin originally described by Zeder, Rudolphi and Bremser
1854	de Filippi	Described <u>Cercaria</u> <u>echinatoides</u> in Italy
1855	La Valette	Described <u>Cercaria</u> <u>echinifera</u>, <u>Cercaria</u> <u>spinifera</u> and <u>Distoma</u> <u>echiniferum</u>
1857	Pagenstecher	Described <u>Cercaria</u> <u>magna</u>, <u>Distoma</u> <u>echinatoides</u> <u>anodontae</u> and <u>D.</u> echiniferum <u>paludinae</u> from Germany
1858 - 1859	Molin	Described adults and hosts in Europe
1858 - 1861	Van Beneden	Described and illustrated larval stages and life cycles
1858	Wedl	Described adults and larvae from Austria
1860	Cobbold	Completed revision of adults and hosts
1873 - 1884	v. Linstow	Described <u>Distoma</u> <u>beleocephalum</u>, <u>D.</u> recurvatum and <u>D.</u> pungens

1887 - 1894	Parona	Published revision of adult echinostomes and hosts from Italy
1890 - 1905	Stossich	Revised knowledge of echinostomes based on the literature and studies of material from Italy
1881 - 1882	Ercolani	Described adults and hosts from Italy
1892	Neumann	Investigated material originally described by Ercolani (1881)
1895 - 1897	Kowalewski	Described adults and hosts from Poland
1895	Railliet	Described and illustrated adults and larvae from Europe
1895	Sonsino	Described the first echinostome cercaria (Cercaria agilis) from Africa
1896	Hassal	Described adults and hosts from the USA
1896 - 1899	Looss	Described adults and hosts from Africa
1899 - 1909	Braun	Completed revisions on adults and hosts from Europe, Asia, Africa and America
1900 - 1908	Ratz	Described as Distoma saginatum, Pegosomum asperum, P. saginatum, P. spiniferum and E. perfoliatum adults from birds in Hungary
1905 - 1925	Nicoll	Descriptions of adults and hosts from Australia and Europe
1908	Garrison	Described human echinostomes from the Philippines
1911 - 1915	Leiper	Described the morphology and biology of adult echinostomes from humans and other mammals in Indo-China
1909 - 1910	Dietz	Published a revision and new systematics of echinostomes arranged in 22 genera
1909	Lühe	Published revision of adult and larval echinostomes
1911 - 1913	Odhner	Revised adults in reptiles, birds and mammals from Egypt
1912 - 1916	Johnston	Described adults, larval stages and hosts from Australia
1912	Soloviev	Described Echinostoma echinatum and E. mesotestius in Turkestan (former USSR)
1913	Gedoelst	Described adults and hosts in Africa
1913 - 1956	Skrjabin & co-workers	Described, illustrated and revised descriptions of adults, larvae and hosts of echinostomes. Two systems were published in 1947 and 1956
1914 - 1962	Cort	Described larval echinostomes (Cercaria reflexae, C. trivolvis and Cercaria sp.) from the USA
1914 - 1940	Linton	Described adults and hosts from North America
1915	Ciurea	Biology and hosts of adults and larval parasites from birds, mammals and humans in Romania
1915	Barker	Described adult echinostomes from rodents in the USA; named Echinoparyphium contiguum, Echinostoma coalitum, E. armigerum, and E. callawayensis
1915 - 1922	Lane	Found adult echinostomes in mammals and humans in Asia
1915	Tanabe	Described adults from Japan

1914 - 1962	Cort & co-workers	Biology and hosts of larval parasites from North America
1917 - 1918	Cawston	Described cercariae (Cercaria catenata and C. arcuata) and hosts from Africa
1917	Muto	Described larval stages from Japan
1917	Ward	Published revision of materials from North America
1918 – 1934	Faust	Morphology, biology and hosts of larval echinostomes from North America, South Africa, and Indo-China
1919 - 1966	Isaitchikov	Morphology, biology and hosts of adult echinostomes from the Ukraine
1920	Johnson	The first monograph on the life cycle and development of larval echinostomes from North America
1920 - 1922	Leon	Morphology and biology of adults in birds and mammals from Romania
1922	Ando	Biology and hosts of adult echinostomes from Japan
1922	Kotlan	Morphology and hosts of adult echinostomes from birds in Hungary
1922	Sewell	Morphology, hosts and systematics of echinostome cercariae from India
1922 - 1951	Travassos & co-workers	Morphology and hosts of adults from Brazil
1922 - 1933	Tubangui	Morphology and life cycle of adults and larval echinostomes from Philippines
1923	Ando & Ozaki	Morphology and hosts of adult worms from mammals in Japan
1923 - 1926	Ando & co-workers	Descriptions of adults from Japan
1923	Miki	Descriptions of adults and larval stages from Japan
1923	Vevers	Morphology of larval stages and hosts from England
1924	Khalil & Abaza	Adults in mammals from Egypt
1924	Lutz	Descriptions of adults and larvae from Brazil
1924 - 1928	Skwortzov	Morphology of larval echinostomes and hosts from the Volga River in Russia
1924 - 1927	Mathias	Morphology and life cycles of larval stages from France
1924	Tsuchimochi	Morphology and life cycles of adults and larval stages from Japan
1925 - 1927	Bittner	Morphology and hosts of adults from Germany
1925	Bolle	Morphology and hosts of adults from Germany
1950 - 1953	Dollfus	Morphology of adults in mammals and birds from Africa and Asia
1925	Houdemer	Adults in wild and domestic animals from Vietnam
1925	Krause	Adults from birds in Germany
1925 - 1938	Stunkard	Adults and larval stages from the United States
1925	Zunker	Adults in birds from Germany

1926	Asada	Adults in birds and mammals from Japan
1926 - 1943	Bhalerao	Adults in reptiles, birds and mammals in India
1926 - 1931	Brown	Morphology and life cycle of cercariae from England
1926 - 1927	Kurova	Adults in birds from Turkestan (Former USSR)
1926	Petrov	Adults in birds from Russia
1926	Seifert	Morphology and hosts of larval stages in Germany
1927	Panova	Adults from birds in the Ukraine
1927	Reich	Morphology and hosts of larval stages in Germany
1927	Semenov	Adults from birds in the western region of the USSR
1927 - 1932	Sprehn	Adults from birds in Germany
1928 - 1934	Dubois	Morphology, hosts and systematics of larval stages from Neuchatel, Switzerland
1928	Ivanitcki	Adults from birds in the Ukraine
1928	McCoy	Life cycle of echinostomes from Missouri, USA
1933	Rusin	Life history of larval echinostomes from Central Europe
1933 - 1975	Yamaguti	Morphology, biology, hosts, life cycle and systematic reorganization of echinostomes; two systems, 1958, 1971
1934	Abdel-Azim	Morphology, biology and hosts of echinostome larvae from Africa
1934	Fallis	Hosts and metacercarial development of echinostome cercariae in the United States
1934	Wessenberg-Lund	Morphology, biology and hosts of echinostome cercariae from Denmark
1936	Miller	Morphology, biology and hosts of echinostome cercariae in North America
1937 - 1943	Beaver	Morphology, biology, life history, and systematics of adult and larval echinostomes from the United States
1937 - 1966	Dubinina	Morphology, biology and hosts of adults from birds in Russia
1938 - 1966	Stunkard	Morphology, biology and hosts of adults in birds in the United States
1938 - 1942	Porter	Echinostome cercariae and metacercariae in South Africa
1939	Hübner	Life cycles of echinostome cercariae and metacercariae from Germany
1939 - 1940	Sandground & co-workers	Adult echinostomes from humans in Indonesia with descriptions of the life cycles
1940 - 1943	Mendheim	New system for the family Echinostomatidae
1941 - 1956	Bashkirova	Completed two revisions of the family Echinostomatidae, published in Skrjabin vol. I, 1947 and vol. XII, 1956
1941- 1949	Johnston & Angel	Morphology, biology, hosts and life cycle development of adult and larval echinostomes in Australia

1948 - 1969	Kasimov	Morphology and hosts of echinostomes in birds from Azerbaidzan (former USSR)
1949 - 1983	Bykhovskaya-Pavlovskaya	Morphology of larval and adult echinostomes from birds and fishes in Russia
1949 - 1989	Kurashvili & co-workers	Morphology, biology and hosts of adult echinostomes from birds in Georgia
1949 - 1960	Dubinina	Morphology, biology and hosts of adults in birds from the Volga River in Russia
1949 - 1985	Ginetsinskaya	Morphology, biology, hosts and life cycles of adult and larval echinostomes from Russia
1949 - 1966	Belopolskaya	Morphology, biology, ecology and hosts of adult and larval echinostomes in birds and invertebrates in the former USSR
1949 - 1983	Lie & co-workers	Life cycle, synergistic and antagonistic interactions, host-parasite relationships and various other studies with adult and larval echinostomes from Asia, Africa and South America
1950 - 1954	Najarian	Life cycles of echinostome cercariae from North America
1951 - 1969	Schaktachtinskaya	Morphology, biology and hosts of adults from birds in Azerbaidzan (former USSR)
1951 - 1963	Shevtzov	Morphology, biology and hosts of adults from birds in the Ukraine
1952	Abdel-Malek	Life cycle of larval echinostomes from Africa
1952	Koga	Life cycle of echinostomes from Japan
1953	Kuntz	Life cycle and hosts of echinostomes from Africa
1953 - 1965	Nevostrueva	Life cycle and hosts of echinostomes from Russia
1953 - 1986	Smogozevskaya	Morphology, biology, life cycles and hosts of adult and larval echinostomids. Monograph on echinostomes from the former USSR in 1976
1954 - 1962	Chiaberashvili	Morphology, biology and life cycles of echinostomes from Georgia
1954	Wikgren	Morphology and biology of larval echinostomes from Finland
1956 - 1961	Leonov	Morphology, biology and hosts of echinostomes from fish-eating birds in the Black Sea area
1956 - 1994	Sudarikov	Morphology, biology and hosts of adult and larval echinostomes from Russia
1957 - 1966	Dzavelidze	Life cycles of echinostomes from Georgia
1957 - 1983	Ryzikov & co-workers	Morphology, biology and hosts of adults in Russia
1958 - 1996	Vassilev & co-workers	Biology, hosts and life cycles of echinostomes from Europe and Asia
1958 - 1960	Supperer	Life cycles and hosts of adult and larval echinostomes from Austria
1958 - 1997	Alishauskaite = (Kiseliene)	Life cycles, morphology and hosts of adult and larval echinostomes from Lithuania and Poland
1958 - 1975	Chernogorenko-Bidulina	Larval echinostomes from the Ukraine
1959 - 1960	Ahmed	Life cycles of larval echinostomes from Germany
1959 - 1976	Karmanova	Life cycles and morphology of adult and larval echinostomes from Russia
1960 - 1980	Odening	Described and illustrated morphology, biology and life cycles of echinostomes from Europe and Asia

1960 - 1966	Nasir	Described and illustrated larval and adult echinostomes and life cycle development in England
1960 - 1990	Zdarska	Described ultrastructure and life cycles of echinostome larvae from Czechoslovakia
1960	Jain	Reported life cycles of larval echinostomes from India
1960 - 1970	Kuprianova-Scakhmatova	Described adult and larval echinostomes and their hosts in Russia and Kirgizia (former USSR)
1961 - 1999	Fried & co-workers	Various studies on the biology, ecology, physiology and life histories of echinostomes in North America; Echinostoma and echinostomiasis in 1990
1961 - 1997	Kiseliene	Morphology, biology, ecology and life histories of larval echinostomes from Lithuania and Europe
1961 - 1964	Khan	Life cycle development of larval echinostomes from England
1962 - 1965	Zajcek	Larval echinostomes from freshwater snails in Czechoslovakia
1962	Wikgren	Larval echinostomes and their hosts in Europe
1963 - 1969	Vaidova	Morphology, biology and hosts of adult echinostomes from fish-eating birds from Azerbaidzan (former USSR)
1964 - 1988	Iskova	Morphology, biology and hosts of adults from birds in the Ukraine; echinostomes from birds in the Caspian and Black Sea regions in 1983
1965 - 1999	Ostrowski de Nunez	Biology, morphology and life history of echinostomes from Argentina
1965 - 1970	Brygoo & co-workers	Life cycle of echinostomes from Madagascar
1966 - 1972	Kosupko	Biology and life cycles of echinostomes from Russia
1966 - 1968	Dzaparidze	Morphology and biology of larval stages from Georgia (former USSR)
1966	Tchertkova	Biology, morphology and hosts of Echinochasmus in Russia
1967	Bisseru	Biology and ecology of larval stages from Africa
1968	Howell	Hosts and morphology of echinostomes from Australia
1968 - 1972	Dönges	Experimental studies on redial generations, their hosts and the morphology of echinostome cercariae from Africa
1968 - 1972	Rysavy & co-workers	Field and experimental studies on the morphology, biology, life cycles and antagonistic interactions of echinostomes from Egypt
1970	Neavalova	Larval echinostomes from freshwater snails in Czechoslovakia
1971	Richard	Developed a new nomenclature for the argentophilic structures of the nervous system of echinostome cercariae
1971	Egizbaeva	Morphology, biology and hosts of adults in Kazakstan (former USSR)
1973 - 1975	Rysavy, Moravec & co-workers	Life cycles and antagonistic interactions of larval echinostomes in Africa
1977 - 1999	Kanev & co-workers	Morphology, biology, life cycles and other studies with adult and larval echinostomes from Europe, Asia, Africa, Australia, North and South America; Check lists of Echinis, Echinostoma, and Echinostomatidae in 1990

1978 - 1990	Michov & co-workers	Protein fractions in adult echinostomes
1979 - 1990	Bayssade-Dufour & co-workers	Described argentophilic structures of echinostome cercariae
1980 - 1992	Christensen & co-workers	Used adult and larval echinostomes as laboratory models for various experimental studies
1980 - 1999	Loker & co-workers	Used adult and larval echinostomes for various immunological studies
1981 - 1990	McDonald	Key to adults echinostomes from waterfowl
1983 - 1986	Shishov & co-workers	Aminergic structures of the nervous system of echinostome rediae, cercariae and metacercariae
1983 - 1998	Mutafova & co-workers	Karyotypes of echinostomes
1983 - 1990	Dragneva & co-workers	Antigen similarities between echinostomes and invertebrate hosts
1983 - 1988	Poljakova & co-workers	Scanning electron microscopy of adult and larval echinostomes
1983 - 1995	Gorchilova & co-workers	Ultrastructural and enzymocytochemical studies with adult and larval trematodes
1983 - 1990	Busta & co-workers	Morphology, biology and life cycles of echinostomes from Europe and Asia
1984 - 1999	Dimitrov & co-workers	Argentophilic structures of echinostome miracidia, rediae and cercariae
1985 - 1990	Nasincova & co-workers	Life cycles, biology and taxonomy of echinostomes from Europe
1987 - 1991	Voltz & co-workers	Comparative studies of enzymes and karyotypes of echinostomes from different populations
1989 - 1999	Grabda-Kazubska & co-workers	Morphology, life cycle, DNA studies of echinostomes from Europe
1989 - 1999	Haas & co-workers	Chemical cues in snail-host finding by echinostome miracidia and cercariae
1989-1999	Fujino & co-workers	Transmission and scanning electron microscopy studies of adult and larval echinostomes
1990 - 1996	Gabrashanska & co-workers	Mineral components of echinostome larvae and their intermediate hosts
1990 - 1999	McCarthy	Biological and ecological studies of echinostome larvae and adults and their hosts
1997 - 1999	Morgan & Blair	DNA analysis of echinostomes from different hosts and geographical regions
1997 - 1999	Toledo & co-workers	Morphology and life cycles of echinostomes from Spain

An interesting history has developed around the echinostomes. "Biblia Naturae" published by Swammerdam (1737/38) is believed to be the first book in which echinostomes were described and illustrated. Significant systematic investigations began several years later when adult echinostomes were collected from naturally infected birds and mammals in Central Europe.

Key books on echinostomes, have been published by Dietz (1909, 1910), Mendheim (1940, 1943), Skrjabin (1947, 1956) and Yamaguti (1958, 1971). Major reviews and primary literature on echinostomes are found in international parasitology journals, i.e., Journal of Parasitology, Parasitology, Advances in Parasitology, Parasitology Research (formerly Zeitschrift für Parasitenkunde), International Journal for Parasitology, Parasite (formerly Annales de Parasitologie Humaine et Comparée), Experimental Parasitology and Journal of Helminthology. Some articles appear in regional parasitology journals such as Acta Parasitologica (formerly Acta Parasitologica Polonica), Parasitologia, Helmintologia, Khelmintologia, Applied Parasitology (formerly Angewante Parasitologie), Korean

Journal of Parasitology and others. Articles on echinostomes also appear in biological journals that are not devoted exclusively to parasitology, such as Invertebrate Biology (formerly Transactions of the American Microscopical Society), Journal of Invertebrate Pathology, Biochemistry, Evolution, Immunology, and Molecular Biology.

Echinostomes are often classified into family, subfamily and genera with names such as Echinis, Echinostoma, Echinostominae, Echinostomatidae and others. A check list of Echinis, Echinostoma and Echinostomatidae was published by Kanev (1990).

2. Life Cycle

All echinostomes possess a complicated life cycle expressed by: (1) alternation of seven generations known as the adult, egg, miracidium, sporocyst, redia, cercaria and metacercaria; and (2) inclusion of three host categories, known as the final, first and second intermediate hosts. Final hosts are vertebrate animals in which the adult worms develop. First intermediate hosts are freshwater and marine snails where sporocyts, rediae and cercariae develop. Second intermediate hosts are invertebrates and some amphibian vertebrates in which metacercariae develop.

The life cycle pattern of echinostomes is quite different. For example, some echinostomes such as Moliniella Odening (1964a) and Neoacanthoparyphium Dzavelidze (1957) (known also as Echinoparyphium) use only one snail species as the first intermediate host. Others, such as Echinostoma Kanev (1994) and Cathaemasia Szidat (1939) use at least five snail species, belonging to five or more genera. Some echinostomes, such as Petasiger (Beaver, 1939), and Paryphostomum (Lie and Basch, 1967b) use only fish as second intermediate hosts, while Hypoderaeum (Wikgren, 1956) and Echinoparyphium (Kanev et al., 1994c) use various invertebrate and vertebrate animals as second intermediate hosts. Some echinostomes e.g., Parorchis (Holliman, 1961), considered also as a philophthalmid do not have second intermediate hosts and their cercariae encyst on plants, shells, stones and other inert objects.

3. Form and Function of Echinostomes

3.1. EGGS

The echinostome egg is ectolecithal and typically oval in shape. Egg size is variable being relatively large in Echinostoma (Kanev et al., 1995 a, b) and in Echinoparyphium (Kanev et al., 1995 b) greater than 100 μm in length, or relatively small, less than 100 μm in length as in Pelmatostomum (Dietz, 1909, 1910) and Sodalis (Lühe, 1909). Most echinostome eggs have a rigid tanned eggshell surface, with an operculum at one end and a nodule-like thickening at the opposite end (Kanev, 1994, Kanev et al., 1994c, 1995 a, b). Infrequently, some echinostomes, (e.g. Cathaemasia Dollfus (1951), Sultanov (1961) have no thickening of the shell, while others such as Aporchis Timon-David (1955) possess a long filament at one pole. Yellow, dark-brown, and silver-white are the most common colours reported for echinostome eggs (Dietz, 1909, 1910, Skrjabin, 1947, 1956,·Yamaguti, 1958, 1971,·Kanev, 1994), Kanev et al., 1994c, 1995 a, b). Morphological features of echinostome eggs, including drawings, light micrographs and scanning electron micrographs have been published by Kanev (1982a, 1994), Lie and Basch (1967b), Kanev et al.

(1994 c,1995 a, b), Krejci and Fried (1994), Dönges (1972).

Fertilization in echinostomes is either self or cross with perhaps the most common and preferred type being cross-fertilization (Nollen, 1983). Formed in the ovary, eggs are released into the ootype where fertilization and early development occur. Surrounded with yolk material, the fertilized eggs are passed into the uterus, where further development, including the formation of the eggshell, occurs. In some echinostomes (e.g., Neoacanthoparyphium, Kanev, 1984) and Protechinostoma (Beaver, 1943) the uterus is short containing one or few eggs; in other echinostomes (e.g., Himasthla, Stunkard, 1960) and Ignavia (Freitas, 1948) the uterus is long, with numerous eggs.

The eggs are released through the female gonopore. Reddy and Fried (1996) investigated egglaying in vitro in nutritive and non-nutritive media. It is suggested that eggs, which are not fertilized or abnormal are expelled through Laurer's canal (Ginetsinskaya and Dobrovolskii, 1962). Development, hatching and infectivity of echinostome eggs have been described by numerous authors (Kanev, 1994, Lie and Basch, 1967b, Kanev et al., 1994c, 1995a, b, Idris and Fried, 1996).

Most echinostome eggs are expelled with the host's faeces. However, eggs of Nephroechinostoma are passed in the urine (Oshmarin and Belous, 1951) and eggs of Balfouria located in the host's abdominal cavity are not released in the feces or urine. The mode of egg release of Balfouria is uncertain. All echinostome eggs are adapted for aquatic environments and are killed if desiccated. Fish and freshwater insects are predators of echinostome eggs (Kanev, 1985).

Egg development in echinostomes is variable and some genera (e.g., Cathaemasia and Parorchis) release eggs with fully developed miracidia (Oshmarin and Belous, 1951, Kanev, 1985) which hatch and escape from the eggs within minutes of egg release. In other echinostomes, such as Echinoparyphium and Hypoderaeum eggs are laid uncleaved (Kanev et al., 1994c; Skrjabin, 1956; Yamaguti, 1971) and miracidia develop within the eggs after 10-20 days of embryonation. In Aporchis, eggs are laid with fully developed miracidia (Skrjabin, 1956) which do not hatch in water, but remain in the egg until it is ingested by the snail intermediate host. These eggs hatch only when ingested by the first intermediate host. Idris and Fried (1996) found that fully developed eggs of Echinostoma caproni, which usually hatch in water, may hatch in the snail gut.

3.2. MIRACIDIA

After hatching from the egg, the echinostome miracidium swims in a straight foreward manner and rotates with the terebratorium telesoped (Kanev, 1985). The swimming behavior varies in different genera. For example, miracidia of the genus Echinostoma swim rapidly forward in a straight line (Kanev, 1985) whereas Echinoparyphium miracidia change direction frequently (Kanev, 1985). As in other digeneans, echinostome miracidia do not feed, but swim and search for a suitable snail host. Kanev (1981b, 1982a, 1985) found that Echinostoma barbosai miracidia had a half-life of 5-7 hr at 18-22°C, whereas Echinoparyphium aconiatum miracidia survived for 8 hr.

Miracidial morphology can be examined live with or without intravital stains to demonstrate glandular and excretory systems. The miracidium is small (about 100 μm in length), transparent, and elongate with eyespots; its body is covered with cilia (Lie and Umathevy, 1965, Lie and Basch, 1967a, Dönges, 1972). Treatment of miracidia with silver nitrate allows for observations on the number and arrangement of the ciliated epidermal plates and the argentophilic papillae.

Dönges (1972) listed the number of epidermal plates for numerous echinostome miracidia, including 6:6:4:2 in Artyfechinostomum, Chaunocephalus, Echinoprypahium, Echinostoma, Hypoderaeum, Paryphostomum and Pegosomum; 6:9:4:2 in Isthmiophora; 6:6:4:3 in Euparyphium; 6:8:4:2 in Acanthoparyphium and Himasthla; 6:6:3:2 and 6:6:4:2 in Echinochsmus. Argentophilic papillae of echinostome miracidia have been described by numerous authors (Kanev, 1985, Dimitrov et al., 1995, Beaver, 1937, Jourdane and Kulo, 1981). Dimitrov et al. (1995) described about 30 papilla-like structures in the miracidia of Echinostoma trivolvis and established a model to show the arrangement of these papillae. Retractile apical papillae followed by a pyriform gland have been described for echinostome miracidia by Lie and Umathevy (1965), Lie and Basch (1967a), Kanev (1985), Kanev et al. (1995b). Two lateral processes each situated posterior to the lateral anterior epidermal plate, with short seta-like structures immediately anterior to each process have been reported for the miracidia. Two dark-brown eye- spots, consisting of two pairs of pigmented bodies are located side by side. Each pigmented body consists of a pair of oval discs and a pair of rods situated posterior to the oval discs (Lie and Umathevy, 1965; Kanev, 1994). The excretory system consists of two flame cells, two excretory ducts, and two excretory pores (Lie and Umathevy, 1965; Lie and Basch, 1967a; Kanev, 1985, 1994; Kanev et al., 1995b). In the miracidia of Echinostoma one flame cell is located on the left in the ventral and posterior half of the body, and a second on the right, in the dorsal and anterior half of the body (Lie and Umathevy, 1965, Kanev, 1994). The excretory pores open between the 3rd and 4th rows of epidermal plates.

Miracidia of echinostomes penetrate the snail host at the mantle edge, foot or other exposed parts of the soft snail body. Kanev (1987) found that miracidia of Echinostoma revolutum shed both their cilia and epidermal plates prior to penetration into the intermediate host, Lymnaea stagnalis. However, it has been suggested that some miracidial - sporocyst transformations in echinostomes could be completed without shedding the epidermal plates as has been reported for other trematodes. How echinostome miracidia find snail hosts has been studied by Haas et al. (1995a) and Haas and Haberl (1997).

Experimental studies suggest that miracidia of Echinostoma and Echinoparyphium show a positive phototaxis and a negative geotaxis (Jeyarasasingham et al., 1972, Kanev, 1985, 1994, Kanev et al., 1994c).

Miracidia of some echinostomes, e.g., Echinoparyphium aconiatum Kanev (1982a), and Neoacanthoparyphium echinatoides Kanev (1984) infect only one snail species, Lymnaea stagnalis and Viviparus viviparus, respectively. Other echinostome miracidia, e.g., Echinostoma echinatum Kanev (1985) and Cathaemasia hians Szidat (1939) infect snails belonging to at least 5 different species.

Kanev and Vassilev (1981) found that miracidia of Echinostoma are immobilized and inactived by certain chemicals from plants (such as Juglan regia and Salix alba) and also from Physa acuta snails which are resistant to infection. Chaetogaster limnaei, freshwater fishes, and insects are predators of miracidia (Kanev, 1985 and Kanev and Vassilev, 1981).

3.3. SPOROCYSTS

Johnson (1920) suggested that Echinostoma revolutum sporocysts occurred infrequently because the miracidia of this species could develop directly into rediae. Experimental studies on E. revolutum showed that all miracidia develop into sporocysts (Beaver, 1937; Kanev, 1994, 1985). No evidence has been found to support direct development of miracidia into rediae in E. revolutum.

Echinostome sporocysts are sac-like structures that are formed after miracidial transformation. Miracidia usually transform at the site of penetration, typically the mantle, foot, head or other exposed parts of the snail host.

Dissection and histological examination of experimentally infected snails showed that sporocysts of Paryphostomum segregatum Lie and Basch (1967b) develop near the site of miracidial entry, i.e., the foot and mantle edge. In other echinostomes, such as Echinostoma rodriguesi Hsu and Basch (1968) and E. revolutum Kanev (1985), sporocysts migrate some distance from the place of entry and develop in the heart cavity or aorta.

Morphologically all known echinostome sporocysts are similar in shape, size, color, and contents. Newly formed sporocysts are about 100 μm long. Lie and Basch (1967b) reported that it took 5 to 7 days for P. segregatum sporocysts to develop and produce the first generation of rediae. A similar period of development (5-8 days) was found for echinostomes in the genus Echinoparyphium (Kanev et al., 1994c) and in Echinostoma (Kanev, 1982a, 1985, 1994, Kanev et al., 1995a, b). Fully developed sporocysts, about 250-500 μm long, with rediae, redial embryos and germinative balls were found 2-4 weeks after exposure of snails to miracidia.

Old, empty, dark-yellow, or grey, shriveled and deformed sporocysts have been found from 3 to 15 wk post exposure (Kanev, 1985, 1994; Kanev et al., 1994c). Young, fully developed sporocysts of Echinostoma revolutum were found in the alimentary tract of Chaetogaster limnaei; sporocysts have also been found infected with microbes and microsporidia (Kanev, 1985).

Encapsulated, inactivated, and expelled echinostome sporocysts resulting from snail host defence mechanisms and immunological reactions in non-compatible snail hosts have been described by Lie and Heyneman (1975, 1976a, b), by Lie et al. (1975b, 1976), and by Kanev (1987).

Indirect and direct antagonism against echinostome sporocysts, including cannibalism and predatory destruction of echinostome sporocysts by echinostome rediae have been described by Lie et al. (1968a, b, 1973, 1975a) and Kanev (1981a).

3.4. REDIA

Echinostome rediae are elongate structures with a mouth, pharynx, sac-like gut, a collar-like tegumentary ring (collar) at the anterior end and two posterior stump-like locomotor appendages (the ambulatory buds).

The first rediae (mother rediae), are produced by the sporocysts. The next redial generations (daughter and grandaughter rediae) are produced by the previous redial generation. The exact number of redial generations is unknown. Some authors, including Skrjabin (1947, 1956) and Yamaguti (1958, 1971, 1975) have suggested that most echinostome species possess two or three redial generations. Results of redial transplantation studies using surgically implanted rediae, under laboratory conditions from one snail to another suggested that the number of redial generations for at least Isthmiophora melis and Echinoparyphium aconiatum is unlimited. Rediae will reproduce successive generations as long as the snail host is alive (Heyneman, 1966; Dönges, 1971; Dönges and Götzelmann, 1975),

Rediae of all generations share similar structures. Some rediae, such as Echinoparyphium rubrum (Kanev et al., 1998) possess an elongate, sausage-like body, while in other echinostome rediae, such as Neoacanthoparyphium echinatoides (Kanev, 1984) the body is sac-like. A well developed strong muscular pharynx and a short sac-like gut are found in the rediae of Echinoparyphium aconiatum (Kanev, 1982a). A strong muscular pharynx, followed by a long, sinuous gut are characteristic of the rediae of Paryphostomum radiatum, and Paryphostomum segregatum (Lie and Basch, 1967b). Rediae with a relatively small pharynx and a long gut have been described for Isthmiophora melis, whereas those of Echinostoma revolutum and E. echinatum (Kanev, 1985, 1994) possess a small pharynx and a short gut. Most echinostome rediae, e.g., Echinostoma and Echinoparyphium possess a collar in the form of an uninterrupted tegumentary ring (Beaver, 1939, Lie and Umathevy, 1965, Lie and Basch, 1967a, Kanev, 1994, Kanev et al., 1994c, 1995a). Some echinostome rediae, e.g., Paryphostomum segregatum (Lie and Basch, 1967b) possess a collar of four independent protrusions, i.e., one ventral, one dorsal and two lateral. Appendages, also called locomotor organs are well developed and conspicuous in Paryphostomum and Isthmiophora and poorly developed in Neoacanthoparyphium. In all echinostome rediae the birth pore is located just behind the collar, on the dorsal side.

Rediae (mother rediae), which develop in sporocysts produce only rediae. Rediae of the next generation (daughter and granddaughter) can produce both rediae and cercariae. Photomicrographs of echinostome rediae containing rediae and cercariae has been published by Kanev (1980).

Echinostome rediae, especially these of Paryphostomum, with a well-developed pharynx, gut and locomotory organs are predators on rediae and sporocysts of echinostome and non-echinostome trematodes, including schistosomes (Lie, 1967, 1969, Lie et al., 1967, 1968a, b, 1970, 1972, 1973, 1975a, Kanev, 1982b, c). Based on this antagonistic and predatory behavior, echinostome rediae have been tested in the laboratory and the field for use as possible control agents of snails infected with Schistosoma (Lie et al., 1970, 1972).

Rediae impregnated with silver nitrate allow for studies on the number and pattern of argentophilic structures on the redial tegument (Dimitrov and Kanev, 1984; Dimitrov et al., 1985; Grabda-Kazubska and Laskowski, 1996). Histochemical methods for impregnating rediae with glyoxylic acid to show aminergic elements of the nervous system have been described by Shishov and Kanev (1986). Antigen similarity between echinostome rediae of Echinostoma aconiatum and the first and second intermediate snail hosts, Lymnaea stagnalis and Physa acuta have been studied by Dragneva and Kanev (1983). The number and morphology of chromosomes in echinostomes was studied by Baršiené (1993), Mutafova et al. (1986),and Mutafova and Kanev (1986). DNA sequencing and phylogenetic analyses of echinostomes based on redial material have been done by Grabda-Kazubska et al. (1998). Light, scanning and transmission electron microscopical observations of echinostome rediae have been published by Krejci and Fried (1994) and Fried and Awatramani (1992). Mineral composition, including identification of Ca, K, N, Fe, Zn, Cu, and other elements in echinostome rediae and their intermediate hosts have been studied by Gabrashanska et al. (1990, 1991). DNA and RNA content of echinostome rediae have been measured by laser technology (Gergova et al., 1996). Hyperparasites, including microbes, microsporidia and metacercariae of echinostome and non-echinostome trematodes, have been found in echinostome rediae by Kanev (1985) and Canning (1995). Echinostome rediae have been found in the digestive tract of Chaetogaster limnaei; presumably they were ingested by the chaetogasters (Kanev, 1985).

3.5. CERCARIA

Echinostome cercariae are distome with a tail and a crown of spines located typically on a collar-like tegumentary flap surrounding the oral sucker, known as the head-collar. General morphology and biology of echinostome cercariae, including drawings and photomicrographs have been published in numerous papers, i.e., Beaver (1939), Lie and Umathevy (1965), Lie and Basch (1967a, b), Yamaguti (1971), Jeyarasasingham et al. (1972), Kanev (1985, 1994), Kanev et al. (1994c, 1995b), and Fried et al. (1998b). In some cercariae, such as Moliniella, Neoacanthoparyphium and Pseudechinoparyphium, the body is large, reaching up to 850 µm in length (Kanev, 1982a, 1984). In cercariae of Echinochasmus and Microparyphium the body is small, about 100 µm in length (Karmanova and Ilyuschina, 1969). In the most common group of echinostome cercariae, including those of Echinostoma, Echinoparyphium and Hypoderaeum the body is about 200 to 500 µm in length (Lie and Umathevy, 1965, Lie and Basch, 1967a, and Kanev et al., 1994c). For most echinostome cercariae, the collar and collar-spines are well developed and clearly visible. In the cercariae of Echinochasmus, the collar-spines are not visible, but later become apparent in the metacercariae and adults (Karmanova and Ilyuschina, 1969). In the cercariae of Cathaemasia, the collar-spines are well developed and clearly visible, but are lost or havely visible in the adult worms (Szidat, 1939). Prepharyngeal bodies, of unknown origin and function, are found in the cercariae of Paryphostomum and Echinostoma histricosum (Lie and Basch, 1967b). The esophagus is poorly developed and difficult to observe in Echinochasmus (Karmanova and Ilyuschina, 1969). A well-developed esophagus, containing numerous refractile bodies arranged in 2-3 rows, has been reported for Cercaria echinata, C. echinatoides and C. Moliniella anceps (Kanev, 1982a). At least five different groups of excretory systems in echinostome cercariae have been described by Sewell (1922). They are as follows: "Echinostoma", "Coronata", "Echinatoides", "Stenostome" and "Mesostome" groups. The number and arrangement of flame cells are present in three main groups. From 10 to 20 flame cells arranged in pairs are reported in Echinochasmus; 20 to 50 cells, arranged by three in Echinostoma and Echinoparyphium; and from 50 to 100 cells arranged by three or four in Moliniella and Pseudechinoparyphium. The main excretory ducts are simple and short, containing from 100 to 500 excretory granules, located between the pharynx and ventral sucker in Hypoderaeum and Echinostoma (Kanev, 1981b; Jeyarasasingham et al., 1972). Large excretory ducts, containing numerous refractile granules and finger like branches or diverticuli at the level of pharynx have been described in Cercaria hoanophila (Szidat, 1939). In Episthmium, Episthochasmus and Echinochasmus the excretory ducts are long, extending from the level of pharynx to the posterior end of body, but contain few (from 10 to 20) excretory granules. Some echinostome cercariae and especially those of the genus Echinostoma possess two types of glands, known as penetration and paraesophageal glands. Althoug well-studied morphologically, the exact function of these glands is still unknown (Lie, 1966, Kanev et al., 1994a; Humphries and Fried, 1996). A third type of gland cell, (cystogenous glands), has been reported for all echinostome cercariae. In some cercarial groups, mainly those of Echinostoma and Paryphostomum, the cystogenous glands contain long stick-like granules arranged symmetrically in a "V" or "X" configuration (Lie and Basch, 1967b; Karmanova and Ilyuschina, 1969). Most echinostome cercariae, including Echinostoma and Echinoparyphium, have been described with cystogenous glands, containing short, "sand"-like granules (Lie and Umathevy, 1965; Lie and Basch, 1967a; Jeyarasasingham et al., 1972; Kanev, 1984; Kanev et al., 1994c, 1995a, b). In most echinostome cercariae, the tail is simple, and cylindrical with a conical tip (Lie and Basch, 1967b, Kanev, 1982a). In Echinostoma the tail is cylindrical with a sharp unilateral finger-like tip (Kanev, 1994, Jeyarasasingham et al., 1972). In some species of Echinochasmus, the tail is cylindrical, with a bilaterally filiform narrowing followed by a fold-like protuberance (Kanev, unpublished results). Compared with the body, the cercarial tail is almost equal in length to the body, as in Echinostoma. The tail is about 5 times longer than the body as in Shiginella. The tail is at least 10 times longer than the body as in Petasiger; rarely, is the tail shorter than the body as in some cercariae of Echinochasmus Lühe (1909), Sewell (1922), Lie and Umathevy (1965), Lie and Basch (1967a), Karmanova and Ilyuschina (1969), Kanev (1982a, 1984, 1994). Light, scanning and transmission electron microscopic observations have been used to investigate the fin-fold structures located on the cercarial tail (Odening, 1964b; Lie and Umathevy, 1965; Kosupko, 1972; Kanev, 1984; Zdarska et al., 1989; Kanev et al., 1993). Tails without finfolds have been found in many echinostome cercariae, including Hypoderaeum, Isthmiophora and Echinoparyphium Kanev et al. (1994c). Two, long, well-developed finfods have been described for cercariae of Moliniella and Neoacanthoparyphium Kanev (1984). Seven finfolds have been described in the cercariae of Echinostoma Kanev et al. (1993). The nervous system of echinostome cercariae, including argentophilic structures and aminergic structures, has been studied by numerous authors, i.e., Richard (1971), Shishov and Kanev (1986), Kanev et al. (1987), Dimitrov (1987), Bayssade-Dufour et al. (1989), Grabda-Kazubska et al. (1991), Dimitrov et al. (1997). Chromosome studies and DNA measurements of echinostome cercariae have been described in numerous papers by Mutafova et al. (1991), Baršiené (1993), Gergova et al. (1996). Microelements of cercariae and their hosts have been studied by Gabrashanska et al. (1990, 1991). Echinostome cercariae have been used for various ultrastructural and histochemical studies. The ultrastructural effects of in vitro trypsin action (Stoitsova and Kanev, 1987), histochemical glycogen, neutral lipids, exogenous glucose (Fried et al., (1998a), histological and histochemical studies (Zdarska and Nasincova, 1985), and scanning and transmission electron microscopy, have been studied extensively by Poljakova-Krusteva and Kanev (1983), Zdarska et al. (1987, 1989). The emergence of echinostome cercariae under different conditions has been studied by McCarthy and Kanev (1992), and Schmidt and Fried (1996). Attraction and chemical cues in snail-host finding by echinostome cercariae have been studied by Evans and Gordon (1983 a, b), Fried and King (1989), Haas (1994), Haas et al. (1995b). Cercarial specifity toward the second intermediate host has been studied under both field and laboratory conditions by Evans and Gordon (1983 a, b), McCarthy (1990), McCarthy and Kanev (1992). Survival and infectivity of echinostome cercariae, including their swimming patterns and their mode of encystment, have been described in numerous papers, Beaver (1943), Lie and Umathevy (1965), Lie and Basch (1967a); Kanev (1981b, 1982a, 1984, 1994), Kanev et al. (1994c, 1995a, b), Fried et al. (1995), and Pechenik and Fried (1995). Some cercariae, such as Parorchis, encyst outside the host. Others, including Moliniella, encyst preferably in freshwater molluscs. Tadpoles and frogs are preferable hosts for Cercaria choanophila, while fishes are the main second intermediate hosts for Echinochasmus and Petasiger. Cercariae of Echinostoma and Echinoparyphium encyst in the kidneys and eye cavities of frogs and water turtles (Kanev, 1994; Kanev et al., 1994c). Chaetogaster limnaei, freshwater fishes, and beetles are predators of echinostome cercariae (Kanev, 1985).

3.6. METACERCARIA

All echinostome cercariae encyst outside of a host, e. g., on surfaces, or in or on second intermediate hosts. Such metacercariae are called cysts or encysted metacercariae. Cysts are different in shape (e. g., spherical for Hypoderaeum and Echinoparyphium; subspherical or oval for Isthmiophora and Echinochasmus; elliptical for Cathaemasia and Himasthla). The diameter is small, up to 100 μm in Echinochasmus, large, up to 400 μm for Moliniella, and mid-size, between 100 and 200 μm, in Echinostoma, Echinoparyphium, and Hypoderaeum. The cyst wall is composed of several layers, about 5 μm thick in Himasthla and up to 50 μm in thickness in Moliniella. The metacercarial body is enclosed inside of the cyst. Structural, ultrastructural and histochemical studies on both cyst walls and the larva within the cyst have been studied by numerous authors, i.e., Gulka and Fried (1979), Gorchilova and Kanev, (1986), Krejci and Fried (1994), Smoluk and Fried (1994), Ursone and Fried (1995b), Fried et al. (1998b). The formation of metacercarial cysts in vivo, in vitro and ectopically has been described by Stein and Basch (1977), and Fried et al. (1997) . In vitro excystation and cultivation of excysted metacercariae to ovigerous adults, in the allantois of chick embryo's, has been described by Chien and Fried (1992) and Fried and Rosa-Brunet (1991). Histochemical studies on hydrolytic enzymes, in excysted metacercariae ,of echinostomes was been studied by Le Flore and Bass (1983). Penetration sites, migration routes, and sites of localization of echinostome cercariae and metacercariae, have been studied by numerous authors, i.e., Beaver (1937, 1939), Wikgren (1956), Dzavelidze (1957), Lie and Umathevy (1965), Lie and Basch (1967a, 1967b), Hsu and Basch (1968), Jeyarasasingham et al. (1972), Kanev (1981b, 1982a, 1984, 1985, 1994), Kanev et al. (1994c, 1995a, b). Some echinostomes such as Hypoderaeum, Echinoparyphium and Echinostoma use the urinary duct for cercarial entry and eventual encystment in the kidney of freshwater snails, frogs and turtles. Some echinostome cercariae use the lacrimal ducts to enter and encyst in the eye cavity of frogs and turtles. Other metacercariae, such as of Isthmiophora and Paryphostomum localize in the skin, fins and scales of fishes. Metacercariae of some echinostomes have been found on the mantle edge of freshwater snails harboring rediae and cercariae and serving as the first intermediate host. Fishes are intermediate host for Echinohasmus and Petasiger with the metacercariae encysting mainly in the gills. Fishes are intermediate hosts for Shiginella with the metacercariae located in the esophagus. Metacercariae of Cathaemasia encyst preferably in the nasal cavity of tadpoles and frogs. Metacercariae of Echinostoma encyst in leeches and planarians. Echinostome metacercariae, including Echinostoma, Hypoderaeum and Echinoparyphium encyst within the rediae of the same echinostome species. Some metacercariae (e.g., Parorchis) are infective immediately after they encyst. Others such as Hypoderaeum and Echinoparyphium need to develop for up to seven days before they are infective. Metacercariae of Moliniella and Pseudechinoparyphium are infective 20 days after encystation.

3.7. ADULTS

3.7.1. Body Shape and Size

After the metacercarial cysts are ingested or the cercariae penetrate the host, the larvae develop into adults in specific sites within the definitive host. Members of the Echinostomatidae live in reptiles, birds and mammals where the adults are parasitic in the intestine, occasionally the bursa of Fabricii, and rarely in the urethra and kidney. In general, all adult echinostomes have a collar with either one or two rows of spines, and spines on the tegument. They have an oral sucker, prepharynx, pharynx, oesophagus, caeca reaching well into the posterior aspect of the body, acetabulum (ventral sucker) well developed and in middle third of the body; a median genital pore, tandem testes, ovary pretesticular and median, and excretory vesicle Y-shaped.

The body size of echinostomes varies considerably. Some, such as Neoacanthoparyphium and Pseudechinoparyphium, are small ranging from 1-3 mm. Others in the genus Aporchis Skrjabin (1956), Yamaguti (1958, 1971) are large, more then 50 mm in length. Body shape of echinostomes also varies considerably. Some are fusiform, tapering at both ends, e.g., Echinoparyphium. Others are long and thin resembling a filament or filiform, e.g., Aporchis. Still others are linguiform, tongue shaped, broad posteriorly tapering to a rounded anterior, e.g., Cathaemasia. Widening or enlargement of the general body shape is also found. The genus Drepanocephalus has an enlargement in the anterior region.

3.7.2. Collar and Collar Spines

Light, scanning, and transmission electron microscopy has been used by numerous authors for the examination and description of collars and collar spines of both adult and larval echinostomes Skrjabin (1956), Yamaguti (1958, 1971), Lie (1963), Lie and Basch (1967b), Poljakova-Krusteva and Kanev (1983), Gorchilova and Kanev, (1986), Grabda-Kazubska and Kiseliene (1991), Kanev and Busta (1992), Ursone and Fried (1995a), and Toledo et al. (1996). Broken and abnormal collar spines have also been examined using the same optical methods in adult and larval echinostomes, Lie (1963), Kanev and Busta (1992).

The collar of echinostomes is useful for determining certain genera in the family Echinostomatidae. Some echinostomes have enlarged protuberances in the forebody but without the typical collar, e.g., Stephanoprora. Other echinostomes, have poorly developed collars, e.g., Pristicola and Microparyphium. Moderately developed collars are found in specimens of the genus Artyfechinostomum and these are considered more developed than the collars in Pristicola and Microparyphium, but not as well developed as the collar in Echinoparyphium. Another different collar type consists of two lobes as in the genus Patagifer (see Skrjabin, 1956, and Yamaguti, 1958, 1971).

Collar spines are also used to identify echinostome genera. Spines vary in size, shape, arrangement, and location. Some spines are cylindrical, large with a sharp anterior and a rounded posterior end, e.g., Cathaemasia; other spines are small, flat and rounded on both ends as in Echinoparyphium. Some echinostomes have spines that are identical in size and shape, e.g., Nephroechinostoma, while others have spines that vary in size from small to large, e.g., Petasiger (Skrjabin, 1956; Yamaguti, 1958, 1971). Some spines are located only on the ventrolateral folds of the collar, e.g., Sodalis. The arrangement of spines varies. Some echinostomes have a single row interrupted in two places, e.g., some species of Aporchis, other echinostomes have the row of spines interrupted in only one place along the dorsal collar region, e.g., Echinochasmus; still others have a complete, uninterrupted row of spines, forming a crown-like structure, e.g., Cleophora. In some cases the crown of spines is lobed in two sections, e.g., Patagifer or the collar has several crowns of spines, e.g., species of Stephanoprora.

3.7.3. Tegument

The tegument of echinostomes has been examined by light, scanning and transmission electron microscopy, and by enzymocytochemical methods. It is typical of trematode teguments (Poljakova-Krusteva and Kanev, 1983; Ursone and Fried, 1995a; Gorchilova and Kanev, 1984, 1992, 1994). The tegument has numerous pits, which increase surface area to enhance absorption and secretion. The tegument consists of an outer membrane, a middle layer, and the basal membrane. The middle layer contains various discoidal and membranous bodies. The outer membrane is trilaminate.

In echinostomes the distal cytoplasmic layer, both apically and basally, is covered with folded plasma membranes and tegumental cells. The cytoplasmic connections of the tegumental cells link them with the overlying syncytium. Enzyme activity is not constant throughout the tegument. The localization of enzymic activity of adenosine triphosphatase, adenosine diphosphatase, acid phosphatase, and succinate dehydrogenase on the tegument has been described by Gorchilova and Kanev (1983, 1984, 1992, 1994). The tegumentary spines vary in size, shape and patterns. The distribution of tegumentary spines has been described by Skrjabin (1956), Yamaguti (1958), Lie (1963), Lie and Umathevy (1965), Lie and Basch (1967a, b); Yamaguti (1971); Kanev (1984, 1994) and Toledo et al. (1996).

3.7.4. Suckers

The suckers of echinostomes fit the overall description of digeneans. Members of the family Echinostomatidae have two suckers, an oral sucker and ventral sucker (acetabulum), and only one genus has no oral sucker. The suckers are cup shaped, uniform in structure, and covered with tegument.

The oral sucker is well developed, located in the anterior most portion of the forebody. The size ratio between the oral sucker and the acetabulum varies depending on the genus. In the genus Singhia the oral sucker is equal in size to the acetabulum. The oral sucker can be up to four times smaller then the acetabulum in the genera Echinostoma and Hypoderaeum or larger then the acetabulum in the genus Pristicola. In some cases the oral sucker is absent as in the genus Pegosomum.

The ventral sucker is found on the ventral body surface. The position on the trematode varies depending on the genus of echinostome. The acetabulum is located in the anterior portion of the forebody in the genus Skrjabinophora and the posterior portion of the forebody in Episthmium. In Singhiatrema the acetabulum is in the middle of the body and in the hind-body in Balfouria and Chaunocephalus. The structural composition of the acetabulum varies from very large with well-developed musculature as in Hypoderaeum to a weakly developed structure as in Balfouria and Mesorchis.

3.7.5. Nervous System

Shishov and Kanev (1986) used histochemical techniques with glioxylic acid and green light fluorescence. They were able to determine the distribution of aminergetic neurons in adult echinostomes. The nervous system of echinostomes, is typical of digeneans as described in Hyman (1951), and Bullock and Horridge (1965). The cerebral ganglia are paired, and are posterior and dorsal to the pharynx. Both anterior and posterior longitudinal nerve cords arise from the cerebral ganglia. The ventral and dorsal longitudinal nerve cords are connected throughout the worm body by cross-commissures. Within the suckers and subtegument is located motor and sensory plexi with nerve endings in the tegument that are well-developed. See Chapter II for an up date of the nervous system of echinostomes.

3.7.6. Sensory Structures

Sensory structures in preadult and adult echinostomes were studied by light, fluorescent, scanning and transmission electron microscopy by Donovick and Fried (1988), Fried et al. (1990), McCarthy and Kanev (1992), Kruse et al. (1992), and Kowalevski (1897). Some of these studies showed nerve fibres terminating in processes just below the tegument. Some of the studies showed differences in the size and shape of tegumentary papillae. In some adult echinostomes, e.g., Echinostoma trivolvis and Echinostoma caproni, variation in the number of cilia, ranging from none, one, to several, on the tegumentary papillae have been reported by Poljakova-Krusteva and Kanev (1983), Fried et al. (1990), and Kruse et al. (1992).

The exact function of the papillae-like structures in adult echinostomes is not known. The location of the papillae, and papillae-like structures in adult echinostomes varies depending on the genus. In the genus Sodalis and Eurycephalus, large papillae-like structures are found at the level of the testes (Ovtcharenko, 1955). Similar structures are located near the genital pore in Hypoderaeum Nestorov et al. (1996). In both cases the function of these structures is unknown.

3.7.7. Digestive System

Most digeneans have a well developed, closed digestive system. The digestive system or gut of Digenea is divided into two sections, the foregut and the caeca. The foregut consists of the mouth, located in the oral sucker; the mouth serves for both ingestion and egestion. The mouth is followed by a prepharynx, a muscular pharynx when present, and the oesophagus. Absorption and secretion occur within the ceca. To aid in absorption some regions of the caeca have microvilli that increase the surface area. Special cells located in the ceca are responsible for secreting substances that aid in digestion.

The digestive system of echinostomes has been well studied and described in many papers by Skrjabin (1947, 1956), Yamaguti (1958, 1971), Lie and Umathevy (1965) Lie and Basch (1967a, b), Kanev (1984), Kanev et al. (1994c). Echinostomes show considerable in the structure of variation the digestive system depending on the genus.

The prepharynx in Hypoderaeum is undeveloped to the extent of being almost imprerceptible. The size of the prepharynx varies. In Acanthoparyphium the prepharynx is short while in some species of Stephanoprora and Velamenophorus the prepharynx is long.

The pharynx, located just posterior to the oral sucker, is often compared in size to the oral sucker for taxonomic purposes. The pharynx of echinostomes varies in size, depending upon the species. The pharynx in the genus Himasthla is smaller than the oral sucker; but is equal in size to the oral sucker in some species of Stephanoprora, and larger than the oral sucker in some species of Echinostoma. Adults in the genus Pegosomum have only a pharynx and the oral sucker is absent.

Oesophagus size and appearance are used to help differentiate genera. A short oesophagus is present in Hypoderaeum, Multispinotrema, and Skrjabinophora, while a long oesophagus occurs in Chaunocephalus, Petasiger, and Parallelotestis. The oesophagus is straight in Eurycephalus, but curved in Pegosomum.

The caecal bifurcation follows the oesophagus and is anterior to the acetabulum. The distance anterior to the acetabulum where the bifurcation occurs can be used to separate several genera within the Echinostomatidae. In Pegosomum and Balfouria the bifurcation occurs considerably anterior to the acetabulum, while the bifurcation is just anterior to the acetabulum in Prinosoma and Echinochasmus.

The caecal length varies and is a characteristic used to separate some genera. A very short caeca that reached the posterior margin of the acetabulum is found in the genus Singhiatrema. Caeca that extend to the posterior margin of the body are present in the genus Mesorchis. Intermediate caecal length occurs in most echinostomes but caecal variation is common.

3.7.8. Male Reproductive System

The male reproductive system of echinostomes is typical of most digeneans with the exception of the dioecious schistosomes. The echinostome, male system usually consists of two testes. The vas efferentia connect the testes to the vas deferens. The vas deferens then lead to the cirrus sac. The cirrus sac contains the seminal vesicle, ejaculatory duct, cirrus, and prostate gland.

The morphology and physiology of the male reproductive system of echinostomes has been well studied by Skrjabin (1947, 1956), Yamaguti (1958, 1971), Grabda-Kazubska and Kiseliene (1991), and Toledo et al. (1996). In the Echinostomatidae most variation within the male system occurs in the shape of the testes and the size and location, of the cirrus sac, shape of the cirrus and location of the genital pore.

The variation in structure and abnormalities in the male system of echinostomes have been studied by Ando and Ozaki (1923) and Kanev et al. (1994b, d). The location of the testes in relation to the body is variable. In the genus Pelmatostomum the testes are located in the anterior portion of the body and in the Echinostoma the testes are located in the posterior portion of the body. In Prinosoma and Mesorchis the testes are located in the middle of the body. The testes may be in tandem as in Moliniella or diagonal and oblique to each other as in Singhia, Neopetasiger and Balfouria. In some genera, Parallelotestis, Caballerotrema, Parorchis, and Singhiatrema the testes are side by side to each another.

The shape of the testes is diverse and can be used to distinguish echinostome genera. The shapes range from smooth and oval in Chaunocephalus to differences in the number and types of lobes as in Artyfechinostomum with 5-6 distinct lobes or Reptilitrema with dendriform lobes. The size of the testes in relation to body size can be used in some cases to help determine genera. Comparatively, large testes are found in Pegosomum and Saakotrema while relatively small testes are found in Balfouria and Chaunocephalus.

The cirrus sac size and it location in relation to the acetabulum is used to separate some genera. Drepancephalus has a small cirrus sac located just anterior to the acetabulum whereas in Pegosomum the cirrus sac is located in the same area as in Drepancephalus but is larger and longer. The cirrus sac in Echinostoma varies but is found in the acetabular region as compared to the cirrus sac in Acanthoparyphium and Aporchis which is found posterior to the acetabulum. The cirrus, located within the cirrus sac, has been studied by scanning electron microscopy in eight species of echinostomes. In this study Nestorov et al. (1996) compared size, shape, and surface texture of the cirrus. Some were found to be spinate as in Isthmiophora, but others had papilla-like structures as in Echinoparyphium.

The genital pore, the opening of the reproductive system, may vary in structure. Either the male and female reproductive systems have independent openings into the genital pore as in Petasiger and Monilifer or the two reproductive systems merge to form a common duct which in turn opens into the genital pore as in Pristicola.

3.7.9 Female Reproductive System

The female reproductive system of echinostomes is typical of digeneans, i.e. single ovary either posterior or anterior to the testes. As the oviduct leaves the ovary it enlarges to form the seminal receptacle which stores the spermatozoa. Vitellaria are paired structures located laterally within the body and connect with the oviduct by the vitelline ducts. The ootype, an enlargement of the oviduct is surrounded by glandular tissue (Mehlis' gland) and is where fertilization usually takes place. The ootype, in turn opens into the uterus in which eggs develop. In mature digeneans, large number of eggs distends the uterus.. The metraterm is the muscular terminal region of the uterus that acts as a receiving chamber for the cirrus, the male copulatory organ.

The single ovary in echinostomes varies in shape and can be used to distinguish genera. The ovary is smooth or round in <u>Acanthoparyphium</u> (Yamaguti, 1971;·Kanev, 1990) and lobed in <u>Euparyphium</u> (Dietz, 1909, 1910; Yamaguti, 1958). In some genera as in <u>Artyfechinosomum</u> (Skrjabin, 1956;·Yamaguti, 1958, 1971) the ovary has an shape of the irregular shape. The size of the ovary as compared to the testes is also helpful in separating genera. In <u>Echinostoma</u> and <u>Echinoparyphium</u> the ovary is smaller then the testes. In the genus <u>Balfouria</u> the ovary is usually larger then the testes. Within the Echinostomatidae the ovary may be located in different regions of the body.

Vitellaria are follicular and are useful to help determine genera and species of echinostomes. A small number of vitellaria occur in <u>Neoacanthoparyphium</u> and extend from the anterior testis to the posterior end of the body. Most echinostome genera have a large number of vitellaria. Vitellaria are usually composed of two groups located laterally in the body. They are classified as being separate, merging, or split. Several genera that demonstrate separate vitellaria are <u>Hypoderaeum</u>, <u>Acanthoparyphium</u>, <u>Cleophora</u>, <u>Singhiatrema</u>, and <u>Parorchis</u>. Merging vitellaria occur when the two lateral groups merge. Where they merge, i.e., behind the first testis, or in the posterior region of the body, helps to distinguish echinostome genera. Those with merging vitellaria include <u>Aporchis</u>, <u>Himasthla</u>, <u>Episthmium</u>, <u>Episthochasmus</u>, <u>Echinochasmus</u> and <u>Pegosomum</u>. Split vitellaria, two groups per side separated by the acetabulum, ovary, uterus and testes are found in several genera including <u>Choanocephalus</u> and <u>Balfouria</u>.

The uterus is usually located in the middle of the body or intracaecally. The size and shape of the uterus varies between genera. The uterus is either branched or relatively smooth. A small uterus with only several eggs is found in <u>Neoacanthoparyphium</u>, while a long uterus with many eggs occurs in <u>Mehrastominae</u>.

Not all echinostomes have a seminal receptacle. For instance, <u>Echinostoma</u> <u>revolutum</u> and <u>E.</u> <u>echinatum</u>, do not have one. The function of the seminal receptacle is done a portion of the uterus in these species. Other species, <u>Echinostoma</u> <u>nudicaudata</u>, and <u>E.</u> <u>pinicaudatum</u> have well developed seminal receptacle.

4. Host

The family Echinostomatidae, has variety of hosts. Developmental stages of echinostomes occur in various species of aquatic snails in numerous snail genera including Physa, Lymnaea, Helisoma, and Bulinus. Many animals serve as definitive hosts for adult echinostomes. Such hosts include fresh water fish, marine fish, reptiles, birds, mammals, and humans (Dietz, 1909, 1910, Mendheim, 1940, 1943, Skrjabin, 1947, 1956, Yamaguti, 1958, 1971).

The sites that echinostomes infect within the final host are diverse. Most species reside in the digestive system. The sites of echinostomes include the oesophagus, duodenum, jejunum, ileum, caecum, cloaca, and bursa Fabricii. The liver, gall bladder and uterus also may serve as sites. Some species even encyst in the musculature of the intestinal wall.

Development to sexual maturity within the final host ranges from 7 to 20 days. Once adults reach maturity they survive for 3 weeks to more than one year depending on the species of echinostome.

5. References

Ando, A. and Ozaki, Y. (1923) Sur aquarte nouvelles especes de Trematodes du genere Echinostoma. Dobuts Zasshi, 29, 1-25.

Baršienė, J. (1993) The Karyotypes of Trematodes. Vilnius "Academia".

Bayssade-Dufour, Ch, Albaret, J. L., Grabda-Kazubska, B., and Chabaud, A. G. (1989) Nomenclature proposeé pour la chétotaxie des cercaires de Plagiorchiata. Annales de Parasitologie Humaine et Comparée, 94, 426-432.

Beaver, P. C. (1937) Experimental studies on Echinostoma revolutum (Froelich) a fluke from birds and mammals. Illinois Biological Monographs, Urbana, Illinois. 15, 1-96.

Beaver, P. C. (1939) The morphology and life history of Petasiger nitidus Linton (Echinostomatidae). Journal of Parasitology, 25, 269-275.

Beaver, P. C. (1943) Studies on Protechinostoma mucronisertulatum n.g. n.n. (Psilostomum reflexae Feldmann, 1941), a trematode (Echinostomatidae) from the sora rail. Journal of Parasitology, 29, 65-70.

Bullock, T. H. and Horridge, G. A. (1965) Structure and Function in the Nervous System, Vol. 1, W. H. Freeman, San Francisco.

Canning, E. (1995) The microsporidian parasites of Plathelminthes: their morphology, development, transmissions and pathogenicity. CIH Miscellaneous publication No 2. London.

Chien, W. Y. and Fried, B. (1992) Cultivation of excysted metacercariae of Echinostoma caproni to ovigerous adults in the allantois of the chick embryo. Journal of Parasitology, 78, 1019-1023.

Dietz, E. (1909) Die Echinostomatiden der Vogel. Zoologische Anzeiger, 34, 180-192.

Dietz, E. (1910) Die Echinostomatiden der Vogel. Zoologische Jahrbucher, 12, 265-512.

Dimitrov, V. (1987) Argentophilic structures of larval stages of some trematodes. Ph. D. Thesis, Bulgarian Academy of Sciences, Sofia.

Dimitrov, V. and Kanev, I. (1984) Argentophilic tstructure of Paryphostomum radiatum (Dujardin, 1845) Dietz, 1909 rediae and cercariae (Trematoda: Echinostomatidae). Khelmintologia, 18, 37-45.

Dimitrov, V., Kanev, I., Busta, J., Nguen Thi Le, and Hd Duy Ngoc. (1985) Argentophilic structures of Echinostoma audyi Lie and Unathevy, 1965 (Trematoda: Echinostomatidae) rediae and cercariae on materials from Czechoslovakia and Vietnam. Khelmintologia, 19, 34-43.

Dimitrov, V., Kanev, I., Fried, B., and Radev, V. (1995) Argentophilic structures of miracidia of Echinostoma trivolvis (Cort, 1914) (Trematoda: Echinostomatidae). Journal of Parasitology, 81 (2), 306-307.

Dimitrov, V., Kanev, I., Fried, B., and Radev, V. (1997) Argentophilic tegumentary papillae of the cercariae of Echinostoma trivolvis (Cort, 1914) (Trematoda: Echinostomatidae). Parasite, 2, 186-194.

Dollfus, R. P. (1951) Miscellanea helminthologica maroccana. I. Quelques trématodes, céstodes et acanthocephales. Archives de l'Institut Pasteur du Maroc, 4, 104-229.

Dönges, G. (1971) The potential number of redial generations in echinostomatids (Trematoda). International Journal for Parasitology, 1, 51-59.

Dönges, J. (1972) Miracidium von Istmiophora melis (Shrank, 1788) (Echinostomatidae). Ökologie und Morphologie. Zeitschrift für Parasitenkunde, 41, 215-230.

Dönges, J. M. and Götzelmann, M. (1975) Echinostome rediae: Effect of pretreatment with cold on transplantation to uninfected snails. Experimental Parasitology, 38, 228-232.

Donovick, R. A. and Fried, B. (1988) Scanning electron microscopy of Echinostoma revolutum and E. liei from domestic chicks. Journal of the Pennsylvania Academy of Sciences, 62, 78-82.

Dragneva, N. and Kanev, I. (1983) Antigen similarity between rediae and cercariae of Echinoparyphium aconiatum Dietz, 1909 (Trematoda: Echinostomatidae) and their intermediate snail host. Khelmintologia, 16, 29-36.

Dzavelidze, G. I. (1957) Cycle of development of a new echinostomatid Echinoparyphium colchicum nov. sp. Tezisi Dokladov Vsesoiuznoi Nauchnoi Konferencii, 105-106,

Evans, N. A. and Gordon, D. M. (1983 a) Experimental observations on the specifity of Echinoparyphium recurvatum toward second intermediate hosts. Zeitschrift für Parasitenkunde, 69, 217-222.

Evans, N. A. and Gordon, D. M. (1983b) Experimental studies on the transmission dynamics of the cercariae of Echinoparyphium recurvatum (Digenea: Echinostomatidae). Parasitology, 87, 167-174.

Freitas, J. F. T. (1948) Echinostomatidae parasito de uretér de ave. Revista Brasileira de Biologia, Rio de Janeiro, 8, 489-492.

Fried, B. and Awatramani, R. (1992) Light and scanning electron microscopical observations of the daughter rediae of Echinostoma trivolvis (Trematoda). Parasitology Research, 78, 257-259.

Fried, B. and King, W. (1989) Attraction of Echinostoma revolutum cercariae to Biomphalaria glabrata dialysate. Journal of Parasitology, 75, 55-57.

Fried, B. and Rosa-Brunet, L. C. (1991) Cultivation of excysted metacercariae of Echinostoma caproni (Trematoda) to ovigerous adult on the chick chorioallantois. Journal of Parasitology, 77, 568-571.

Fried, B., Irwin, S. W. B., and Lowry, S. F. (1990) Scanning electron microscopy of Echinostoma trivolvis and E. caproni (Trematoda) adults from experimental infections in the Golden Hamster. Journal of Natural History, 24, 433-440.

Fried, B., Idris, N., and Oshawa, T. (1995) Experimental infection of juvenile Biomphalaria glabrata with cercariae of Echinostoma trivolvis. Journal of Parasitology, 81, 308-310.

Fried, B., Schmidt, K. A., and Sorensen, R. E. (1997) In vivo and ectopic encystment of Echinostoma revolutum and chemical excystation of the metacercariae. Journal of Parasitology, 83, 251-254.

Fried, B., Eyster, L. S., and Pechenik, J. A. (1998a) Histochemical glycogen and neutral lipid in Echinostoma trivolvis cercariae and effects of exogenous glucose on cercarial longevity. Journal of Helminthology, 72, 83-85.

Fried, B., Frazer, B., and Kanev, I. (1998b) Comparative observations on cercariae and metacercariae of Echinostoma trivolvis and Echynoparyphium sp.. Journal of Parasitology, 84, 623-626.

Gabrashanska, M., Kanev, I. , and Damijanova, A. (1990) Mineral content of four paraitic species from class Trematoda and their freshwater snail hosts. In: Proceedings of the fourth International Symposium on Molybdenum, Vanadium and other trace elements (Jena), 440-446.

Gabrashanka, M., Damyanova, A., and Kanev, I. (1991) Mineral compoition of Echinostoma revolutum (Froelich, 1802) and its snail hosts Lymnaea stagnalis. Khelmintologia, 31, 3-7.

Gergova, S., Kanev, I., Naumova, E., Michailova, A., Panaiotova, M., Nestorov, M., and Gabev, E. (1996) Measurements of DNA and RNA amount in the rediae and cercariae of digenean trematodes Echinostoma revolutum and Pseudechinoparyphium echinatum. In: Proceedings of seventh European Mltiquiioquium of Parasitology, (EMOP - VII), Parma, Italy, Bg2.10, 159.

Ginetsinskaya, T. A. and Dobrovolskii, A. A. (1962) Fauna larval trematodes in freshwater snails from Volga. II. Echinostome cercariae (Echinostomatidae). Trudi Astrakhanskoi Zapovednika, Astrakhan, 6, 45-91.

Gorchilova, L. and Kanev, I. (1983) Studies on the echinostomatids (Trematoda) in Bulgaria. VII. The ultrastructure of the tegument of Echinostoma audyi Lie et Umathevy, 1965. Khelmintologia, 15, 3-11.

Gorchilova, L. and Kanev, I. (1984) Enzymocytochemical characteristics of the tegument and intestinal wall of marites from the genus Echinostoma with 37 collar spines. Khelmintologia, 18, 31-36.

Gorchilova, L. and Kanev, I. (1986) Electron-microscopy studies on the Echinostoma revolutum (v Froelich, 1802), E. echinatum (Zeder, 1803) and Echinoparyphiym aconiatum (Dietz, 1909) (Trematoda: Echinostomatidae) metacercariae. Khelmintologia, 22, 3-8.

Gorchilova, L. and Kanev, I. (1992) Structural and functional characteristics of the tegument and intestinal wall in Hypoderaeum conoideum (Bloch, 1782) (Trematoda: Echinostomatidae). Comptes rendus de l'Academie Bulgare des Sciences, 45, 133-135.

Gorchilova, L. and Kanev, I. (1994) Echinostoma echinatum (Zeder, 1803): Ultrastructure and enzymocytochemical characteristic of the tegument and of the gut wall. Helmintologia, 31, 133-138.

Grabda-Kazubska, B. and Kiseliene, V. (1991) The life cycle of Echinoparyphium mordwilkoi Skrjabin, 1915 (Trematoda: Echinostomatidae). Acta Parasitologica Polonica, 36, 167-173.

Grabda-Kazubska, B. and Laskowski, Z. (1996) On the morphology and chaetotaxy of rediae and cercariae of Isthmiophora melis (Schrank, 1788) (Trematoda, Echinostomatidae). Acta Parasitologica, 41, 7-12.

Grabda-Kazubska, B., Borsuk, P., Laskowski, Z., and Moné, H. (1998) A phylogenetic analysis of trematodes of the genus Echinoparyphium and related genera based on sequencing of internal transcribed spacer region of rDNA. Acta Parasitologica, 43, 116-121.

Grabda-Kazubska, B., Niewiadomska, K., Kanev, I., and Bayssade-Dufour, Ch. (1991) Trematodes nervous system. Annales des Parasitologie Humaine et Comparee, 66, 24-31.

Gulka, G. and Fried, B. (1979) Histochemical and ultrastructural studies on the metacercarial cyst of Echinostoma revolutum (Trematoda). International Journal for Parasitology, 9, 57-59.

Haas, W. (1994) Physiological analyses of host-finding behaviour in trematode cercariae: adaptations for transmission success. Parasitology, 109, 515-521.

Haas, W. and Haberl, B. (1997) Host recognition by trematode miracidia and cercariae, in: Fried, B. and Graczyk, T. K., (eds.), Advances in Trematode Biology, CRC Press, 197-227.

Haas, W., Haberl, B., Kable, M., and Körner, M. (1995a) Snail-host-finding by miracidia and cercariae: chemical host cues. Parasitology Today, 11, 468-472.

Haas, W., Körner, M., Hutterer, E., Wegner, M., and Haberl, B. (1995b) Finding and recognition of the snail intermediate hosts by 3 species of echinostome cercariae. Parasitology, 110, 133-142.

Heyneman, D. (1966) Successful infection with larval echinostomes surgically implanted into the body cavity of the normal snail host. Experimental Parasitology, 18, 220-223.

Holliman, R. B. (1961) Larval trematodes from the Apalachee Bay area, Florida, with a check list of known marine cercariae arranged in a key to their subfamilies. Tulane Studies in Zoology, New Orleans, 9, 1-74.

Hsu, K. C., Lie, K. J., and Basch, P. F. (1968) The life history of Echinostoma rodriguesi sp. n. Journal of Parasitology, 54, 333-339.

Humphries, J. E. and Fried, B. (1996) Histological and histochemical studies on the paraoesophageal glands in cercariae and metacercariae of Echinostoma revolutum and E. trivolvis. Journal of Helminthology, 70, 299-301.

Hyman, L. H. (1951) The Invertebrates: Platyhelminthes and Rhynchocoela - The Acoelomate Bilateria, Vol. II, McGraw-Hill, New York.

Idris, N. and Fried, B. (1996) Development, hatching, and infectivity of Echinostoma caproni (Trematoda) eggs, and histologic and histochemical observations on the miracidia. Parasitology Research, 82, 136-142.

Jeyarasasingham, U., Heyneman, D., Lim, H-K, and Mansour, N. (1972) Life cycle of a new echinostome from Egypt, Echinostoma liei sp. nov. (Trematoda: Echinostomatidae). Parasitology, 65, 203-222.

Johnson, J. C. (1920) The life cycle of Echinostoma revolutum (Froelich). University of California. Publication in Zoology, 19, 335-388.

Jourdane, J. and Kulo, S. D. (1981) Etude experimentale du cycle biologique de Echinostoma togoensis n. sp. parasite a l'etat larvaire de Biomphalaria pfefferi. Annales Parasitologie Humaine et Comparee, 56, 477-488.

Kanev, I. (1980) Studies on the species belonging to echinostomes (Trematoda) in Bulgaria. IV. On the development and ecology of Echinostoma audyi Lie and Umathevy, 1965. Khelmintologia, 9, 39-50.

Kanev, I. (1981a) Ecology of echinostomatids found in Bulgaria. I. On the possibility of a superinvasion of snails with larval stages of one and the same species of genus Echinostoma. Khelmintologia, 11, 42-55.

Kanev, I. (1981b) Studies on the species composition of the echinostomatids (Trematoda) in Bulgaria. VI. Establishment of an echinostome with the features of Echinostoma barbosai Lie et Basch, 1966. Khelmintologia, 12, 36-47.

Kanev, I. (1982a) Comparative studies on the morphology and biology of Echinostoma nudicaudatum Nasir, 1960, Cercaria deficipinatum Khan, 1960 and Echinoparyphium aconiatum Dietz, 1909 (Trematoda: Echinostomatidae). Khelmintologia, 14, 29-43.

Kanev, I. (1982b) Ecology of the echinostomatids (Trematoda) found in Bulgaria. II. Antagonism between the larval stages of two trematode species in the body of one and the same intermediate host. Khelmintologia, 13, 45-52.

Kanev, I. (1982c) Ecology of the echinostomatids (Trematoda) found in Bulgaria. III. Antagonism between Echinostoma audyi and Echinostoma lindoense in the body of Lymnaea stagnalis snails. Khelmintologia, 13, 53-60.

Kanev, I. (1984) Studies on the species composition of family Echinostomatidae (Trematoda) in Bulgaria. VII. On the morphology and biology of Neoacanthoparyphium echinatoides (Fillipi, 1854) Odening, 1962., in ed. I. Vassilev. Faune, taxonomy and ecology of helminths on birds, Publishing House of Bulgarian Academy of Sciences, Sofia, 122-134.

Kanev, I. (1985) On the morphology, biology, ecology and taxonomy of Echinostoma revolutum group (Trematoda: Echinostomatidae: Echinostoma). Dissertation for scientific degree "Doctor of biological sciences". Bulgarian Academy of Sciences, Sofia.

Kanev, I. (1987) Encapsulation and inactivation of Echinostoma revolutum (v. Froelich, 1802) (Trematoda: Echinostomatidae) sporocysts from the tissue and haemolymph of Lymnaea stagnalis snails. Khelmintologia, 24, 26-31.

Kanev, I. (1990) Check list of the parasitic worms of Echinis, Echinostoma, Echinostomatidae with references for its remaining, replacement and reclassify. Publishing House of Bulgarian Academy of Sciences, Sofia.

Kanev, I. (1994) Life cycle, delimitation and redescription of Echinostoma revolutum (Froelich, 1802) (Trematoda: Echinostomatidae). Systematic Parasitology, 28, 125-144.

Kanev, I. and Busta, J. (1992) Variation and abnormality in collar spination of adult and larval echinostomes (Trematoda). Acta Parasitologica Polonica, 37, 47-49.

Kanev, I. and Vassilev, I. (1981) Ecology of echinostomatids (Trematoda) found in Bulgaria. IV. Immobilisation of miracidia from some plant and snail substrates. Biology, 378-391.

Kanev, I., Vassilev, I., Bayssade-Dufour, Ch., Albaret, J. L., and Casone, J. (1987) Chetotaxie cercarienne d'Echinostoma revolutum (Froelich, 1802) et E. echinatum (Zeder, 1803) (Trematoda: Echinostomatidae) Annales des Parasitologie Humaine et Comparee, 62, 222-234.

Kanev, I., Eizenhut, U., Ostrowski de Nunez, M., Manga-Gonzalez, M. Y., Tzolov, D., and Radev, V. (1993) Redescription of the tail and fin folds of Echinostoma revolutum cercariae from its type locality (Trematoda: Echinostomatidae). Annales des Parasitologie Humaine et Comparee, 68, 125-127.

Kanev, I., Eisenhut, U., Ostrowski de Nunez, M., Manga-Gonzalez, M. Y., and Radev, V. (1994a) Penetration and paraoesophageal gland cells in Echinostoma revolutum cercariae from its type locality. Helminthologia, 30, 131-133.

Kanev, I., Kirev, T., Dimitrov, V., and Radev, V. (1994b) Structural and functional abnormalities in the female reproductive system of Hypoderaeum conoideum (Trematoda: Echinostomatidae). Acta Parasitologica Polonica, 39, 103-105.

Kanev, I., McCarthy, A., Radev, V., and Dimitrov, V. (1994c) Dimorphism and abnormality in the male reproductive system of four digenean parasite species (Trematoda). Acta Parasitologica Polonica, 39, 107-109.

Kanev, I., Radev, V., Vassilev, I. Dimitrov, V., and Minchella, D. (1994d) The life cycle of Echinoparyphium cinctum (Rudolphi, 1803) (Trematoda: Echinostomatidae) with re-examination and identification of its allied species from Europe and Asia. Helminthologia (Kosice), 31, 73-82.

Kanev, I., Dimitrov, V., Radev, V., and Fried, B. (1995a) Redescription of Echinostoma jurini (Skvortzov, 1924) (Trematoda: Echinostomatidae) with a discussion of its identity and characteristics. Annales des Naturhistorisches Museum, Wien, 97 B, 37-53.

Kanev, I., Fried, B., Dimitrov, V., and Radev, V. (1995b) Redescription of Echinostoma trivolvis (Cort, 1914) (Trematoda: Echinostomatidae) with a discussion of its identity and characteristics. Systematic Parasitology, 32, 61-70.

Kanev, I., Sorensen, B., Sterner, S., and Fried, B. (1998) The identification and characteristics of Echinoparyphium rubrum (Trematoda: Echinostomatidae) based on experimental evidence of the life cycle. Acta Parasitologica, 43, 181-188.

Karmanova, E. M. and Ilyuschina, T. L. (1969) The life cycle of Echinochasmus coaxatus. Trudy Gel'mintologicheskoi Laboratorii AN SSSR, 20, 60-70.

Kosupko, G. A. (1972) Studies on morphology and biology of Echinostoma revolutum (Froelich, 1802) and E. miyagawai Ishii, 1932 (Trematoda: Echinostomatidae) on experimental material. Ph. D. Thesis, VIGIS, Moscow.

Kowalevski, M. (1897) O przedstswicielach rodzaja Echinostomum (Rud.1809) u kaczki I kury, oraz slow kilka w kwestyi synonimiki. Kosmos. Lwow, 2, 554-565.

Krejci, K. G. and Fried, B. (1994) Light and scanning electron microscopic observations of the eggs, daughter rediae, cercariae and encysted metacercariae of Echinostoma trivolvis and E. caproni. Parasitology Research, 80, 42-47.

Kruse, D. M., Fried, B., and Hosier, D. W. (1992) The expulsion of Echinostoma trivolvis (Trematoda) from ICR mice: scanning electron microscopy of the worms. Parasitology Research, 78, 74-77.

Le Flore, W. and Bass, H. (1983) Observations on morphology and hydrolytic enzime histochemistry of excysted metacercariae of Himasthla rhigedana (Trematoda: Echinostomatidae). International Journal for Parasitology, 13, 179-183.

Lie, K. J. (1963) The life history of Echinostoma malayanum Leiper, 1911. Tropical and Geographical Medicine, 15, 17-24.

Lie, K. J. (1966) Studies on Echinostomatidae (Trematoda) in Malaya. XIV. Body gland cells in cercariae of Echinostoma audyi Lie and Umathevy and E. lindoense Sanground and Bonne. Journal of Parasitology, 52, 1049-1051.

Lie, K. J. (1967) Antagonistic interaction between Schistosoma mansoi sporocysts and echinostoma rediae on the snail Australorbis glabratus. Nature (London), 211, 1213-1215.

Lie, K. J. (1969) Role of immature rediae in antagonism of Paryphostomum segregatum to Schistosoma mansoi and larval development in degenerated sporocysts. Zeitschrift für Parasitenkunde , 32, 316-323.

Lie, K. J. and Basch, P. F. (1967a) The life history of Echinostoma paraensei sp.n. (Trematoda). Journal of Parasitology, 53, 1192-1199.

Lie, K. J. and Basch, P. F. (1967b) The life history of Paryphostomum segregatum Dietz, 1909. Journal of Parasitology, 53, 280-286.

Lie, K. J. and Heyneman, D. (1975) Studies on resistance in sails: A specific tissue reaction of Echinostoma lidoense in Biomphalaria glabrata snails. International Journal for Parasitology, 5, 621-626.

Lie, K. J. and Heyneman, D. (1976a) Studies on resistance in snails. 5. Tissue reactions to Echinostoma lidoense sporocysts in sensitized resensitized Biomphalaria glabrata. Journal of Parasitology, 62, 292-297.

Lie, K. J. and Heyneman, D. (1976b) Studies on resistance in sails. 6. Escape of Echinostoma lindoense sporocysts from encapsulation in the snail heart and subsequent loss of the host's ability to resist infection by the same parasite. Journal of Parasitology, 62, 298-302.

Lie, K. J. and Umathevy, T. (1965) Studies on Echinostomatidae (Trematoda) in Malaya. VIII. The life history of Echinostoma audyi sp.n. Journal of Parasitology, 51, 781-788.

Lie, K. J., Basch, P. F., and Heyneman, D. (1968a) Antagonism between two species of echinostomes (Paryphostomum segregatum and Echinostoma lindoense) in the snail Biomphalaria glabrata. Zeitschrift für Parasitenkunde, 30, 117.

Lie, K. J., Basch, P. F., and Heyneman, D. (1968b) Direct and indirect antagonism between Paryphostomum segregatum and Echinostoma paraensei in the snail Biomphalaria glabrata. Zeitschrift für Parasitenkunde, 31, 101.

Lie, K. J., Basch, P. F., and Hoffman, A. (1967) Antagonosm between Paryphostomum segregatum and Echinostoma barbosai in the snail Biomphalaria straminea. Journal of Parasitology, 53, 1265.

Lie, K. J., Basch, P. F., and Umathevy, T. (1965) Antagonism between two species of larval trematodes in the same snail. Nature (London), 206, 422-425.

Lie, K. J., Heyneman, D., and Jeong, K. H. (1976) Studies on resistance in sails. 7. Evedence of interference with the defence reaction in Biomphalaria glabrata by trematode larvae. Journal of Parasitology., 62, 608-615.

Lie, K. J., Heyneman, D., and Kostanian, N. (1975a) Failure of Echinostoma lindoense to reinfect snails already harboring that species. International Journal for Parasitology, 5, 483-486.

Lie, K. J., Heyneman, D., and Lim, H. K. (1975b) Studies on resistance in sails. Specific resiatance induced by irradiated miracidia of Echinostoma lindoense in Biomphalaria glabrata snails. International Journal for Parasitology, 5, 627-631.

Lie, K. J., Kwo, E. H., and Ow-Yang, C. K. (1970) A field trail to test the possible control of Schistosoma spindale by means of interspecific trematoda antagonism. South Asian Journal of Tropical Medicine and Public Health, 1, 19.

Lie, K. J., Kwo, E. H., and Ow-Yang, C. K. (1972) Further field trail to control Schistosoma spindale by trematode antagonism. South Asian Journal of Tropical Medicine and Public Health, 2, 237-243.

Lie, K. J., Lim, H. K., and Ow-Yang, C. K. (1973) Antagonism between Echinostoma audyi and Echinostoma hystricosum in the snail Lymnaea rubiginosa with a discussion on pattarns of trematoda interaction. South Asian Journal of Tropical Medicine and Public Health, 4, 504-508.

Lühe, M. (1909) Parasitische Plattwürmer. I. Trematoden. Süsswasserfauna Deutschland, Heft 17, 215 pp.

McCarthy, A. (1990) The influence of second intermediate host dispersion pattern upon the transmission of cercariae of Echinoparyphium recurvatum (Digenea: Echinostomatidae). Parasitology, 101, 43-47.

McCarthy, A. and Kanev, I. (1992) Pseudechinoparyphium echinatum (Digenea: Echinostomatidae): Experimental observations on cercarial specificity toward second intermediate hosts. Parasitology, 100, 423-428.

Mendheim, H. (1940) Beitrage zur Systematik and Biologie der Familie Echinostomatidae (Trematoda). Nova Acta Leopoldina, 8, 489-688.

Mendheim, H. (1943) Beitrage zur Systematik und Biologie der Familie Echinostomatidae. Archiv für Naturgeschifte, 12, 175-302.

Mutafova, T. and Kanev, I. (1986) On the Echinostoma revolutum (v. Froelich, 1802) and Echinostoma echinatum (Zeder, 1803) karyotype (Trematoda: Echinostomatidae). Khelmintologia, 22, 37-42.

Mutafova, T., Kanev, I., and Angelova, R. (1986) On the Hypoderaeum conoideum (Bloch, 1782) karyotype (Trematoda: Echinostomatidae). Khelmintologia, 22, 33-36.

Mutafova, T., Kanev, I., and Eisenhut, U. (1991) Karyological studies of Isthmiophora melis (v. Schrank, 1788) (Trematoda: Echinostomatidae) from its type locality. Journal of Helminthology, 65, 255-258.

Nestorov, M., Kanev, I., Fried, B., Nollen, P., Dezfuli, B., and Radev, V. (1996) Surface morphology of genital atrium of eighteen digenean trematode species. In: Proceedings of seventh European Multiquollquium of Parasitology, (EMOP-VII), Parma, Italy, A6.21, 92.

Nollen, P. M. (1983) Patterns of sexual reproduction among parasitic platyhelminths, Parasitology, 86, 99-120.

Odening, K. (1964a) Die Entwicklungzyclen einiger Trematoden-arten des Blesshuhns Fulica atra L. im Raum Berlin. Biologische Rundschau, 2, 129-131.

Odening, K. (1964b) What is Cercaria spinifera La Valette? Some remarks on the species identity and biology of some echinostome cercariae. In: "Parasitic worms and Aquatic conditions". Proceedings of Symposium, Prague, 91-97.

Oshmarin, P. G. and Belous, E. V. (1951) The importance of localization of helminths in the construction of their systematics as a pattern for new echinostomes from the kidney of the eagle. Dokladi AN USSR, n.s., an. 19, 77, 165-168.

Ovtcharenko, D. A. (1955) Eurycephalus dogieli g.n. sp.n. New trematode from vipi. Doklady AN USSR, 104, 157-159.

Pechenik, J. A. and Fried, B. (1995) Effect of temprature on survival and infectivity of Echinostoma trivolvis cercariae: a test of the energy limitation hypothesis. Parasitology, 111, 373-378.

Poljakova-Krusteva, O. and Kanev, I. (1983) Studies on echinostomatids (Trematoda) in Bulgaria. VIII. Scanning electron microscopic study of Echinostoma lindoense Sandground et Bonne, 1940 (Trematoda: Echinostomatidae). Khelmintologia, 15, 63-71.

Reddy, A. and Fried, B. (1996) Egg-laying in vitro of Echinostoma caproni (Trematoda) in nutritive and non-nutritive media. Parasitology Research, 82, 475-476.

Richard, J. (1971) La chétotaxie des cercaires. Valeur systématique et philétique. Mémoires du Museum National d'Histoire Naturelle, Zoologie, 67, 1-179.

Schmidt, K. A. and Fried, B. (1996) Emergence of cercariae of Echinostoma trivolvis from Helisoma trivolvis under different condititions. Journal of Parasitology, 82, 674-676.

Sewell, R. S. (1922) Cercariae indicae. Indian Journal of Medical Research, 10, 370.

Shishov, B. and Kanev, I. (1986) Aminergic elements in the nervous sytem of Echinotomatids and Philophthalmids. Parasitologia, 22, 46-54.

Skrjabin, K.I. (1947) Trematodes of animals and man. Essentials of trematodology. Vol.I Izdatelstvo AN SSSR. Moscow

Skrjabin, K.I. (1956) Trematodes of animal and man. Essentials of trematodology, Vol. 12. Izdatelstvo AN SSR, Moscow

Smoluk, J. L. and Fried, B. (1994) Light microscopic observations of Echinostoma trivolvis (Trematoda) metacercariae during in vitro excystation. Parasitology Research, 80, 435-438.

Stein, P. and Basch, P. (1977) Metacercarial cyst formation in vitro of Echinostoma paraensei. Journal of Parasitology, 63, 1031-1040.

Stoitsova, S. and Kanev, I. (1987) Ultrastructural effects of in vitro trypsin action on two trematode cercariae. In: Proceedings of fifth National Conference of Parasitology, 1-3 X (Varna), 183.

Stunkard, H. W. (1960) Problems of generic and specific determination in digenetic trematodes with special reference to the genus Microphallus Ward, 1901. Library of Dr Eduardo Caballero y Caballero, Instituto de Biologia, Universidad Natural de Mexico, 299-309.

Sultanov, M. A. (1961) New species of trematodes from birds in the Uzbekistan SSR. Dokladi Akademii Nauk Uzbek SSR, 11, 58-62.

Swammerdam, J. (1737/38) Bijbel der Ntuure. Leide, Amsterdam.

Szidat, L. (1939) Beiträge zum Aufbau eines natürlichn Systems der Trematoden. I. Die Entwicklung von Echinocercaria choanophila U. Szidat zu Cathaemasia hians und die Abteilung der Fasciolidae von der Echinostomidae. Zetschrift für Parazitenkunde, 11, 239-283.

Timon-David, J. (1955) Trématodes des Groénlands de l'ile de Rion. Annales des Parasitologie Humaine et Comparee, 30, 446-476.

Toledo, R., Muñoz-Antoli, C., Perez, M., and Esteban, J. G. (1996) Redescription of the adult stage of Hypoderaeum conoideum (Bloch, 1782) (Trematoda: Echinostomatidae) and new record in Spain. Research and Reviews in Parasitology, 56, 195-201.

Ursone, R. L. and Fried, B. (1995a) Light and scanning electron microscopy of Echinostoma caproni (Trematoda) during maturation in ICR mice. Parasitology Research, 81, 45-51.

Ursone, R. L. and Fried, B. (1995b) Light microscopic observations of Echinostoma caproni metacercariae during in vitro excystation. Journal of Helminthology, 69, 253-257.

Wikgren, B. J. (1956) Studies on Finnish larval flukes with a list of known Finnish adult flukes (Trematoda: Malacocotylea). Acta Zoologica Fennca, 91, 1-106.

Yamaguti, S. (1958) Systema Helminthum. Vol.I. The Digenetic Trematodes of Vertebrates, Part I and II. Interscience Publishers, London and New York.

Yamaguti, S. (1971) Synopsis of Digenetic Trematodes of Vertebrates. Keigaku Publishing, Tokyo

Yamaguti, S. (1975) A synoptical review of life cycle histories of digenetic trematodes of vertebrates. Keigaku Publishing, Tokyo

Zdarska, Z. and Nasincova, V. (1985) Histological and histochemical studies of the cercaria of Echinostoma revolutum. Folia Parasitologica, 32, 341-347.

Zdarska, Z., Nasincova, V., and Valkounova, J. (1989) Ultrastructure of the tail of Echinostoma revolutum cercaria. Folia Parasitologica, 36, 239-242.

Zdarska, Z., Nasincova, V., Sterba, J., and Valkounova, J. (1987) Ultrastructure of a new type of sensory ending in Echinostoma revolutum cercaria (Trematoda: Echinostomatidae). Folia Parasitologica, 34, 311-315.

THE SYSTEMATICS OF THE ECHINOSTOMES

A. KOSTADINOVA[1] and D.I. GIBSON[2]

[1] *Department of Biodiversity, Central Laboratory of General Ecology, Bulgarian Academy of Sciences, 2 Gagarin Street, 1113 Sofia, Bulgaria*

[2] *Department of Zoology, The Natural History Museum, Cromwell Road, London, SW7 5BD, UK*

Contents

1. Introduction

"For anyone who thinks all the world's systematic problems are solved or easily solvable, a journey through these discussions in the echinostome literature would be exceedingly educational." (Roberts and Janovy, 2000)

The family Echinostomatidae Looss, 1899 is a rather heterogeneous group of cosmopolitan, hermaphroditic digeneans that parasitize, as adults, numerous vertebrate hosts of all classes. This group exhibits a substantial taxonomic diversity (91 nominal genera are described), which is associated with the broad range of final hosts and a wide geographical distribution. Adult echinostomes are predominantly found in birds, but also parasitize mammals and

31

B. Fried and T.K. Graczyk (eds.),
Echinostomes as Experimental Models for Biological Research, 31–57.
© 2000 *Kluwer Academic Publishers. Printed in the Netherlands.*

occasionally reptiles and fishes. Their typical site of location is the intestine, but there are examples from the bile duct, gall bladder, ureters and urinary tubules of the kidneys, cloaca and bursa Fabricii of birds. The main distinguishing feature of the family Echinostomatidae is the presence of a circumoral head collar armed with one or two ventrally interrupted crowns of large spines. The pattern of the crown armature is an essential diagnostic character at both suprageneric and generic level, and in most cases the size of the spines is used at the species level. Consequently, illustrations of the collar spines, after cutting the entire collar at its base and mounting in glycerine, is desirable in order to study these characters accurately and consistently.

Although a number of echinostome species had been described within the collective genus *Echinostoma* Rudolphi, 1809, as a result of the work of C.A. Rudolphi and others during the 19th and early 20th centuries (e.g. Looss, 1896, 1899; Diesing, 1850; Dujardin, 1845; Creplin, 1829, 1837; Fuhrmann, 1904; etc), it was Dietz (1909, 1910) who first applied comparative morphology in establishing the first classification of the Echinostomatidae. Dietz subdivided this heterogeneous group into 22 natural genera, thus providing a basic pattern for further studies on the group. Odhner (1910) distinguished three subfamilies, the Echinostomatinae Looss, 1899, Himasthlinae Odhner, 1910 and Echinochasminae Odhner, 1910. Later, Travassos (1923) erected the subfamily Chaunocephalinae and Mendheim (1940, 1943) proposed a system comprising nine subfamilies and 27 genera.

Accumulation of additional data, gained from studies on both the comparative morphology of the adults and life-histories, was reflected in further modification to the classification of the group, indicating its complexity and heterogeneity. Thus Skrjabin and Bashkirova (1956) accepted 11 subfamilies and 40 genera, while Yamaguti (1958, 1971) recognized as valid 12 subfamilies and 50 genera. Table 1 presents the more recent and widely used classification systems of the Echinostomatidae, as well as that which we advocate as a result of a recent comparative morphological study based on the examination of type and newly collected material and a critical evaluation of the published data (Kostadinova, in prep.).

The criteria generally utilized at suprageneric levels are: the subdivision of the body into regions; the position of the internal seminal vesicle; the site within the final host; the degree of development, structure and shape of the collar; the arrangement of the collar spines in a double or single row; the presence/absence of a dorsal interruption of the spine crown; and the position of the testes and vitellarium. At the generic level, the following characters traditionally used for identification are generally considered useful: the degree of development and morphology of the collar; the number, shape, arrangement and relative size of the collar spines; the morphology of the male terminal genitalia (position of cirrus-sac, structure of internal seminal vesicle, development of pars prostatica, size and armament of cirrus); the position of the ovary and testes; the location and structure of the vitellarium; the nature of the tegumental armament; and the presence of a uroproct. We also found useful certain ratios, such as body shape, determined by the maximum body width as a proportion of body length (BW%), and the relative length of the forebody (FO%), uterine field (U%) and post-testicular field (T%) as a proportion of body length.

Figure 1 provides an estimate of the species richness of genera within the Echinostomatidae based on our data, compared to the list of species published by Yamaguti (1971). Although the inventory of echinostomes is still far from complete (new additions are expected, especially from South America, Africa and Australia) and the generic composition of the two data-sets differ (see Table 1), a clear pattern emerges. The distribution of taxonomic diversity, expressed as the number of species in individual genera, is very uneven within the Echinostomatidae and is characterized by a large number of genera with low species diversity, one consistently species-rich genus (*Echinostoma*) and a few intermediate genera. In both data-sets the less species-rich genera (consisting of 1-10 species) comprise a similar proportion of all taxa [76% (39 genera, as listed by Yamaguti) and 73% (32 genera in our data-set)], and these include a relatively large number of monotypic genera (14 and 7, respectively) and virtually all monotypic subfamilies.

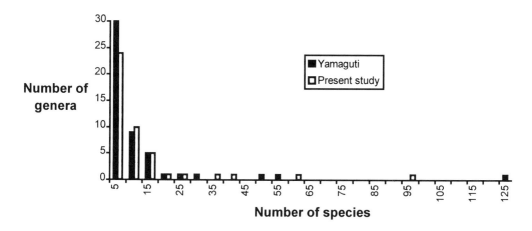

Figure 1. The distribution of species diversity among echinostomatid genera based on our data and Yamaguti (1971).

This picture is congruent with the conclusion of Dial and Marzluff (1989) that the overdominance of one taxon is a common, non-random feature of distributions of taxonomic diversity. Although the taxonomic biases appear to affect the picture, which is more obvious on the left side of the distribution (Fig. 1), few conclusions can be drawn from the common pattern of unequal diversity within the Echinostomatidae. The list of the species-rich genera (consisting of more than 20 species) is almost identical and the number of species within each genus represents a fairly similar proportion of the total number of species in both data-sets. These are *Echinostoma* (24% and 19.6% of species in Yamaguti's and the present data, respectively), *Echinoparyphium* (10.2% and 8.5%), *Echinochasmus* (9.6% and 12.8%), *Stephanoprora* (5.9% and 7.0%) and *Himasthla* (4.1% and 4.9%), the 'dominant' genus being historically the 'oldest'. A common feature of these genera is their cosmopolitan distribution and utilisation of a wide range of hosts (both intermediate and final), which has largely contributed to the observed diversity. Another probable explanation for the observed pattern, however, is that these taxa have not been subjected to comprehensive revisions. In fact the last revision of the Echinostomatidae at the generic and specific levels was almost 50 years ago (Skrjabin and Bashkirova, 1956) and it was based only on data from published descriptions. Table 2 lists the diagnostic morphological characters of the largest genera within the Echinostomatidae; the main characteristics utilized for their determination at suprageneric level are presented schematically in Figure 2.

Of the species-rich genera of the Echinostomatidae discussed above, *Echinostoma* and *Echinoparyphium* have become rather complex and include a large number of morphologically similar species. Thus, more than 120 and 60 nominal species have been described within *Echinostoma* and *Echinoparyphium*, respectively. One could speculate that this pattern represents another digenean example of species flocks, since two large groups of morphologically similar species are present in both *Echinostoma* and *Echinoparyphium* (with 37/45 and 43/45 collar spines, respectively). However, a critical examination of the literature suggests that this may not be the case. Both genera are characterized by a long history of inadequate descriptions and/or poor specific diagnoses, extensive synonymy and other nomenclatural problems. Loss of the type-material, including that of *E. revolutum*, the type-species of *Echinostoma*, its inaccessibility in many cases or simply its non-existence, further complicate the situation; all these obstacles make a comprehensive revision of these two groups a life-time's work. Recently, however, species-groups, recognized in relation to the number of spines on collar and other morphological and life-cycle features, such as the '*revolutum*' group, encompassing species of *Echinostoma* which possess 37 collar spines, and the '*recurvatum*' group within the genus *Echinoparyphium*, have been accepted.

We shall focus on one of the latter groups in order to examine the various approaches to species discrimination, since the species involved are among the most extensively studied echinostomes. An attempt will be made to identify the problems in recent taxonomic treatments which have resulted in the current state of uncertainly rather than present an exhaustive review of all of the studies which have dealt with some aspects of systematics of the group.

2. *Echinostoma*: the systematics of the 37-collar-spined '*revolutum*' species group

2.1. CURRENT STATUS

The systematics of the echinostomes within the '*revolutum*' group has long been problematical due to both the interspecific homogeneity of characters used to differentiate species, which are largely related to adult morphology, and the poor differential diagnoses of newly established taxa. As a result, a large number of species has been described within the group for which no reliable morphological characters enabling species discrimination exist. Beaver (1937) first suggested the hypothesis of polymorphism of the species which he identified as *E. revolutum,* the type-species of *Echinostoma*. He considered nine species synonymous with *E. revolutum* and suggested that the remaining 11 species were so inadequately described or distinguished on the basis of such variable characters that they should be regarded as 'syn. inq.'. Since the publication of Beaver's monograph, 19 other species possessing 37 collar spines have been added to the '*revolutum*' group. Although the practice of describing new species without a critical study of the species already established in the group has persisted, a sufficient body of data has been accumulated from studies on the life-histories of a number of species, and especially on more detailed morphological studies of cercariae, which are useful for discrimination at the specific level.

Table 1. Recent classifications of the Echinostomatidae.

Skrjabin & Bashkirova (1956)	Yamaguti (1971)	Kostadinova (in prep.)
Echinostomatinae Looss, 1899	**Echinostomatinae Looss, 1899**	**Echinostomatinae Looss, 1899**
Echinostoma Rudolphi, 1809	*Echinostoma* Rudolphi, 1809	*Echinostoma* Rudolphi, 1809
Baschkirovitrema Skrjabin, 1944	*Artyfechinostomum* Lane, 1915	*Bashkirovitrema* Skrjabin, 1944
Dietziella Skrjabin & Bashkirova, 1956	*Dietziella* Skrjabin & Bashkirova, 1956	*Drepanocephalus* Dietz, 1909
Drepanocephalus Dietz, 1909	*Drepanocephalus* Dietz, 1909	*Echinodollfusia* Skrjabin & Bashkirova, 1956
Echinodollfusia Skrjabin & Bashkirova, 1956	*Echinoparyphium* Dietz, 1909	*Echinoparyphium* Dietz, 1909
Echinoparyphium Dietz, 1909	*Euparyphium* Dietz, 1909	*Euparyphium* Dietz, 1909
Euparyphium Dietz, 1909	*Hypoderaeum* Dietz, 1909	*Hypoderaeum* Dietz, 1909
Ignavia Teixeira de Freitas, 1948	*Isthmiophora* Lühe, 1909	*Isthmiophora* Lühe, 1909
Longicollia Bykhowskaya-Pavlovskaya, 1953	*Metechinostoma* Petrochenko &	*Longicollia* Bykhowskaya-Pavlovskaya, 1953
Moliniella Hubner, 1939	Khrustaleva, 1963	*Lyperorchis* Travassos, 1921
Nephrostomum Dietz, 1909	*Neoacanthoparyphium* Yamaguti, 1958	*Moliniella* Hubner, 1939
Parallelotestis Belopolskaya, 1954	*Neoechinostoma* Agraval, 1963	*Neoacanthoparyphium* Yamaguti, 1958
Paryphostomum Dietz, 1909	*Nephrostomum* Dietz, 1909	*Pameileenia* Wright & Smithers, 1956
Patagifer Dietz, 1909	*Pameileenia* Wright & Smithers, 1956	*Parallelotestis* Belopolskaya, 1954
Petasiger Dietz, 1909	*Parallelotestis* Belopolskaya, 1954	*Paryphostomum* Dietz, 1909
Prionosoma Dietz, 1909	*Parechinostomum* Dietz, 1909	*Petasiger* Dietz, 1909
Allechinostomatinae Sudarikov, 1950	*Paryphostomum* Dietz, 1909	*Prionosoma* Dietz, 1909
Allechinostomum Odhner, 1910	*Petasiger* Dietz, 1909	*Prionosomoides* Freitas & Dobbin, 1967
Sobolevistoma Sudarikov, 1950	*Prionosoma* Dietz, 1909	*Singhia* Yamaguti, 1958
Stephanoprora Odhner, 1902	*Prionosomoides* Freitas & Dobbin, 1967	**Chaunocephalinae Travassos, 1923**
Chaunocephalinae Travassos, 1923	*Protechinostoma* Beaver, 1943	*Chaunocephalus* Dietz, 1909
Chaunocephalus Dietz, 1909	*Pseudechinochasmus* Verma, 1936	*Balfouria* Leiper, 1908
Balfouria Leiper, 1908	*Pseudechinostomum* Odhner, 1910	**Echinochasminae Odhner, 1910**
Echinochasminae Odhner, 1910	*Skrjabinophora* Bashkirova, 1941	*Echinochasmus* Dietz, 1909
Echinochasmus Dietz, 1909	*Testisacculus* Bhalerao, 1927	*Dissurus* Verma, 1936
Mesorchis Dietz, 1909	**Chaunocephalinae Travassos, 1923**	*Mehrastomum* Saksena, 1959
Saakotrema Skrjabin & Bashkirova, 1956	*Chaunocephalus* Dietz, 1909	*Microparyphium* Dietz, 1909
Velamenophorus Mendheim, 1940	**Echinochasminae Odhner, 1910**	*Pulchrosomoides* Freitas & Lent, 1937
Eurycephalinae Skrjabin & Bashkirova, 1956	*Echinochasmus* Dietz, 1909	*Saakotrema* Skrjabin & Bashkirova, 1956
Eurycephalus Ovcharenko, 1955	*Allechinostomum* Odhner, 1910	*Stephanoprora* Odhner, 1902
Himasthlinae Odhner, 1910	*Beaverostomum* Gupta, 1963	*Uroproctepisthmium* Fischthal & Kuntz, 1976
Himasthla Dietz, 1909	*Episthmium* Lühe, 1909	**Himasthlinae Odhner, 1910**
Acanthoparyphium Dietz, 1909	*Microparyphium* Dietz, 1909	*Himasthla* Dietz, 1909
Aporchis Stossich, 1905	*Patagifer* Dietz, 1909	*Acanthoparyphium* Dietz, 1909
Artyfechinostomum Lane, 1915	*Stephanoprora* Odhner, 1902	*Aporchis* Stossich, 1905
Cloeophora Dietz, 1909	**Himasthlinae Odhner, 1910**	*Artyfechinostomum* Lane, 1915
Pelmatostomum Dietz, 1909	*Himasthla* Dietz, 1909	*Caballerotrema* Prudhoe, 1960
Reptiliotrema Bashkirova, 1941	*Acanthoparyphium* Dietz, 1909	*Cloeophora* Dietz, 1909
Hypoderaeinae Skrjabin & Bashkirova, 1956	*Aporchis* Stossich, 1905	*Curtuteria* Reimer, 1963
Hypoderaeum Dietz, 1909	*Bashkirovitrema* Skrjabin, 1944	**Ignaviinae Yamaguti, 1958**
Multispinotrema Skrjabin & Bashkirova, 1956	*Cloeophora* Dietz, 1909	*Ignavia* Teixeira de Freitas, 1948
Skrjabinophora Bashkirova, 1941	*Curtuteria* Reimer, 1963	**Nephrostominae Mendheim, 1943**
Microparyphiinae Mendheim, 1943	*Dissurus* Verma, 1936	*Nephrostomum* Dietz, 1909
Microparyphium Dietz, 1909	*Echinodollfusia* Skrjabin & Bashkirova,	*Patagifer* Dietz, 1909
Nephroechinostomatinae Oshmarin & Belous,	1956	**Pegosominae Odhner, 1910**
1951	*Himasthloides* Alekseev, 1965	*Pegosomum* Ratz, 1903
Nephroechinostoma Oshmarin & Belous, 1951	*Longicollia* Bykhowskaya-Pavlovskaya,	**Pelmatostominae Yamaguti, 1958**
Pegosomatinae Odhner, 1910	1953	*Pelmatostomum* Dietz, 1909
Pegosomum Ratz, 1903	**Ignaviinae Yamaguti, 1958**	**N. subfam.**
Sodalinae Skrjabin & Schulz, 1937	*Ignavia* Teixeira de Freitas, 1948	*Ruffetrema* Saxena & Singh, 1982
Sodalis Kowalewsky, 1902	**Mehrastominae Saksena, 1959**	**Singhiatreminae Simha, 1962**
	Mehrastomum Saksena, 1959	*Singhiatrema* Simha, 1954
	Pegosominae Odhner, 1910	**Sodalinae Skrjabin & Schulz, 1937**
	Pegosomum Ratz, 1903	*Sodalis* Kowalewsky, 1902
	Pelmatostominae Yamaguti, 1958	
	Pelmatostomum Dietz, 1909	
	Saakotrematinae Yamaguti, 1971	
	Saakotrema Skrjabin & Bashkirova, 1956	
	Singhiatrematinae Simha, 1962	
	Singhiatrema Simha, 1954	
	Singhiinae Yamaguti, 1958	
	Singhia Yamaguti, 1958	
	Caballerotrema Prudhoe, 1960	
	Sodalinae Skrjabin & Schulz, 1937	
	Sodalis Kowalewsky, 1902	

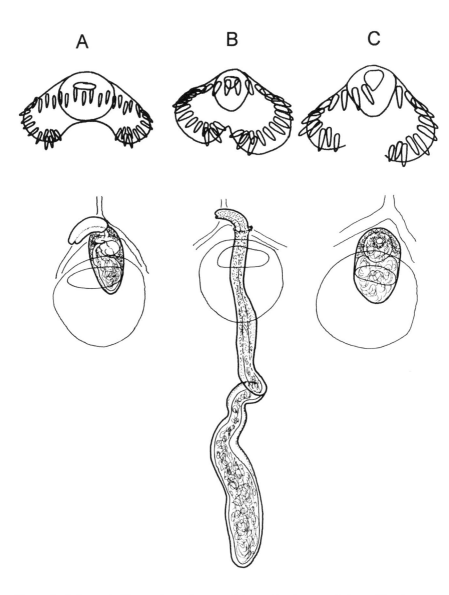

Figure 2. Schematic illustration of some of the main characteristics utilized for the determination of the major subfamilies of the Echinostomatidae. A. Echinostomatinae. B. Himasthlinae. C. Echinochasminae.

Perhaps among the most influential studies were those of Lie and colleagues, who described the life-cycle stages of *E. lindoense* (see Lie, 1964) and *E. audyi* based on material from Malaysia and of two new echinostome species from Brazil, *E. barbosai* and *E. paraensei* (see Lie and Umathevy, 1965; Lie and Basch, 1966, 1967). It is worth noting that these authors were the first to identify almost perfectly, despite the limited technical methods used at that time, the number and position of the tail finfolds in the cercariae of the '*revolutum*' group, and to apply a comparative approach in studying the paraoesophageal gland-cell outlets and chaetotaxy patterns of cercariae in an attempt to distinguish these closely related species (Lie, 1966a,b) (the last two still serve as the only morphological criteria used for species separation).

Our knowledge of the systematics of echinostomes advanced with the more recent publications of Prof. I. Kanev and his co-authors (for references see Kanev, 1994; Kostadinova, 1999; Kostadinova et al., in press (a,b)), whose conclusions were based on examination of life-cycles of echinosomes in experimental conditions. In the conclusions of his thesis, Kanev (1985) considered valid only four species of the '*revolutum*' group: *E. revolutum, E. echinatum, E. trivolvis* and *E. caproni*. Subsequently, he (1994) re-described *E. revolutum* based on material from Europe and also described the life-cycles and cercarial chaetotaxy of the North American species *E. trivolvis* [syn. *E. revolutum* (*sensu* Beaver,1937)] (see Kanev et al., 1995a; Dimitrov et al., 1997) and of another Euro-Asian species, *E. jurini* (see Kanev et al., 1995b; Dimitrov and Kanev, 1992). The synonyms and comparative data on the diagnostic features, as used by Kanev and colleagues, for the species currently recognized within the '*revolutum*' group, are presented in Table 3. We have added to the list three species (*E. parvocirrus, E. miyagawai* and *E. friedi*), which have been distinguished on the basis of experimentally obtained data on their life-cycles (Nassi and Dupouy, 1988; Kostadinova et al., in press (a); Toledo et al., in press), and *E. paraensei,* a Brazilian species which had been considered a synonym of *E. caproni* (see Kanev, 1985), following evidence from isoenzyme and molecular studies (Sloss et al., 1995; Morgan and Blair, 1995, 1998b) that proved its distinct species status.

2.2. SPECIES DIFFERENTIATION: TAXONOMIC APPRAISAL OF OLD AND NEW EVIDENCE

As shown in Table 3, recent data (Morgan and Blair, 1998b; Kostadinova et al., in press (a); Toledo et al., in press) tend to support the opinion that species diversity within the '*revolutum*' group is much higher than suggested by Kanev (1994) for the four species *E. revolutum, E. echinatum, E. caproni* and *E. trivolvis* . In order to test this prediction, we will examine the evidence provided from studies at different levels, organismal to molecular, in relation to characteristics that were specifically identified as useful in the species discrimination of *Echinostoma* spp., as well as other features generally accepted as criteria for species differentiation. The specific question we address is whether convincing evidence exists to support the present conceptions of the taxonomy of the group.

Table 2. Characteristic features of the species-rich genera of the Echinostomatidae.

Feature	Echinostoma	Echinoparyphium	Echinochasmus	Stephanoprora	Himasthla
Body size	Medium to large[1]	Small to medium	Small	Small to large	Medium to large
Body shape	Elongate (BW = 10-20%)	Elongate to elongate-oval (BW = 13-33%)	Elongate-oval (BW = 17-50%)	Elongate (BW = 7-20%)	Notably elongate and slender (BW = 2.5-14%)
Forebody	Very short (FO = 10-20%)	Long (FO = 20-40%)	Long (FO = 25-50%)	Short (FO = 15-26%)	Extremely short (FO = 3-10%)
Head collar:					
Ventral ridge	Present	Present	Absent	Absent	Present
Dorsal interruption	Absent	Absent	Present	Present	Absent
Spines	31-55	29-45	20-24 or 30-34	22 (26 type only)	29, 31 or 34-40 (type only)
Lateral spines	Single row	Double row	Single row	Single row	Single row
Dorsal spines	Double row	Double row	Single row	Single row	Single row
Angle spines	2 × 5	2 × 4	2 × 2-3	2 × 2	2 × 4-5
Testes	In 3rd quarter of body	In mid-hindbody	In mid-hindbody	Equatorial or pre-equatorial	Fairly close to posterior extremity
Cirrus-sac	Elongate-oval	Elongate-oval	Elongate-oval	Elongate-oval	Very long and slender
Internal seminal vesicle	With saccular posterior and tubular anterior portion	Simple or bipartite-saccular, or twisted tubular	Bipartite-saccular	Bipartite-saccular	Tubular, with dilated posterior part
Cirrus	Strong, tubular, unspined	Long, tubular, unspined	Short, unspined	Small, unspined	Long, muscular, spined
Uterus	Long to very long (U = 20-45%)	Very short to short (U = 3-20%)	Very short (U = 0-10%)	Moderately long	Very long (U = 40-80%)
Vitelline fields	Between VS[2] and posterior extremity	Not reaching VS anteriorly	Anterior limits between VS and pharynx	Anterior limits at anterior testis	Anterior limits variable, up to mid-level of cirrus-sac
First intermediate hosts	Freshwater pulmonate and prosobranch gastropods	Freshwater pulmonate and prosobranch gastropods	Freshwater prosobranch gastropods	Prosobranch (occasionally pulmonate) gastropods	Marine prosobranch gastropods
Second intermediate hosts	Freshwater gastropods, bivalves, planarians and tadpoles	Freshwater gastropods, bivalves and tadpoles	Freshwater teleosts, occasionally snails and tadpoles	Freshwater, brackish-water and marine teleosts	Marine bivalves, prosobranch gastropods, annelids
Final hosts	Wide range of birds and mammals	Wide range of birds and mammals	Wide range of fish-eating birds and mammals	Wide range of fish-eating birds, mammals and reptiles	Birds[3], incidentally in fish and mammals
Distribution	Cosmopolitan	Cosmopolitan	Cosmopolitan	Cosmopolitan	Cosmopolitan

[1] Small (<5 mm), medium (5-10 mm); large (> 10 mm).

[2] VS = ventral sucker.

[3] Charadriiformes, Anseriiformes and Ciconiiformes

Table 3. Features of the currently recognised species within the 'revolutum' group.

Species/Character	Synonyms	Geographical distribution	Final hosts	First intermediate hosts	Cercaria: No. of pores of penetration glands	Cercaria: No. of pores of para-oesophageal glands (total)
E. revolutum Frölich, 1802	E. paraulum Dietz, 1909 E. audyi Lie & Umathevy, 1965 E. ivaniosi Mohandas, 1973	Europe, Asia	Birds only	Lymnaea stagnalis, other Lymnaea spp.	4	16-20
E. echinatum (Zeder, 1803)	E. lindoense Sandground & Bonne, 1940 E. barbosai Lie & Basch, 1966 Cercaria spinifera La Valette, 1855	Europe, Asia, South America	Birds and mammals	Lymnaea spp., Planorbarius spp., Planorbis spp., Anisus spp., Gyraulus spp., Biomphalaria spp.	6	60-64
E. jurini (Skvortsov, 1924)	E. bolschewense Nasincova, 1991 E. sisjakowi (Skvortsov, 1935) E. orlovi Romashov, 1966	Europe, Asia	Mammals only	Viviparus contectus, V. viviparus	6	8-10
E. caproni Richard, 1964*	E. liei Jeyarasasingam et al., 1972 E. togoensis Jourdane & Kulo, 1981	Africa, Madagascar	Birds and mammals	Biomphalaria spp, Bulinus spp.	8?**	0
E. trivolvis (Cort, 1914)	E. rodriguesi Hsu et al., 1968* E. armigerum Barker & Irvine in Barker, 1915 E. coalitum Barker & Beaver in Barker, 1915 E. callawayensis Barker & Noll in Barker, 1915 E. multispinosum Vigueras, 1944 Echinoparyphium contiguum Barker & Barston in Barker, 1915	North America	Birds and mammals	Helisoma trivolvis	6	4-6
E. paraensei Lie & Basch, 1967	-	Brazil	Mammals only	Biomphalaria glabrata, Physa gyrina	6-8	0
E. parvocirrus Nassi & Dupouy, 1988	-	Guadeloupe	Birds only	Biomphalaria glabrata	6	24
E. miyagawai Ishii, 1932	-	Europe, Asia	Birds and mammals	Planorbis planorbis, Anisus vortex, Planorbarius corneus, Lymnaea spp.	6	42-46
E. friedi Toledo et al., in press	-	Europe	Birds and mammals	Lymnaea peregra, L. corvus, Gyraulus chinensis	6	46-50

* Sensu Kanev (1985); described from material ex Physa rivalis in Brazil (Hsu et al., 1968).
** Queried because this feature is described for E. liei only.

2.2.1. Morphology

Investigations on morphology far outnumber those on other topics. The following are comments on some of the main criteria which have been utilized in recent years in echinostome systematics.

The paraoesophageal gland-cell outlets in the cercariae have been considered the most reliable feature in distinguishing the species of the '*revolutum*' group (Kanev, 1985, 1994; Kanev et al., 1993, 1995a,b), despite the fact that the consistency of observations after the application of vital staining techniques is rather low, as stressed by Lie (1966a) and confirmed by other authors (Vasilev and Kanev, 1979; Kanev et al., 1993; Vasilev et al., 1982; Kanev and Odening, 1983; Morgan and Blair, 1998b), and this introduces significant artifactual variation. Descriptions of the number and distribution of the para-oesophageal gland-cell outlets are available for all species of the group, and it appears from the existing data (see Table 3) that the species exhibit differences with respect to this characteristic of their cercariae (with the exception of *E. caproni* and *E. paraensei*, which lack paraoesophageal gland-cell outlets). However, difficulty encountered in the observation of paraoesophageal gland-cells and outlets forms a serious barrier to considering this feature reliable for species identification, since inconsistencies and discrepancies appear in the descriptions if more than one sample (isolate) of a given species is studied (Kostadinova et al., in press (a)). Thus variable patterns have been reported for *E. echinatum* (see references in Table 4 and Kostadinova et al., in press (a), for a detailed comparison); however, at least some of this variation appears to be introduced by synonymy. Thus, for example, Kanev (1994) considered *E. revolutum* of Našincová (1986) (described with a total of 30-36 para-oesophageal glands) synonymous with *E. echinatum*, thus extending the range of variation in the number and position of paraoesophageal gland-cells in this species. The position of *E. barbosai* (originally described with 43 paraoesophageal gland-cells) is also unclear, since this species was also placed in synonymy with *E. echinatum* (see Mutafova and Kanev, 1987). These data, as well as recent descriptions of the cercariae of *E. parvocirrus, E. miyagawai* and *E. friedi* (see Table 3), suggest that, if we consider the pattern of cercarial paraoeseophageal gland-cell outlets a distinguishing feature, the species diversity of the group possessing more than 20 gland-cell outlets and referred to as *E. echinatum* (see Kanev, 1994; Kanev et al., 1995a,b) is much greater than previously believed.

Patterns of the distribution of sensilla in cercariae is another larval characteristic which is considered reliable for species discrimination within the '*revolutum'* group (Kanev, 1994; Kanev et al., 1995a,b; Dimitrov et al., 1997). Although these authors have not compared in detail the sensillary distribution patterns in the species they studied (*E. revolutum, E. echinatum, E. trivolvis* and *E. jurini*), they have suggested that 'argentophilic structures in these species differ in number and position' (see e.g. Kanev et al., 1995a,b). Data on cercarial chaetotaxy are available for all of the species of the '*revolutum'* group listed in Table 3, although they are rather unsatisfactory for some species (only drawings of the sensilla are presented: see Lie, 1966b; Lie and Basch, 1966, 1967; Hsu et al., 1968; Kosupko, 1969, 1971a; Nezvalová, 1970), and comparison reveals pattern variation on different scales (see Table 4).

Table 4. Criteria used for species differentiation and evidence used in the evaluation of the present systematic arrangement of the 'revolutum' group.

Characteristics	Evidence	Species	Reference
Morphology			
(i) Cercarial paraoesophageal gland-cell outlets	* Differences between species of *Echinostoma* associated with the number and distribution of para-oesophageal gland-cell outlets	*E. revolutum, E. trivolvis, E. jurini, E. caproni, E. echinatum*	Lie (1966a); Kanev (1994); Kanev et al. (1995a,b)
		E. parvocirrus	Nassi and Dupouy (1988)
		E. miyagawai	Kosupko (1969, 1971a); Kostadinova et al. (in press(a))
		E. paraensei	Lie and Basch (1967)
		E. friedi	Toledo et al. (in press)
	* Variable patterns within species	*E. echinatum*	Vasilev and Kanev (1979); Kanev (1981); Vasilev and Kanev (1981); Vasilev et al. (1982); Kanev (1994); Kanev et al. (1995a,b), see Kostadinova et al. (in press (a))
(ii) Patterns of cercarial chaetotaxy	* Differences between species of *Echinostoma* in relation to the number and distribution of sensilla	*E. revolutum, E. trivolvis, E. jurini, E. caproni, E. echinatum, E. parvocirrus*	Kanev et al. (1987); Dimitrov and Kanev (1992); Dimitrov et al. (1997)
		E. miyagawai	Kosupko (1969, 1971a); Kostadinova (1999)
		E. friedi	Toledo et al. (in press)
		E. revolutum	Dimitrov et al. (1985); Kanev et al. (1987); Nezvalova (1970) - see Kostadinova (1999)
	* Within species pattern variation	*E. echinatum*	Kanev et al. (1987)
		E. jurini	Dimitrov and Kanev (1992)
		E. trivolvis	Fried and Fujino (1987); Dimitrov et al. (1997)
	* Differences in descriptions of species considered synonymous	*E. revolutum – E. audyi*	Lie (1966b); Kanev et al. (1987) - see Kostadinova (1999)
		E. echinatum – E. lindoense	Lie (1966b); Kanev et al. (1987) - see Kostadinova (1999)
	* Almost identical patterns in currently recognised species	*E. caproni – E. liei*	Richard and Brygoo (1978)
		E. caproni, E. jurini	Richard (1971); Richard and Brygoo (1978); Dimitrov and Kanev (1992) - see Kostadinova (1999)

Table 4. Continued.

(iii) Adult morphology	* 'Morphological features of adult worms are especially unstable for species diagnosis'	All 37 collar-spined *Echinostoma* spp.	Kanev (1985)
	* Minute size of medio-dorsal spines is a unique characteristic feature of *E. paraensei*	*E. paraensei*	Lie and Basch (1967)
	* Differences in testes shape and distribution of vitelline follicles	*E. revolutum – E. miyagawai*	Kosupko (1971a, 1972)
	* Dimorphism in testes shape	*E. revolutum*	Kanev et al. (1994)
	* Morphometric separation of species based on the relative size of collar spines	*E. miyagawai – E. revolutum* of Našincová (1986)	Kostadinova et al. (in press (a,b))
	* Statistically significant differences in growth rates between species	*E. paraensei – E. caproni*	Meece and Nollen (1996)
		E. trivolvis – E. revolutum	Humphries et al. (1997); Fried et al. (1997)
		E. miyagawai – E. revolutum	Kosupko (1971a, 1972)
		E. miyagawai – E. revolutum of Našincová (1986)	Kostadinova et al. (in press (b))
	* Different patterns of allometric growth	*E. miyagawai – E. revolutum* of Našincová (1986)	Kostadinova et al. (in press (b))
Life-cycle			
(i) Snail host susceptibility	* Species variability in susceptibility within the genus *Biomphalaria* (exp. system)	*E. caproni*	Jeyarasasingam et al. (1972); Christensen et al. (1980)
	* Strain differences in susceptibility of snails (*Lymnaea stagnalis* – natural populations; *Biomphalaria*, *Helisoma* – exp. systems)	*E. revolutum, E. caproni, E. trivolvis*	Kanev (1987); Jeyarasasingam et al. (1972); Christensen et al. (1980); Huffman and Fried (1990); Jourdane and Kulo (1981); Fried et al. (1987)
	* Within-strain differences in infection rates associated with snail size (*B. glabrata* exp. system)	*E. caproni*	Jeyarasasingam et al. (1972); Christensen et al. (1980); Jourdane and Kulo (1981)
	* Non-susceptibility could be strongly selected for through 3-4 snail generations (*B. glabrata* exp. system)	*E. caproni*	Langand et al. (1998)
(ii) Final host – parasite compatibility	* Species show variable recovery rates in different mammal experimental hosts (mice, hamsters)	*E. caproni, E. trivolvis*	Meece and Nollen (1996)
	* Low worm recovery in chicks irrespective of their age (1-day-old, 21-day-old)	*E. revolutum, E. trivolvis*	Fried et al. (1997); Humphries et al. (1997)
	* Worm recovery and worm expulsion patterns are infection-dose dependent (mice, hamsters)	*E. caproni*	Barus et al. (1974); Christensen et al. (1981); Odaibo et al. (1988); Yao et al. (1991)
	* Different mice strains show different compatibility	*E. caproni*	Kaufman and Fried (1994)

Table 4. Continued.

Behaviour			
(i) Distribution along the intestine	* Species exhibit characteristic location along the intestine	E. revolutum	Kosupko (1971a, 1972); Fried et al. (1997); Sorenesn et al. (1997) Lie and Basch (1967); Nollen (1996a,b)
		E. paraensei	Kosupko (1971a); Kostadinova et al. (in press (a))
		E. miyagawai	Meece and Nollen (1996)
	* Age-related migration of worms along the intestine	E. caproni	Kosupko (1971a,b); Kostadinova et al. (in press (a))
		E. miyagawai	Nollen (1996a, 1997)
	* Overlapping distribution in concurrent infections	E. caproni – E. paraensei E. caproni – E. trivolvis E. revolutum – E. trivolvis	Fried et al. (1997)
(ii) Mating behaviour	* Non-restricted mating pattern with consistently higher self- than cross-insemination rates (single an multiple worm infections)	E. trivolvis, E. paraensei, E. caproni	Nollen (1990, 1993, 1996b)
	* Interspecies mating in concurrent infections	E. caproni – E. trivolvis	Nollen (1997)
Chromosomes	* Significant differences in chromosome morphology	E. revolutum, E. echinatum, E. miyagawai, Echinostoma sp. (ex Physa)	Baršienė and Kiselienė (1991); Baršienė (1993)
	* Substantial variation between isolates from different species of Lymnaea (L. stagnalis, L. auricularia, L. ovata)	E. revolutum	Baršienė and Kiselienė (1991)
Isoenzymes	* Fixed allelic variation between currently recognised species	E. caproni, E. trivolvis, E. paraensei	Kristensen and Fried (1991); Ross et al. (1989); Sloss et al. (1995)
	* Differences in PGM and PGI patterns between laboratory isolates of species considered synonymous	E. caproni group	Voltz et al. (1985, 1986, 1988)
	* Isolates maintained in laboratory show loss of heterozygocity	E. caproni	Voltz et al. (1988)
	* Hybrid breakdown between populations of species considered synonymous	E. caproni group	Trouvé et al. (1998)
DNA	* Low sequence divergence in rDNA ITS regions between currently recognised species	E. revolutum, E. caproni, E. paraensei, E. trivolvis	Morgan and Blair (1995); Sorensen et al. (1998)
	* Ribosomal DNA sequence variation between isolates is similar to interspecific variation within the 'revolutum' group	E. revolutum, E. trivolvis	Sorensen et al. (1998)
	* Intraspecific variations in CO1 and ND1 mitochondrial genes	E. caproni, E. revolutum, E. paraensei	Morgan and Blair (1998a,b)
Geographical distribution	* New 'discoveries'	E. revolutum (North America and Australia)	Sorensen et al. (1997); Morgan and Blair (1998b)
		E. paraensei (Australia)	Morgan and Blair (1998b)

A recent attempt at a comparative analysis of the cercarial sensory arrangements reported in the species of the '*revolutum*' group (Kostadinova, 1999) may provide an explanation for the observed variability. Thus a comparison of the chaetotaxy of *E. revolutum* by Kanev et al. (1987) and two descriptions of *E. audyi* (syn. of *E. revolutum sensu* Kanev, 1985, 1994) (Dimitrov et al., 1985; Lie, 1966b) revealed that the sensory pattern in *E. revolutum* has been incompletely described and that substantial differences exist between the three descriptions in 9 of the 16 recognized groups on the head collar and in all groups along the body (Kostadinova, 1999). An additional problem with respect to the use of the chaetotaxy in species discrimination arises from comparison of the other species within the '*revolutum*' group (*E. echinatum*) for which more than one description of sensory pattern exist. Kanev et al. (1987) described two different patterns of sensillary distribution for *E. echinatum* (as *E. echinatum* and *Echinostoma* sp. (?) *echinatum*) and also recorded variations in the chaetotaxy of this species in cercariae from different molluscan hosts. Although these authors suggested that their description of *E. echinatum* differs from that of *E. lindoense* (syn. of *E. echinatum sensu* Kanev, 1985, 1994) only in the number of the tail sensilla, we have found additional differences in five groups on the head collar and three groups along the body, which indicate that the sensory pattern of *E. echinatum* was incompletely described. These results cast doubt on the synonymy of *E. lindoense* and *E. echinatum* suggested by Kanev (1985, 1994). Surprisingly, examination of the descriptions of *E. jurini* and *E. caproni* suggests that the chaetotaxy patterns in these two species are almost identical, which implies that they are indistinguishable in this respect (i.e. either they are synonymous or chaetotaxy is in this instance an unreliable taxonomic technique).

Toledo et al. (in press), in an analysis of a larger data-set, concluded that a combination of four clusters of sensilla (CIIIV$_1$, CIIIV$_2$, CIVDL and UV$_b$) can be useful for taxonomic studies within the '*revolutum*' group. They have suggested that the pattern (with respect to these clusters) observed in *E. echinatum* by Kanev et al. (1987) has not been described in any other species of *Echinostoma* and that, although the general pattern of *E. revolutum* as presented by Kanev et al. (1987) is consistent with that of the species belonging to the '*revolutum*' group, its identity is dubious. Toledo et al. (in press) compared the published descriptions of cercarial chaetotaxy irrespective of synonymy and revealed five species groupings on the basis of the identity or substantial similarity of the four clusters: (i) *E. lindoense* of Lie (1966b), *E. miyagawai* and *E. paraulum* of Nezvalová (1970); (ii) *E. caproni*, *E. barbosai*, *E. parvocirrus* and *E. liei*; (iii) *E. jurini*, *E. paraensei* and *E. rodriguesi*; (iv) *E. revolutum*, *E. trivolvis*, *E. friedi* and *E. audyi* of Dimitrov et al. (1985); and (v) *E. audyi*. This picture is rather different from the current concept of the group (see Table 3).

Overall, assuming that the procedure for the impregnation of the sensilla produces consistent results and despite of the different nomenclatures used for their interpretation, too much variation of pattern exists within the '*revolutum*' group. On the other hand, 12 of 27 groups of cercarial sensilla exhibit either constancy, intraspecific variability or overlapping ranges among the *Echinostoma* spp. possessing 37 collar spines, which suggests that they may be more suitable for comparative studies at the generic or suprageneric level (Kostadinova, 1999).

Kanev (1985) stated in the conclusions of his thesis on *Echinostoma* spp. from the '*revolutum*' group that 'Morphological features of adult worms are especially unstable for species diagnosis'. In subsequent publications based on this study, he suggested that, since adult *E. revolutum* cannot be identified (distinguished from the sympatric *E. echinatum*) using morphological criteria, 'the echinostomes with 37 collar spines from naturally infected

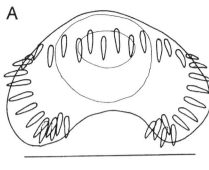

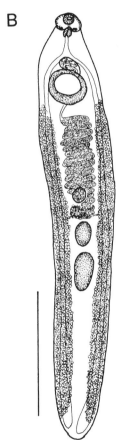

Figure 3. *Echinostoma revolutum* 18 d.p.i. from experimental infection of pigeon. Cercarial source: *Lymnaea peregra*, Bulgaria. A. Collar. B. Entire worm. Scale-bars: A, 500 µm; b, 3 mm.

birds in Europe and Asia should be referred to members of a '*revolutum*' group' (see Kanev, 1994) and that 'Practically, all 37-spined adults of *Echinostoma* spp. in mammals in Europe might be identical with *E. jurini*' (see Kanev et al., 1995b). Although these conclusions were based on studies of a number of *Echinostoma* spp., they appear to overestimate the degree of morphological variability in the absence of experimental support. Kanev (1985, 1994) did not conduct a detailed morphological study of those species which he re-described after completing the life-cycle in experimental conditions, and, because of the preconception that adult morphology is of limited utility for species discrimination, did not present enough accurate information on the variability of adult worms which would enable adequate comparison with subsequent descriptions. In this respect, the recent descriptions of *E. revolutum, E. jurini* and *E. trivolvis* (see Kanev, 1994; Kanev et al., 1995ab) have not improved on those of earlier workers in the field (e.g. Lie et Basch, 1966, 1967; Hsu et al., 1968) in that they add very little to our knowledge of the adult morphometric variability in terms of age and development. In addition, the pooling of data from worms of different age further increases variation and makes comparison difficult.

On the whole, the species of the '*revolutum*' group are characterized by a substantial interspecific homogeneity of the morphological characters of the adult stage which tend to be used for species discrimination in other groups of echinostomes. Only a few attempts have been made to evaluate morphological variability within this group. The first quantitative study (Beaver, 1937), which focused on morphometric characteristics of worms obtained experimentally from a variety of final hosts, has revealed broad intraspecific variation in the adult morphology of a species identified as "*E. revolutum*" (syn. of *E. trivolvis*). Beaver reported substantial morphometric variation related to different types of hosts (bird or mammal), suggested that "*E. revolutum*" is a polymorphic species with a cosmopolitan distribution and placed in synonymy *Echinostoma* spp. which exhibited overlapping ranges of morphometric features. For a long period his conclusions were widely accepted and almost all specimens recovered in natural and experimental infections were referred to as to *E. revolutum*.

Kosupko (1971a, 1972) first described differences in growth rates and variations in some qualitative morphological features which she used to distinguish two Palaearctic species, *E. revolutum* and *E. miyagawai*, in an experimental study on their life-cycles. She clearly stated that adult *E. revolutum* and *E. miyagawai* differ in the shape of the testes (elongate-oval and smooth *vs* spherical and indented) and the position of vitellarium (extra-caecal and non-overlapping in the post-testicular field *vs* confluent). Although Kanev (1985, 1994) placed *E. miyagawai* of Kosupko (1972) in synonymy with *E. echinatum*, he did not consider these features as discriminating between the latter and *E. revolutum*, and subsequently described testicular shape as dimorphic (i.e. 'two testicular morphic forms, smooth and lobed') in *E. revolutum* (Kanev et al., 1994). Our re-examination of Kanev's voucher material of *E. revolutum*, however, revealed that the specimens represent two distinct forms, which differ from each other and from the re-description of *E. revolutum* based upon them (Kanev, 1994) not only in metrical characters but also with respect to testicular shape and the distribution of vitelline follicles (see Kostadinova et al., in press (a), for details). Unfortunately, these data indicate that the high heterogeneity of adult morphology in *E. revolutum*, as suggested by Kanev (1985, 1994) and (Kanev et al., 1994), is probably a result of the examination of composite material.

Recently, we completed the life-cycle of *E. revolutum* in the laboratory in order to assess the morphological variability in the adult stage (Kostadinova et al., unpublished data). Preliminary examination of abundant material suggests a significant homogeneity of the metrical features of adult worms (at different ages, 13, 14 and 18 p.i.) and constancy in the shape of the testes and distribution of the vitellarium, as described for this species by Kosupko (1971a,b, 1972; see also Fig. 3). This homogeneity is what we expected and had already observed in experimentally raised adults of *E. miyagawai* (see Kostadinova et al., in press (a,b)), since the worms originate from a single source and are, therefore, genetically identical. It is likely that morphometric separation of further species is possible if appropriate methods are applied, e.g. it has been shown in comparative studies that species differ in their growth rates in the final host [*E. paraensei vs E. caproni*, *E. trivolvis vs E. revolutum*, *E. miyagawai vs E. revolutum* of Našincova (1986); see Table 4 for references]. We have also revealed significant differences in 22 and 23 (of 35) metrical characters between the corresponding age subsets of the latter two species, which were related to the different allometric growth patterns that also contribute to the different testicular shape. Application of cluster and discriminant analyses showed unambiguous separation of the two samples with respect to both age and species, and identified the variables that yielded a 100% accurate identification (see Kostadinova et al., in press (b), for details). Further analysis of patterns of intraspecific variation based on accurate and well-documented morphological data with reference to the age and host will be required in order to enable a re-evaluation of the species and an understanding the relationships within the '*revolutum*' group.

2.2.2. Life-cycle

Host-specificity has been considered the most important feature discriminating the species within the '*revolutum*' group (Kanev, 1985, 1994; Kanev et al., 1995a,b). Although the speciation model assuming isolation at the stage of the most primitive host in the life-cycle is somewhat uncertain (Bray, 1991), specificity towards the first intermediate host may serve as a characteristic feature at the specific level, since the degree of compatibility between parasite and snail seem to be largely determined by genetic factors (Rollinson and Southgate, 1987). There are two basic requirements for snail susceptibility studies, i.e. the determination of the natural host of the parasite, which is usually a result of a field study, and subsequent recognition of the potential intermediate hosts in laboratory experiments and field studies. Available data on the *Echinostoma* spp./first intermediate host relationships from experimental and/or field studies are limited. Typically, the identification of the first intermediate host (see Table 3) comes from the original description of the species (field collection), from some recent re-descriptions (field and laboratory infections, e.g. *E. revolutum, E. trivolvis, E. miyagawai*) and a few other records for some of the species (e.g. *E. revolutum, E. jurini, E. trivolvis, E. paraensei, E. caproni*). With few exceptions (see e.g. Kosupko, 1972; Jeyarasasingam et al., 1972; Lie and Basch, 1967, 1968; Toledo et al., in press) these records lack quantitative data in support to the parasite/host specificity and present only lists of mollusc species tested and/or positive following infection experiments. As already discussed (see Kostadinova et al., in press (a), for details), the list of the 12 lymnaeid species which serve as the first intermediate hosts of *E. revolutum* presented by Kanev (1994) and the statement that this species 'does not develop in some 100 other prosobranch and pulmonate snails' were not based on experimental results, where only three lymnaeid and two planorbid species were tested. The snail-host data presented in the re-description of *E. revolutum*, as well as the suggestion that *E. echinatum* develops in lymnaeid and planorbid snails (Kanev et al., 1995a,b), appear to be a result of the 'lumping' of published intermediate host-records following the synonymy of these two species suggested by Kanev (1985, 1994).

There are also a series of apparently inexplicable records of *Echinostoma*/molluscan relationships. These include *E. rodriguesi*, whose natural first intermediate host is *Physa rivalis*. Kanev (1985) listed it as a synonym of *E. trivolvis*; the latter, however, develops in *Helisoma trivolvis* only. *P. rivalis* was successfully infected in the laboratory with *E. paraensei*, whose first intermediate host is *B. glabrata*, and Lutz (1924) reported larval stages of three echinostome species possessing 37 collar spines (*E. revolutum, E. erraticum* Lutz, 1924 and *E. nephrocystis* Lutz, 1924) from the same host from Brazil. Although it is believed that there is only one African species of the '*revolutum*' group, *E. caproni*, and that it develops in *Biomphalaria* spp. (see e.g. Panaiotova et al., 1996), Bisseru (1967) infected *Bulinus truncatus, B. tropicus, Planorbarius dufouri* and *B. glabrata* from Zimbabwe with a species he identified as *E. revolutum*, and Wright et al. (1979) studied '*E. revolutum*' from *B. senegalensis* from the Gambia.

The finding of cercariae of the '*revolutum*' group in physid snails in Europe (see Našincova, 1991; Baršiene, 1993) is another unusual case; the latter author suggested that the karyotype of *Echinostoma* from *Physa fontinalis* is rather distinct as compared to the other three *Echinostoma* spp. studied by her (*E. revolutum, E. echinatum, E, miyagawai*). Finally, Morgan and Blair (1998b) identified *E. revolutum* metacercariae from Australia on the basis of mtDNA sequences, but failed to find the first intermediate lymnaeid snail host, despite of extensive sampling, and suggested that *E. revolutum* may complete its life-cycle using a planorbid snail as the first host.

This picture of the parasite/molluscan host relationship within the group is too incomplete to provide criteria for species differentiation, particularly in view of the results of more detailed studies on snail host susceptibility. The laboratory experiments on the *Biomphalaria* spp. – *E. caproni* and *Helisoma trivolvis* – *E. trivolvis* experimental systems (see Table 4) have revealed variability in snail susceptibility at different scales. In view of these species, strain and age-related variations, and the finding that non-susceptibility could be selected for, resulting in large differences among snails from the same strain (Lagrand et al., 1998), we suggest that careful experimental design is required in studying *Echinostoma* spp./first intermediate host compatibility and a quantitative approach needed for data evaluation and publication.

It is much more difficult to interpret the differences between echinostomes of the '*revolutum*' group with respect of their definitive host-range. As shown in Table 3, the specificity to the final host, although considered important for species discrimination (Kanev, 1985, 1994; Kanev et al., 1995a,b), is only roughly defined as the ability to infect avian (*E. revolutum, E. parvocirrus*) or mammalian hosts (*E. jurini* and *E. paraensei*) or both (*E. echinatum, E. caproni, E. trivolvis, E. miyagawai* and *E. friedi*). This division is based only on experimental life-history studies and therefore provides only a limited approximation of the situation in nature. Unfortunately, due to the bias that adult morphology is unsuitable for species identification, no recent attempt has been made to compare the morphology of adult worms from experimental and natural infections. Thus a substantial amount of information from documented records and/or museum material from a wide range of natural definitive hosts of echinostomes is unavailable. As in the case of the *Echinostoma* spp./first intermediate host relationship, with few exceptions, we have failed to reveal quantitative evidence in support of conclusions on the host-range of the adult stage. Recent detailed experimental studies on *Echinostoma* spp./final host compatibility indicate that this relationship is difficult to test experimentally (see Table 4). Thus Fried et al. (1997) discussed the low worm recovery of the 'avian species' *E. revolutum* and *E. trivolvis* in experimental infections of chicks and suggested that the significance of infectivity and worm recovery data in studies on echinostomes in chicks must be viewed with caution. Variable patterns of compatibility have also been found in another species, *E. caproni*, which matures in both mammals and birds, but has been mostly used as a mammalian/echinostome model in experimental studies. These findings underline the need for caution in interpreting the parasite-host relationships from the existing published data.

2.2.3. Behaviour

Distribution along the intestine was suggested by Kosupko (1971a, 1972) as a species-specific feature in *E. revolutum* and *E. miyagawai*. She documented age-related migration along the intestine in *E. miyagawai* and demonstrated on experimental material that the two species differed in the permanent location of adult worms in the host intestine (rectum and caeca in *E. revolutum* vs ileum in *E. miyagawai*). We have confirmed the distributional pattern of *E. miyagawai* in more detail (Kostadinova et. al., in press (a)), and, as shown in Table 4, the characteristic niche in the intestine of the final host was considered stable in recent studies on *E. revolutum* and in two other species, *E. caproni* and *E. paraensei*. The observed patterns of overlapping distribution in concurrent infections, however, suggest that interspecific mating is possible, although probably rare, as revealed in the laboratory for *E. caproni* and *E. trivolvis* by Nollen (1997). The experimental studies on the mating behaviour of echinostomes (see Table 4 for references) have shown consistently higher self- than cross-insemination rates (e.g. 58% vs 14% in *E. trivolvis*, respectively; 75% vs 21% in *E. paraensei*); this may have a significant effect on the population structure and species differentiation of echinostomes.

2.2.4. Genetic characteristics (karyology, isoenzymes and molecular studies)

Although chromosome morphology has only been briefly applied in attempts to discriminate species of the '*revolutum*' group, the significant differences found between three species (*E. revolutum*, *E. miyagawai* and *E. echinatum*) by Baršienė and Kiselienė (1991) tend to support their distinct status. However, the considerable intraspecific differences of *E. revolutum* associated with the first intermediate host (*Lymnaea* spp.) suggest that variation at this level might be a common feature which has not been noted because of the lack of data.

Isoenzyme studies, although largely focused on the '*E. caproni*' group, have provided evidence for the distinct species status of *E. caproni*, *E. trivolvis* and *E. paraensei*, but also detected intraspecific variability within *E. caproni* (see Table 4). These data, coupled with the observation of pre-mating reproductive isolation and low hybrid fecundity between two geographically distant strains of *E. caproni,* suggest a considerable degree of genetic differentiation among the African echinostomes currently treated as strains of *E. caproni* (Trouvé et al., 1996, 1998).

DNA sequence analysis provides tools for the quantitative assessment of genetic heterogeneity and it is widely accepted that ribosomal DNA has proved useful in species discrimination (see Chapter 13). Morgan and Blair (1995) distinguished five *Echinostoma* spp. using ITS sequence data. They observed, however, rather low inter-specific variability among rDNA sequences within the '*revolutum*' group, and Sorensen et al. (1998) have shown rDNA sequence variation between isolates of *E. revolutum* and *E. trivolvis* comparable to that observed between different species, which questions the use of ITS regions for distinguishing between species of this genus. Additional examination of mtDNA CO1 and ND1 sequences revealed that sequence information obtained from the more rapidly evolving mitochondrial genome (the ND1 gene in particular) is useful for studying the relationships within the '*revolutum*' group (Morgan and Blair, 1998a,b), although intraspecific variation was observed.

2.2.5. Geographical distribution

Kanev (1985) suggested that the four species considered valid in his revision of the '*revolutum*' group exhibit generally non-overlapping distribution patterns on a continental scale, with only two sympatric combinations (*E. revolutum* and *E. echinatum* in Europe and Asia and *E. echinatum* (recorded as its synonym *E. lindoense* in Brazil by Lie (1968)), *E. trivolvis* (syn. *E. rodriguesi*) and *E. caproni* (syn. *E. paraensei*) in South America. Recent studies have confirmed the existence of three additional species in Europe, *E. jurini* (?syn. *E. bolschevense*), *E. miyagawai* and *E. friedi* (see Kanev et al., 1995b; Našincová, 1991; Kostadinova et al., in press (a); Toledo et al., in press). Recently, *E. revolutum* has been recorded in North America (Sorensen et al., 1997, 1998), the distribution of *E. trivolvis* was considered restricted to North America (Kanev et al., 1995a) and the validity of the Brazilian species *E. paraensei* was restored (see above). *E. revolutum* and *E. paraensei* were recorded in Australia based on evidence from molecular studies, and one as yet unidentified species, closely related to *E. revolutum,* was found in New Zealand (Morgan and Blair, 1998b).

Unfortunately, it is difficult to comment upon the zoogeographical distribution patterns of the species belonging to the '*revolutum*' group from the limited number of isolated records which are based on completed life-cycles. It appears, however, that both morphological and molecular evidence indicate that the number and distribution of the *Echinostoma* spp. with 37 collar spines deviate from the initial concept (Kanev, 1985) upon which species discrimination was based.

A return to the 'old' literature may be helpful in evaluating new 'discoveries' and may assist in formulating testable questions on the diversity and distribution of echinostomes. Thus with respect to the Australian species we found that Lie and Basch (1967) referred to a close resemblance between the cercaria of *E. paraensei* and an unnamed cercaria described by Johnston and Muirhead (1949) from Australia. As shown in Table 4, this species possesses a unique (according to the original description) morphological feature, i.e. considerable difference in the size of collar spines, which should make the subsequent identification of the Australian material rather straightforward.

Johnston and Angel (1941) described, from the planorbid snails *Amerianna* spp., a cercaria of *E. revolutum* which possesses 39 or 41 paraoesophageal gland-cells (estimated from their figure 1.3). This identification was not considered valid by Kanev (1994). As shown in Table 3, this material is closer to *E. miyagawai* in relation to the paraoesophageal glands. Our examination of the voucher material of *E. revolutum* from the study of Morgan and Blair (PMeta-2, Queensland Museum No. G213944, see Morgan and Blair, 1998b) has revealed that the morphology of the adult form agrees better with the original description of *E. robustum* Yamaguti, 1935 than with either *E. revolutum* or *E. miyagawai* (after Kanev, 1994 and Kostadinova et al. in press (b)). The Australian isolate may, therefore, prove an appropriate example to study the extent of morphological/genetic differentiation among *Echinostoma* spp. A study of the morphology of the cercarial stage of this isolate could greatly clarify the situation. We would suggest, however, a more extensive sampling of all probable molluscan hosts (i.e. not restricted to lymnaeid snails only) in this case.

The study by Lutz (1924) indicates that the diversity of Brazilian species of the '*revolutum*' group is probably higher than currently accepted. He experimentally obtained adults, which he identified as *E. revolutum* and *E. mendax* Dietz, 1909, and described *E. nephrocystis* Lutz, 1924 (all from *Physa* sp.) as well as *E. erraticum* Lutz, 1924 (based on experimental material from *Spirulina* sp.) and two other species, *E. microrchis* Lutz, 1924 and *E. neglectum* Lutz, 1924 (based on adults from naturally infected birds). Although the detailed description of his material in the re-examination by Kohn and Fernandes (1975) indicates that the material of *E. nephrocystis*, *E. microrchis* and *E. neglectum* represents a collection of more than one species, the forms possessing 37 collar spines and identified as *E. erraticum*, *E. nephrocystis*, *E. microrchis* and *E. neglectum* exhibit significant morphological differences.

The composition of the Euro-Asian species of the '*revolutum*' group is far from certain. Although we have included *E. echinatum* in the list of currently recognized species (see Table 3), we would like to suggest that it cannot be considered valid. This species has not been justified in a taxonomic publication, and our recent re-examination of the voucher specimens of *E. echinatum* from the experimental studies of Kanev and colleagues (Kanev, 1985; 1994; Kanev et al., 1987) has shown that a number of specimens were wrongly identified and include *Echinostoma sarcinum* Dietz, 1909 (possessing 47 spines) as well as members of different genera (*Hypoderaeum* and *Echinoparyphium*) (see Kostadinova, 1995 and Kostadinova et al., in press (a) for a description of this material).

2.2.6. Conclusions

We have attempted to examine as many of the criteria which might be useful for species differentiation as the existing data permit. This review of the literature suggests to us that:

(i) There is variation between, but also within, recognized species with respect to larval morphological characteristics, chromosomes, isoenzymes and DNA sequences;

(ii) A more detailed study of the morphological characteristics of the adult stage may still provide useful information for species discrimination;

(iii) Whenever studied in detail (mainly in *E. caproni*), variations in parasite/host compatibility suggest that this relationship is difficult to interpret, hence it is desirable that conclusions on host specificity be based on carefully designed experiments and/or extensive field collections;

(iv) Overall, detected variation is rarely quantified.

3. Suggestions for an integrated approach for investigating the taxonomic diversity in echinostomes

Application of an appropriate species concept is of vital importance in evaluating the taxonomic diversity and patterns of speciation in parasitic worms; however, the biological species concept has many limitations for parasites (see Lymbery, 1992; Lymbery and Thompson, 1996; Meeus et al., 1998; and references therein). Problems with the biological species concept with respect to *Echinostoma* are related to the dispersal patterns of the species on two levels, that of local populations and across broader geographical regions.

• Dispersal at a local scale (or microspacial dispersal): the clonal multiplication of the infective stages within the first intermediate host followed by aggregated (clumped) distribution of metacercariae within the population of the second intermediate host results in an aggregated distribution of similar genotypes within the final host. The probable outcome of this characteristic multiplication/dispersion pattern is a restricted cross-fertilisation due to the mating of siblings. This, coupled with the relatively high rate of self-insemination provides prerequisites for clonal isolation, leading to a high rate of populational differentiation and the possibility of sympatric speciation.

• Dispersal on a wider scale (or macrospacial dispersal): the final hosts are highly vagile and wide-ranging, and hence tend to reduce clonal isloation by extensive gene flow between populations, making allopatric speciation more likely.

With the exception of Beaver (1937) no appropriate hypothesis on the species boundaries within the 'revolutum' group has been formulated. He (rightly) used a single species as a null hypothesis, and tested it by means of examining a large set of morphological data obtained experimentally. The fact that Kanev et al. (1995a) considered the species studied by Beaver synonymous with *E. trivolvis* does not, however, invalidate his conclusions. The species boundaries within the 'revolutum' group, as a result of allopatric speciation (assumed by Kanev, 1985, 1994), are not consistently supported by the data on various aspects of echinostome morphology and biology, nor by recent geographical records based on sequence data (see Table 4). However, a greater diversity of species, presumably resulting from sympatric speciation, is not unequivocally supported by molecular data, since molecular methods do not allow unambiguous differentiation of species. Nevertheless, the pioneer molecular studies have been based on laboratory strains and thus natural genetic variation has not been encountered (Rollinson and Southgate, 1987). Recent findings of Morgan and Blair (1998b) and Sorensen et al. (1998) have shown higher variation in natural populations.

Although much progress has been made recently in gaining information on various aspects of echinostome biology and systematics, it seems that further advancement in our understanding of the phylogeny and relationships, hence the likely mode of speciation of the 'revolutum' group, would result from an integrated approach to the problem of species diversity within this group. The finding of more sympatric species combinations than expected underlines the need for more detailed analyses at multiple levels (e.g. morphological and molecular differentiation, transmission patterns, etc.) and different scales (among species and among populations). It is apparent that studies are needed at the level of local populations, designed to identify whether the degree of intraspecific variation exhibited within presently recognized species of *Echinostoma* is comparable to that between species.

While our outlook at the systematics of the 'revolutum' group might appear somewhat gloomy, we believe that this group provides useful a model for studying patterns of digenean speciation by gaining answers to questions such as: What is the real range of morphological/genetic variability of echinostome populations under natural conditions? What is the degree of morphological/genetic differentiation of the echinostome species in sympatric and allopatric situations?

In conclusion, we would like to suggest the following avenues of research, which we believe would advance our understanding of the taxonomic boundaries within the 'revolutum' group:

• *A comparative approach*: Surprisingly, we do still need a comparable series of studies on echinostome morphology. We lack knowledge of the variability of the morphological characteristics of both larval and adult stages of echinostomes, which are needed in order to assess their value in taxonomy.

• *Re-evaluation of existing taxa*: Many (if not all) species currently recognized within the 'revolutum' group need to be re-examined with respect to accurate and comparable documentation of intraspecific morphological variation.

• *Integrated approach*: Molecular analysis can greatly advance our understanding of the phylogenetic relationships of the species within the group, especially in conjunction with detailed phenotypical characterization of the studied isolates.

• *Diversity*: Studies on the genetic and morphological diversity of natural echinostome populations.

References

Baršienė, J. (1993) *The karyotypes of trematodes*, Vil'nyus, Academia. (In Russian).

Baršienė, J. and Kiselienė, V. (1991) Karyological studies of trematodes within the genus *Echinostoma*, *Acta Parasitologica Polonica* **36**, 23-29.

Barus, V., Moravec, F. and Rysavy, B. (1974) Antagonistic interaction between *Echinostoma revolutum* and *Echinoparyphium recurvatum* (Trematoda) in the definitive host, *Folia Parasitologia* **21**, 155-159.

Beaver, P.C. (1937) Experimental studies on *Echinostoma revolutum* (Frölich) a fluke from birds and mammals, *Illinois Biological Monographs* **15**, 96 pp.

Bisseru, B. (1967) Stages in the development of larval echinostomes recovered from schistosome transmitting molluscs in Central Africa, *Journal of Helminthology* **41**, 89-108.

Bray, R.A. (1991) Species and speciation in parasitic helminths. *Second International School. Parasite Host Environment,* Publishing House of the Bulgarian Academy of Sciences, Sofia, pp. 174-184.

Christensen, N.O., Frandsen, E. and Roushdy, M.Z. (1980) The influence of environmental conditions and parasite-intermediate host-related factors on the transmission of *Echinostoma liei*, *Zeitschrift für Parasitenkunde* **64**, 47-63.

Christensen, N.O., Nydal, R., Frandsen, F. and Nasen, P. (1981) Homologous immunotolerance and decreased resistance to *Schistosoma mansoni* in *Echinostoma revolutum*-infected mice, *Journal of Parasitology* **67**, 164-166.

Creplin, F.C.H. (1829) *Novae observationes de entozois*, Berlin.

Creplin, F.C.H. (1837) *Distoma, Allgemeine Encyclopädie der Wissenschaften und Künste* (Ersch & Grüber) **29**, 309-329.

Dial, K.P. and Marzluff, J.M. (1989) Nonrandom diversification within taxonomic assemblages, *Systematic Zoology* **38**, 26-37.

Diesing, C.M. (1850) *Systema helminthum*. Vol. 1. Vienna, Wilhelmum Braumüller.

Dietz, E. (1909) Die Echinostomiden der Vogel, *Zoologisch Anzeiger* **34**, 180-192.

Dietz, E. (1910) Die Echinostomiden der Vogel, *Zoologisch Jahrbücher*, Suppl. **12**, 265-512.

Dimitrov, V. and Kanev, I. (1992) Chaetotaxy of the cercariae of *Echinostoma jurini* (Skvortsov, 1924) (Trematoda: Echinostomatidae), *Khelmintologiya* **32**, 11-18.

Dimitrov, V., Kanev, I., Fried, B. and Radev, V. (1997) Sensillae [sic] of the cercariae of *Echinostoma trivolvis* (Cort, 1914) (Trematoda: Echinostomatidae), *Parasite* **2**, 153-158.

Dimitrov, V., Kanev, I., Busta, J., Le, N.T. and Ngo, H.Z. (1985) Argentophilic structures of *Echinostoma audyi* Lie et Umathevy, 1965 (Trematoda: Echinostomatidae) rediae and cercariae on materials from Czechoslovakia and Vietnam, *Khelmintologiya* **19**, 34-43 (In Bulgarian).

Dujardin, F. (1845) *Histoire naturelle des helminthes ou vers intestinaux*, Librairie Encyclopédique de Roret, Paris.

Fried, B. and Huffman, J.E. (1996) The biology of the intestinal trematode *Echinostoma caproni*, *Advances in Parasitology* **38**, 312-368.

Fried, B. and Fujino, T. (1987) Argentophilic and scanning electron microscopic observations of the tegumentary papillae of *Echinostoma revolutum* (Trematoda) cercariae, *Journal of Parasitology* **73**, 1169-1174.

Fried, B., Scheuermann, S. and Moore, J. (1987) Infectivity of *Echinostoma revolutum* miracidia for laboratory-raised pulmonate snails, *Journal of Parasitology* **73**, 1047-1048.

Fried, B., Mueller, T.J. and Frazer, B.A. (1997) Observations on *Echinostoma revolutum* and *Echinostoma trivolvis* in single and concurrent infections in domestic chicks, *International Journal for Parasitology* 27, 1319-1322.

Fuhrmann, O. (1904) Neue Trematoden, *Zentralbatt für Bakteriologie, Parasitenkunde, Infektionskrankheiten (und Hygiene)*, *Abt. Original* **37**, 58-64.

Huffman, J.E. and Fried, B. (1990) *Echinostoma* and echinostomiasis, *Advances in Parasitology* **29**, 215-269.

Humphries, J.E., Reddy, A. and Fried, B. (1997) Infectivity and growth of *Echinostosma revolutum* (Frölich, 1802) in the domestic chick, *International Journal for Parasitology* **27**, 129-130.

Hsu, K.C., Lie, K.J. and Basch, P.F. (1968) The life history of *Echinostoma rodriguesi* sp. n. (Trematoda: Echinostomatidae). *Journal of Parasitology* **54**, 333-339.

Jeyarasasingam, U., Heyneman, D, Lim, H.K. and Mansour, H. (1972) Life cycle of a new echinostome from Egypt, *Echinostoma liei* sp. nov. (Trematoda: Echinostomatidae), *Parasitology* **65**, 203-222.

Johnston, T.H. and Muirhead, N.C. (1949) Larval trematodes from Australian freshwater molluscs. Part XIV, *Transactions of the Royal Society of South Australia* **73**, 102-108.

Johnston, T.H. and Angel, L.M. (1941) The life history of *Echinostoma revolutum* in South Australia, *Transactions of the Royal Society of South Australia* **65**, 317-322.

Jourdane, J. and Kulo, S.D. (1981) Étude experimentale du cycle biologique de *Echinostoma togoensis* n. sp., parasite à l'état larvaire de *Biomphalaria pfeifferi* au Togo, *Annales de Parasitologie Humaine et Comparée* **56**, 477-488.

Kanev, I. (1981) Studies on the species composition of the echinostomatids (Trematoda) in Bulgaria. VI. Establishment of an echinostome with the features of *Echinostoma barbosai* Lie et Basch, 1966, *Khelmintologiya* **12**, 36-47 (In Bulgarian).

Kanev, I. (1985) [*On the morphology, biology, ecology and taxonomy of* Echinostoma revolutum *group (Trematoda: Echinostomatidae:* Echinostoma)]. Thesis, Bulgarian Academy of Sciences, Sofia. (In Bulgarian).

Kanev, I. (1987) Encapsulation and inactivation of *Echinostoma revolutum* (Frölich, 1802) sporocysts (Trematoda: Echinostomatidae) by tissues and hemolymph of snails *Lymnaea stagnalis* (L.) (Mollusca: Gastropoda). *Khelmintologiya* **24**, 26-31 [In Bulgarian].

Kanev, I. (1994) Life-cycle, delimitation and redescription of *Echinostoma revolutum* (Frölich, 1802) (Trematoda: Echinostomatidae), *Systematic Parasitology* **28**, 125-144.

Kanev, I. and Odening, K. (1983) Further studies on *Cercaria spinifera* La Valette, 1855 in Central Europe. *Khelmintologiya* **15**, 24-34.

Kanev, I., Vassilev, I., Bayssade-Dufour, C., Albaret, J.L. and Cassone, J. (1987) Chétotaxie cercarienne d'*Echinostoma revolutum* (Frölich, 1802) et *E. echinatum* (Zeder, 1803) (Trematoda, Echinostomatidae), *Annales de Parasitologie Humaine et Comparée* **62**, 222-234.

Kanev, I., Eisenhut, U., Ostrowski de Nunez, M., Manga-Gonzalez, M.Y. and Radev, V. (1993) Penetration and paraoesophageal gland cells in *Echinostoma revolutum* cercariae from its type locality, *Helminthologia* **30**, 131-133.

Kanev, I., McCarthy, A., Radev, V. and Dimitrov, V. (1994) Dimorphism and abnormality in the male reproductive system of four digenean species (Trematoda), *Acta Parasitologica* **39**, 107-109.

Kanev, I., Dimitrov, V., Radev, V. and Fried, B. (1995a) Redescription of *Echinostoma trivolvis* (Cort, 1914) with a discussion of its identity, *Systematic Parasitology* **32**, 61-70.

Kanev, I., Fried, B., Dimitrov, V. and Radev, V. (1995b) Redescription of *Echinostoma jurini* (Skrjabin, 1924) with a discussion of its identity and characteristics, *Annalen des Naturhistorischen Museums in Wien* **97B**, 37-53.

Kaufman, A.R. and Fried, B. (1994) Infectivity, growth, distribution and fecundity of a six versus twenty-five metacercarial inoculum of *Echinostoma caproni* in ICR mice, *Journal of Helminthology* **68**, 203-206.

Kiselienė, V. K. (1983) The echinostomatids of the muskrat in Lithuania and some aspects of their development, *Acta Parasitologica Lituanica* **20**, 77-88 [In Russian].

Kohn, A. and Fernandes, B.M.M. (1975) Sobre as especies do genero *Echinostoma* Rudolphi, 1809 decritas por Adolpho Lutz em 1924, *Memorias do Instituto Oswaldo Cruz* **73**, 77-89.

Kostadinova, A. (1995) *Echinostoma echinatum* (Zeder, 1803) *sensu* Kanev (Digenea: Echinostomatidae): a note of caution, *Systematic Parasitology* **32**, 23-26.

Kostadinova, A. (1999) Cercarial chaetotaxy of *Echinostoma miyagawai* Ishii, 1932 (Digenea: Echinostomatidae), with a review of the sensory patterns in the '*revolutum*' group, *Systematic Parasitology*, in press.

Kostadinova, A. (in preparation) Family Echinostomatidae. In: Jones, A., Gibson, D.I. & Bray, R.A. (Eds) *Keys to the trematode parasites of vertebrates*. CABI, Wallingford.

Kostadinova, A., Gibson, D.I., Biserkov, V. and Chipev, N. (in press, a) Re-validation of *Echinostoma miyagawai* Ishii, 1932 (Digenea: Echinostomatidae) on the basis of the experimental completion of its life-cycle, *Systematic Parasitology*.

Kostadinova, A., Gibson, D.I., Biserkov, V. and Ivanova, R. (in press, b) A quantitative approach to the evaluation of the morphological variability of two echinostomes, *Echinostoma miyagawai* Ishii, 1932 and *E. revolutum* (Frölich, 1802), from Europe, *Systematic Parasitology*.

Kosupko, G.A. (1969) The morphological peculiarities of *Echinostoma revolutum* and *E. miyagawai* cercariae, *Trudy Vsesoyuznogo Instituta Gel'mintologii im K.I. Skrjabina* **15**, 159-165 (In Russian).

Kosupko, G.A. (1971a) New data in the bioecology and morphology of *Echinostoma revolutum* and *E. miyagawai* (Trematoda: Echinostomatidae), *Byulleten' Vsesoyuznogo Instituta Gel'mintologii im. K.I. Skrjabina* **5**, 43-49 (In Russian).

Kosupko, G.A. (1971b) [Criteria of the species *Echinostoma revolutum*, demonstrated on experimental material.] In: *Sbornik rabot po gel'mintologii posvyashchen 90-letiyu so dnya rozhdeniya akademika K.I. Skrjabina*. 'Kolos', Moscow, pp. 167-175 (In Russian).

Kosupko, G.A. (1972) [*Morphology and biology of* Echinostoma revolutum *Frölich, 1802 and* Echinostoma miyagawai *Ishii, 1932 (Trematoda: Echinostomatidae) studied on experimental material.*] Thesis, Moscow. (In Russian).

Kristensen, A.R. and Fried, B. (1991) A comparison of *Echinostoma caproni* and *Echinostoma trivolvis* (Trematoda: Echinostomatidae) adults using isoelectrofocusing, *Journal of Parasitology* **77**, 496-498.

Langand, J. Jourdane, J. Coustau, C. Delay, B. and Morand, S. (1998) Cost of resistance, expressed as a delayed maturity, detected in the host-parasite system *Biomphalaria glabrata/Echinostoma caproni*, *Heredity* **80**, 320-325.

Lie, K.J. (1964) Studies on Echinostomatidae (Trematoda) in Malaya. VII. The life history of *Echinostoma lindoense* Sandground and Bonne, 1940, *Tropical and Geographical Medicine*. *Amsterdam* **16**, 72-81.

Lie, K.J. (1966a) Studies on Echinostomatidae (Trematoda) in Malaya. XIV. Body gland cells in cercariae of *Echinostoma audyi* Lie & Umathevy, and *E. lindoense* Sandground and Bonne, *Journal of Parasitology* **52**, 1049-1051.

Lie, K.J. (1966b) Studies on Echinostomatidae (Trematoda) in Malaya. XIII. Integumentary papillae on six species of echinostome cercariae, *Journal of Parasitology* **52**, 1041-1049.

Lie, K.J. (1968) Further studies on the life history of *Echinostoma lindoense* Sandground and Bonne, 1940 (Trematoda: Echinostomatidae) with a report of its occurrence in Brazil, *Proceedings of the Helminthological Society of Washington* **35**, 74-77.

Lie, K.J. and Basch, P.F. (1966b) Life history of *Echinostoma barbosai* sp. n. (Trematoda: Echinostomatidae), *Journal of Parasitology* **52**, 1052-1057.

Lie, K.J. and Basch, P.F. (1967) The life history of *Echinostoma paraensei* sp. n. (Trematoda: Echinostomatidae), *Journal of Parasitology* **53**, 1192-1199.

Lie, K.J. and Umathevy, T. (1965) Studies on Echinostomatidae (Trematoda) in Malaya. VIII. The life history of Echinostoma audyi sp. n., *Journal of Parasitology* **51**, 781-788.

Looss, A. (1896) Recherche sur la faune parasitiare de l'Egypte. Premiére partie, *Mémoires de l'Institute Egyptien* **3**, 1-252.

Looss, A. (1899) Weitere Beiträge zur Kenntnis der Trematoden-Fauna Aegyptens zugleich Versuch einer natürlichen Gliederung des Genus *Distomum* Retzius, *Zoologische Jahrbücher, Abt. Systematik* **12**, 521-784.

Lutz, A. (1924) Estudos sobre a evolucao dos Endotrematodes brazileiros, *Memorias do Instituto Oswaldo Cruz* **17**, 55-93.

Lymbery, A.J. (1992) Interbreeding, monophyly and the genetic yardstick: species concepts in parasites, *Parasitology Today* **8**, 208-211.

Lymbery, A.J. and Thompson, R.C.A. (1996) Species of *Echinococcus*: pattern and process, *Parasitology Today* **12**, 486-491.

Meece, J.K. and Nollen, P.M. (1996) A comparison of the adult and miracidial stages of *Echinostoma paraensei* and *E. caproni*, *International Journal for Parasitology* **26**, 37-43.

Meeus, T. de, Michalakis, Y. and Renaud, F. (1998) Santa Rosalia revisited: or why are there so many kinds of parasites in 'The garden of Earthy delights'? *Parasitology Today* **14**, 10-14.

Mendheim, H. (1940) Beiträge zur Systematik und Biologie der familie Echinostomatidae (Trematoda), *Nova Acta Leopoldina* **8**, 489-588.

Mendheim, H. (1943) Beiträge zur Systematik und Biologie der Familie Echinostomatidae, *Archiv für Naturgeschichte* **12**, 175-302.

Morgan, J.A.T. and Blair, D. (1995) Nuclear rDNA ITS sequence variation in the trematode genus *Echinostoma*: an aid to establishing relationships within the 37-collar-spine group, *Parasitology* **111**, 609-615.

Morgan, J.A.T. and Blair, D. (1998a) Relative merits of nuclear ribosomal internal transcribed spacers and mitochondrial CO1 and ND1 genes for distinguishing among *Echinostoma* species (Trematoda), *Parasitology* **116**, 289-297.

Morgan, J.A.T. and Blair, D. (1998b) Mitochondrial ND1 gene sequences used to identify echinostome isolates from Australia and New Zealand, *International Journal for Parasitology* **28**, 493-502.

Mutafova, T. and Kanev, I. (1987) Karyotype and chromosome analysis of collar spined parasitic worms (Trematoda: Echinostomatidae). In: *Fifth National Conference of Parasitology, Varna, Bulgaria*, Abstracts, p. 185.

Našincová, V. (1986) Contribution to the distribution and the life history of *Echinostoma revolutum* in Central Europe, *Věstník Československé Společnosti Zoologické* **50**, 70-80.

Našincová, V. (1991) The life cycle of *Echinostoma bolschewense* (Kotova, 1939) (Trematoda: Echinostomatidae), *Folia Parasitologica* **38**, 143-154.

Nassi, H. and Dupouy, J. (1988) Étude experimentale du cycle biologique d'*Echinostoma parvocirrus* n. sp. (Trematoda: Echinostomatidae), parasite larvaire de *Biomphalaria glabrata* en Guadeloupe, *Annales de Parasitologie Humaine et Comparée* **63**, 103-118.

Nezvalová, J. (1970) Beiträg zur kenntnis der cerkarien aus Südmähren, *Spisy Přírodovědecké Fakulty Universita J.E. Purkyně v Brně* **515**, 217-252 (In Czech).

Nollen, P.M. (1990) *Echinostoma caproni*: mating behavior and the timing of development and movement of reproductive cells, *Journal of Parasitology* **76**, 784-789.

Nollen, P.M. (1993) *Echinostoma trivolvis*: mating behavior of adults raised in hamsters, *Parasitology Research* **79**, 130-132.

Nollen, P.M. (1996a) Mating behaviour of *Echinostoma caproni* and *E. paraensei* in concurrent infections in mice, *Journal of Helminthology* **70**, 133-136.

Nollen, P.M. (1996b) The mating behaviour of *Echinostoma paraensei* grown in mice, *Journal of Helminthology* **70**, 43-45.

Nollen, P.M. (1997) Mating behaviour of *Echinostoma caproni* and *E. trivolvis* in concurrent infections in hamsters, *International Journal for Parasitology* **27**, 71-75.

Odaibo, A.B., Christensen, N. O. and Ukoli, F. M. A. (1988) Establishment, survival, and fecundity in *Echinostoma caproni* (Trematoda) infections in NMRI mice, *Proceedings of the Helminthological Society of Washington* **55**, 265-269.

Odhner, T. (1910) No. 23A. Nordostafrikanische Trematoden, grosstenteils vom Weissen Nil. I. Fascioliden. In: *Results of the Swedish Zoological Expedition to Egypt and the White Nile 1901 under the direction of L.A.Jägerskiöld*. K.W. Appelbergs Boktryckeri, Uppsala, 169 pp.

Panaiotova, M, Kanev, I., Fried, B. and Sorensen, R. (1996) Snail host in the life cycle of eight echinostome species possessing head collar with crown of 37-spines arranged around the oral sucker. [*Abstracts of the VIIth European Multicoloquium of Parasitology*], *Parassitologia* **38**, p. 76.

Richard, J. (1971) La chétotaxie des cercaires. Valeur systématique et phylétique. *Mémoires du Muséum National d'Histoire Naturelle, Serie A, Zoologie* **67**, 1-177.

Richard, J. and Brygoo, E.R. (1978) Cycle évolutif du trématode *Echinostoma caproni* Richard, 1964 (Echinostomatoidea). *Annales de Parasitologie Humaine et Comparée* **53**, 265-275.

Roberts, L.S. and Janovy, J. Jr, *Gerald D. Schmidt and Larry S. Roberts' Foundations of Parasitology* (6th Edition), McGraw Hill, Boston.

Rollinson, D. and Southgate, V.R. (1987) The genus *Schistosoma*: a taxonomic appraisal. In: *The biology of schistosomes, from genes to latrines.* (Rollinson, D. and Simpson, A.J.G. Eds). Academic Press, London.

Ross, G.C., Fried, B. and Southgate, V.R. (1989) *Echinostoma revolutum* and *E. liei*: observations on enzymes and pigments, *Journal of Natural History* **23**, 5, 977-981.

Skrjabin, K.I. and Bashkirova, E.Y. (1956) Family Echinostomatidae, *Osnovy Trematodologii* **12**, 53-930. (In Russian).

Sloss, B., Meece, J., Romano, M. and Nollen, P. (1995) The genetic relationships between *Echinostoma caproni, E. paraensei*, and *E. trivolvis* as determined by electrophoresis, *Journal of Helminthology* **69**, 243-246.

Sorensen, R.E., Kanev, I., Fried, B. and Minchella, D.J. (1997) The occurrence and identification of *Echinostoma revolutum* from North American *Lymnaea elodes* snails, *Journal of Parasitology* **83**, 169-170.

Sorensen, R.E., Curtis, J. and Minchella, D.J. (1998) Intraspecific variation in the rDNA ITS loci of 37-collar-spined echinostomes from North America: implications for sequence-based diagnoses and phylogenetics. *Journal of Parasitology* **84**, 992-997.

Toledo, R., Muñoz-Antolí, C. and Esteban, J.G. (in press) Life-cycle of *Echinostoma friedi* n. sp. (Trematoda: Echinostomatidae) in Spain and a discussion on the relationships within the '*revolutum*' group based on cercarial chaetotaxy, *Systematic Parasitology*.

Travassos, L. (1923) Informacoes sobre a fauna helminthologica de Matto Grosso, *Fohla Medica* **3**, 187-192.

Trouvé, S., Renaud. F., Durand, P. and Jourdane, J. (1996) Selfing and outcrossing in a parasitic hermaphrodite helminth (Trematoda, Echinostomatidae), *Heredity* **77**, 1-8.

Trouvé, S., Renaud, F., Durand, P. and Jourdane, J. (1998) Experimental evidence of hybrid breakdown between genetically distinct populations of *Echinostoma caproni, Parasitology* **117**, 2, 133-135.

Vasilev, I. and Kanev, I. (1979) Study on the species belonging of echinostomes (Trematoda) in Bulgaria. III. Determination of two species of echinostome cercariae with 37 collar spines, *Khelmintologiya* **8**, 6-23 (In Bulgarian).

Vasilev, I. and Kanev, I. (1981) Studies on echinostomatids (Trematoda) species composition in Bulgaria. V. On the development and ecology of *Echinostoma lindoense* Sandground et Bonne, 1940, *Khelmintologiya* **11**, 39-50 (In Bulgarian).

Vasilev , I., Kanev, I., Svietlicovski, M. and Bušta, J. (1982) Establishment of an echinostome with the features of *Echinostoma lindoense* Sandground et Bonne, 1940 (Echinostomatidae: Trematoda) in Poland and Czechoslovakia, *Khelmintologiya* **13**, 12-21 (In Bulgarian).

Voltz, A. Richard, J. Pesson, B. and Jourdane, J. (1985) Etude enzymatique comparée d'*Echinostoma caproni* Richard,1964 et *E. togoensis* Jourdane et Kulo, 1981 (Echinostomatidae, Trematoda), *Comptes Rendus de l'Academie des Sciences*, III (Sciences de la Vie) **301**, 33-36.

Voltz, A. Richard, J. Pesson, B. and Jourdane, J. (1986) Etude chimiotaxonomique du genre *Echinostoma*: comparaison d'une souche isolée du Cameroun (E. sp.) a deux especes Africaines (*E. caproni* et *E. togoensis*), *Annales de Parasitologie Humaine et Comparée* **61**, 617-623.

Voltz, A. Richard, J. Pesson, B. and Jourdane, J. (1988) Isoenzyme analysis of *Echinostoma liei*: comparison and hybridization with other African species, *Experimental Parasitology* **66**, 13-17.

Wright, C.A., Rollinson, D. and Goll, P.H. (1979) Parasites in *Bulinus senegalensis* (Mollusca: Planorbidae) and their detection, *Parasitology* **79**, 95-105.

Yamaguti, S. (1958) *Systema helminthum*. Vol. 1. *The digenetic trematodes of vertebrates,* Interscience Publishers, New York.

Yamaguti, S. (1971) *Synopsis of digenetic trematodes of vertebrates*. Vols I and II, Keigaku Publishing Co, Tokyo.

Yao, G., Huffman, J.E. and Fried, B. (1991) The effects of crowding on adults of *Echinostoma caproni* in experimentally infected golden hamsters, *Journal of Helminthology* **65**, 248-254.

ECHINOSTOMES IN VETERINARY AND WILDLIFE PARASITOLOGY

JANE E. HUFFMAN
Fish and Wildlife Microbiology Laboratory, Department of Biological Sciences, East Stroudsburg University, East Stroudsburg, Pennsylvania 18301, USA

1. **Introduction**
2. **Echinostomiasis and Epidemiology**
   2.1. INFECTIVITY, GROWTH AND DISTRIBUTION OF ECHINOSTOMES IN THE RODENT - AVIAN MODEL
   2.1.1. *Infectivity*
   2.1.2. *Growth*
   2.1.3. *Distribution*
   2.2. GEOGRAPHIC DISTRIBUTION
3. **Wildlife, Domestic Animals and Experimental Hosts Infected with Echinostomes**
   3.1. NATURAL AND EXPERIMENTAL INFECTIONS IN WARM BLOODED ANIMALS - AVIAN AND MAMMALIAN
4. **Pathological Effects of Echinostomes in Wildlife, Domestic Animals and Experimental Hosts**
   4.1. DIAGNOSIS
   4.2. CLINICAL EFFECTS
   4.3. GROSS AND HISTOPATHOLOGICAL EFFECTS
   4.4. INFLAMMATORY CELL MECHANISMS
   4.5. CONTROL AND TREATMENT
5. **Effects of Experimental Echinostome Infections in Rodents, Mallard Ducks and Domestic Chicks**
   5.1. SUPERIMPOSED AND CONCURRENT INFECTIONS
   5.2. FERTILITY AND FECUNDITY IN ANIMALS INFECTED WITH ECHINOSTOMES
   5.3. NUTRITIONAL DEFICIENCIES AND ECHINOSTOME EFFECTS
6. **References**

B. Fried and T.K. Graczyk (eds.),
Echinostomes as Experimental Models for Biological Research, 59–82.
© 2000 *Kluwer Academic Publishers. Printed in the Netherlands.*

1. Introduction

This chapter assesses the current status of research concerned with echinostomes in veterinary and wildlife parasitology. The pathology of echinostomiasis represents a complex and diverse set of reactions. The complexity is dependent on a wide variety of factors including, characteristics of echinostome species and the nature of the host's response. The use of the rodent/domestic chick model has provided basic knowledge of echinostomiasis including its etiology, distribution, epizootiology, signs, diagnosis, treatment and control. This chapter deals with the subject of nutritional influences on parasite-host relationships along with the effects of crowding and concurrent infections in animal models. Through the use of animal models fertility and fecundity effects in animals infected with echinostomes can be applied to infection in wildlife and domestic animals. A partial list of echinostomes found in domestic, wild and experimental animals is given.

2. Echinostomiasis and Epidemiology

2.1. INFECTIVITY, GROWTH AND DISTRIBUTION OF ECHINOSTOMES IN THE RODENT-AVIAN MODEL

2.1.1. *Infectivity*
The establishment, survival time and distribution of echinostomes may vary from one host to another. Establishment, survival time and egg production may vary independently of one another in generating the overall reproductive success of the echinostome.

The growth and development of *Echinostoma trivolvis* in natural and experimentally infected avian and mammalian hosts was studied by Beaver (1937). Senger (1954) studied growth, development and survival of *E. trivolvis* in domestic chicks and rats. Infectivity of *E. trivolvis* in domestic chicks has been reported (Senger, 1954; Fried and Weaver, 1969; Fried and Butler, 1978; Fried and Alenick, 1981; Fried, 1984). Fried *et al.* (1988) reported on the infectivity, growth, and development of *E. caproni* in the domestic chick. Kim and Fried (1989) reported on gross and histopathological effects of *E. caproni* in an experimental avian model. Humphries *et al.* (1997) studied the infectivity and growth of *E. revolutum* in the domestic chick. Fried *et al.* (1997) showed that worm body area and organ size can be used to distinguish adults of *E. revolutum* from *E. trivolvis* in domestic chicks. Factors such as age, size of cyst inoculum, pretreatment of metacercarial cysts and host - gut emptying time influence the infectivity of *E. trivolvis* in experimentally infected chicks (Fried *et al.*, 1997). Huffman and Fried (1990) summarized findings of average worm recoveries for *E. trivolvis* in experimentally infected chicks.

Host-parasite differences between species of echinostomes and their host will continue to contribute to a better understanding of the biosystematics of the

37-collar-spined echinostome group. Some of these differences may help elucidate species distinctions when echinostomes are recovered in the wild from naturally infected hosts in both single and multiply infected avian hosts. Franco *et al.* (1986) reported 100% infection of hamsters with *E. trivolvis* and worm recovery averaged 38%. Hamsters have remained infected for up to 123 days post-infection (Mabus *et al.*, 1988). Hosier and Fried (1986) reported that ICR and Swiss Webster (SW) mice can be infected with metacercariae of *E. trivolvis*. The worm burden of SW mice was reduced at 3 weeks post-infection and eliminated by 4 weeks. In ICR mice the worm burden was reduced at 2 weeks and eliminated by 3 weeks. Christensen *et al.* (1981) reported that inbred albino mice exposed to more than 15 encysted metacercariae of *E. caproni* retained consistent worm numbers for up to 70 days. Hosier and Fried (1986) reported a recovery rate of 88% of *E. trivolvis* in mice. Christensen *et al.* (1981) and Sirag *et al.* (1980) reported worm recoveries as high as 94% and 100%, respectively, of *E. caproni* in mice. Recovery rates for *E. paraensei* were reported by Meece and Nollen (1996). The worm recovery rate from mice was 15% (110/950). This low recovery rate may be due to elimination of adults after 14 days postinfection. The worm recovery rates from hamsters was 37% (74/200). The life span for *E. paraensei* has been reported by Lie and Basch (1967) to be 150 days. The life span for *E. revolutum* (referred to as *E. audyi*) has been reported by Lie and Umathevy (1965) to be from 3 to 8 weeks.

2.1.2. *Growth*

There is considerable information on the competition between intestinal helminths (Keymer, 1982). Some information is available on intraspecific crowding with intestinal trematodes. Rankin (1937) noted the relationship between number and size of trematodes within a host and suggested that crowding may be the factor involved. Other reports on crowding with intestinal trematodes include those by Willey (1941), Fried and Nelson (1978), Mohandas and Nadakal (1978), Fried and Freeborne (1984), Franco *et al.* (1988) and Yao et al. (1991). Mohandas and Nadakal (1978) have shown retarded growth of *Echinostoma malayanum* at high worm densities. The growth of *E. caproni* in the domestic chick was studied by Fried *et al.* (1988). Franco *et al.* (1988) studied the effect of crowding on adults of *E. trivolvis* in golden hamsters. Their study reported that in hamsters infected with *E. trivolvis*, increased dosage levels and thus infrapopulation size, influences several aspects of the worm's development in the host intestine. Huffman *et al.* (1988) investigated single and concurrent infections of the golden hamster with *E. trivolvis* and *E. caproni*. The establishment, survival, and fecundity of *E. caproni* infections in NMRI mice were reported by Odaibo *et al.* (1988). Yao *et al.* (1991) examined infectivity, growth, development and prepatency of E. caproni in golden hamsters infected with different worm burdens. To investigate growth and development, worm body area and gonadal area were determined. Meece and Nollen (1996) compared growth of *E. paraensei* and *E. caproni* in mice. *Echinostoma caproni* adults were significantly smaller than those of *E. paraensei* from day 11 through the duration of the experiment. *Echinostoma paraensei* adults reached a maximum body area of 35.2 mm at 39 days, while *E. caproni* reached their maximum body area of 8.7 mm at 28 days. Similar results were found for *E. paraensei* and *E. caproni* grown in hamsters. Lie and Umathevy (1965) reported that *E. revolutum* at 33 days old and grown in ducklings were smaller than those grown in pigeons. Body area for *E. revolutum* ranged from 4.8 mm to 16.6 mm (Lie and Umathevy, 1965). Lie and Nasemary (1973) reported that 24 specimens of *E. ilocanum* from a naturally infected *Rattus argentiventer* ranged in body area from 3.3 mm to 14.9 mm.

2.1.3. *Distribution*

Reviews by Huffman and Fried (1990) and Fried and Huffman (1996) report the distribution of *E. trivolvis* and *E. caproni* in rodents and domestic chickens. *Echinostoma trivolvis* occupies numerous sites in the intestine of the domestic chicken including the ileum, rectum, cloaca, cecum and the bursa of Fabricius (Beaver, 1937; Senger, 1954; Fried and Weaver, 1969; Fried and Butler, 1978; Fried and Alenick, 1981). In my laboratory we have examined singly three species of echinostomes (*E. caproni*, *E. trivolvis*, and *E. revolutum*) in domestic chickens. The chickens were exposed to the echinostomes as day-old chicks and worms were recovered from the intestine on day postinfection. Location of the recovered worms in the intestines of the infected varied according to echinostome species. *Echinostoma caproni* was located in the mid third of the intestine, between the gizzard and the cloaca. Both *E. trivolvis* and *E. revolutum* were found more posteriad in the intestine near the cloaca, but typically, *E. revolutum* was found more posteriad than *E. trivolvis*. Humphries *et al.* (1997) and Fried *et al.* (1997) reported that *E. revolutum* occurred mainly in the posterior aspect of the intestine. Mucha *et al.* (1990) reported that *E. trivolvis* in experimentally infected mallard ducks (*Anas platyrhynchos*) was located in the cloaca and lower ileum. Lie and Basch (1967) reported that *E. paraensei* was generally located in the anterior half of the small intestine of white rats. The location of *E. paraensei* in mice and hamsters was reported by Meece and Nollen (1996). *Echinostoma paraensei* adults were recovered through out the length of the small intestine until day 17 postinfection and then were found clustered at the pylorus - duodenum interface with some worms extending into the stomach. After 20 days, worms were again distributed through out the intestine. Lie and Umathevy (1965) reported that *E. revolutum* was located in the rectum if birds. Unpublished studies in my laboratory showed the distribution of *E. caproni* and *E. trivolvis* in the golden hamster over a 9 wk period. Worm distribution in the small intestine is shown in Figure 1. *E. caproni* was mainly located in the lower third of the small intestine. *E. trivolvis* was distributed throughout the small intestine. Hosier and Fried (1991) examined the distribution of *E. caproni* in female ICR mice. From 1 to 8 wk postinfection most worms localized in the jejunum and ileum; after eight weeks most worms were in the jejunum and duodenum indicating an anteriad shift as the worms aged. Odaibo *et al.* (1988, 1989) reported the distribution of *E. caproni* in NMRI mice. Yao *et al.* (1991) studied the effects of crowding on adults of E. caproni in experimentally infected golden hamsters. In heavy infections worms were more widely distributed in the hamster then in lighter infections.

2.2. GEOGRAPHIC DISTRIBUTION

Echinostomes are commonly found in birds (waterfowl) and mammals and their distribution is ubiquitous. Echinostome distribution is mainly dependent upon the presence of suitable molluscan hosts. Geographic distribution for individual species is given in Chapter 2.

3. Wildlife, Domestic Animals and Experimental Hosts Infected with Echinostomes

3.1. NATURAL AND EXPERIMENTAL INFECTIONS IN WARM BLOODED ANIMALS - AVIAN AND MAMMALIAN

Echinostome adults infect a large number of different warm-blooded hosts. A selective list of echinostome infections in domestic and wild animals listed here comprise only a partial number those that have been found. Skryabin *et al.* (1979) described the echinostomatids of domestic and wild birds discovered by various workers in Russia.

Mammalian hosts for *E. trivolvis* are listed in Table 1. Table 2 lists the avian hosts infected with *E. trivolvis*. Avian and mammalian hosts of *E. caproni* are listed in Table 3. The avian hosts infected with *E. revolutum* are found in Table 5. Table 6 lists the avian and mammalian hosts for selected species of echinostomes.

TABLE 1. Mammalian hosts infected with *Echinostoma trivolvis*

Common name	Species	Reference
Experimental Hosts		
Dog	*Canis familiaris*	Beaver (1937)
Guinea pig	*Cavia porcellus*	Beaver (1937)
Cat	*Felis catus*	Beaver (1937)
Golden hamster	*Mesocriceus auratus*	Franco *et al.* (1986)
House mouse	*Mus musculus*	Beaver (1937)
Rabbit	*Oryctolagus cuniculus*	Beaver (1937)
Norway rat	*Rattus norvegius*	Beaver (1937)
Pig	*Sus scrofa*	Beaver (1937)
ICR mouse	*Mus musculus*	Hosier and Fried (1986)
Swiss Webster mice	*Mus musculus*	Hosier and Fried (1986)
C3H mice	*Mus musculus*	Fujino and Fried, (1993)
C3H/HeN	*Mus musculus*	Fujino *et al.* (1996)
Athymic nude mice	*Mus musculus*	Bindseil and Christensen (1984)
SCID mice	*Mus musculus*	Fujino *et al.* (1998)
Natural Hosts		
Muskrat	*Ondatra zibethica*	Leidy (1988)
Red fox	*Vulpes vulpes*	Smith (1978)
Opossum	*Didelphis virginiana*	Alden (1995)

TABLE 2. Avian hosts infected with *Echinostoma trivolvis*

Common name	Species	Reference
Experimental Hosts		
Domestic chick	*Gallus gallus*	Fried (1984)
Mallard duck	*Anas platyrhynchos*	Mucha *et al.* (1990)
Domestic pigeon	*Columba livia*	Beaver (1937)
Natural Hosts		
Common pintail	*Anas acuta*	Gower (1937)
American widgeon	*Anas americana*	Gower (1937)
Domestic duck	*Anas borchas*	Gower (1937)
	Anas brasiliensis	Beaver (1937)
Cinnamon teal	*Anas cyanoptera*	Wilkinson et al. (1977)
Blue-winged teal	*Anas discor*	Shaw and Kogan (1980)
Eurasian widgeon	*Anas penelope*	Gower (1937)
Mallard	*Anas p. platyrhynchos*	Gower (1937)
Motted duck	*Anas p. fulvicula*	Kinsella and Forrester (1972)
Black duck	*Anas rubripes*	Gower (1937)
Australian black duck	*Anas superciliosa*	Beaver (1937)
Emden goose	*Anser* sp.	Griffiths *et al.* (1976)
Domestic goose	*Anser* sp.	Griffiths *et al.* (1976)
Greytag goose	*Anser anser*	Beaver (1937)
Magpie goose	*Anseranus semipalmata*	Beaver (1937)
Lesser scaup	*Aythya affinis*	Gower (1937)
Redhead	*Aythya americana*	Gower (1937)
Ring-necked duck	*Aythya collaris*	Gower (1937)
Greater scaup	*Aythya marila*	Gower (1937)
Great-horned owl	*Bubo virginianus*	Ramalingam and Samuel (1978)
Rough-legged hawk	*Buteo lagopus*	Beaver (1937)
Muscovy	*Cairina moschata*	Gower (1937)
Partridge		Rim (1982)
Common scoter	*Oidemia nigra*	Gower (1937)
American flamingo	*Phoenicopterus sp.*	Threlfall (1980)
Common grackle	*Quiscalus quiscula*	Stanley and Rabalais (1971)
Mourning dove	*Zenaida macroura*	Barrows and Hayes (1977)

TABLE 3. Mammalian hosts infected with *Echinostoma caproni*

Common name	Species	Reference
Experimental Hosts		
NMRI mice	*Mus musculus*	Odaibo *et al.* (1988)
SS mice	*Mus musculus*	Christensen *et al.*, (1985)
SVS mice	*Mus musculus*	Christensen *et al.*, (1986)
CBA mice	*Mus musculus*	Baek *et al.* (1996)
RAG mice	*Mus musculus*	Frazer *et al.* (1999)
BALB/cABOM mice	*Mus musculus*	Bindseil and Hau (1991)
ICR mice	*Mus musculus*	Fried and Sousa (1990)
Swiss Webster mice	*Mus musculus*	Faust (1996)
House mice	*Mus musculus*	Jeyarasasingam *et al.* (1972)
Jirds	*Meriones unquiculatus*	Mahler *et al.* (1995)
Golden hamster	*Mesocriceus auratus*	Huffman *et al.* (1988)
White rats	*Rattus rattus*	Hansen *et al.* (1991)
Rabbit	*Oryctolagus cuniculus*	Moravec *et al.* (1974)
Natural Hosts		
Rat	*Rattus* sp.	Jeyarasasingam *et al.* (1972)
Egyptian giant shrew	*Crocidura olivieri*	Jeyarasasingam *et al.* (1972)

TABLE 4. Avian hosts infected with *Echinostoma caproni*

Common name	Species	Reference
Experimental Hosts		
Domestic chicken	*Gallus gallus*	Fried et al. (1988)
Pigeon	*Columba livia*	Jeyarasasingam *et al.* (1972)
Finch	*Lonchura striata*	Jeyarasasingam *et al.* (1972)
Domestic duckling	*Anas bochas*	Jeyarasasingam *et al.* (1972)
Falcon	*Falco newtoni*	Richard and Brygoo (1978)

TABLE 5. Avian hosts infected with *Echinostoma revolutum*

Common name	Species	Reference
Experimental Hosts		
Pigeon	*Columba livia*	Lie and Umathevy (1965)
Domestic duck	*Anas borchas*	Lie and Umathevy (1965)
Little cuckoo dove	*Macropygia ruficeps*	Lie and Umathevy (1965)
Black-headed munia	*Lonchura ferruginosa*	Lie and Umathevy (1965)
Domestic chicken	*Gallus gallus*	Nasincova (1986)
Spotted munia	*Lonchura punctulata*	Lie and Umathevy (1965)
Java sparrow	*Padda oryzivora*	Lie and Umathevy (1965)
Natural Hosts		
Unknown - probably birds		

TABLE 6. Avian and mammalian hosts for selected species of *Echinostomes*

Common name	*Echinostome*	Reference
Experimental Hosts		
Domestic chicken	*E. revolutum*	Humpries et al. (1997)
Natural Hosts	(North America)	
Unknown - probably birds		
Experimental Hosts		
Golden hamster	*E. paraensei*	Meece and Nollen (1996)
White mice		Meece and Nollen (1996)
White rat		Lie and Basch (1967)
Natural Hosts		
Unknown - probably rodents		
Experimental Hosts		
Rats	*E. hortense*	Saito (1984)
Natural Hosts		
Humans		
Experimental Hosts		
White rats	*E. ilocanum*	Lee and Nasemary (1973)
Rats		
Dogs		
Rice field rats		

4. Pathological Effects of Echinostomes in Wildlife, Domestic Animals and Experimental Hosts

The *Echinostoma* - rodent model has proved to be an excellent one for studying intestinal trematode relationships as well as the clinical course of infection (Huffman *et al.* 1986; Huffman and Fried, 1990; Fried and Huffman, 1996). Work with echinostomes provides the advantages of a relatively large worm size in hosts that are easily maintained in the laboratory.

4.1. DIAGNOSIS

A diagnosis of acute echinostomiasis is usually made by a postmortem examination of the affected animal. In diagnosing a chronic infection examination of fecal samples for eggs can be made. Echinostome eggs do not usually float in the concentrations of sugar solutions ordinarily used. Sedimentation techniques are preferred because of the high specific gravity of the eggs (Benbrook and Sloss, 1961). It is best to concentrate the eggs by washing the feces through sieves to remove coarse debris and centrifuge at 2000 x g for 5 min or allow the washings to settle. The eggs will then be found in the sediments. Echinostome eggs vary in size. They are operculate and amber in color. The average length of an *E. trivolvis* egg is about 100 μm and the average width is about 70 μm. See Figure 8 in Huffman and Fried (1990) for a photomicrograph of a typical echinostome egg. The finding of eggs in the feces should be correlated with clinical signs and environmental factors. Commercially available techniques such as Fecalyzer (EVSCO Pharmaceuticals, Buena, NJ 08310) can be used to detect echinostome eggs.

The length of time it takes for an echinostome to become ovigerous depends on the host, worm species and the number of worms present. Table 7 depicts the fecundity results from experimental hosts. The distribution and fecundity of two allopatric species of *Echinostoma* in the golden hamster has been investigated in my laboratory. In this study two groups of golden hamsters each were fed 25 cysts of either *E. caproni* or *E. trivolvis*. To determine the number of eggs/gram of feces a modification of the technique from Brindley and Dobson (1981) was used. Another useful technique employs the PARACOUNT - EPG kit (Olympic Equine Products, Issaquah, WA 98027) which can be used for detection of echinostome eggs. Fecal egg counts per worm are shown in Figures 2 and 3. *E. trivolvis* produced significantly more eggs than did *E. caproni*, with fecal egg counts from hamsters infected with *E. trivolvis* being approximately double that of hamsters infected with *E. caproni* throughout the study. The number of eggs produced by either *E. caproni* or *E. trivolvis* in the golden hamster is related to the age of the infection.

4.2. CLINICAL EFFECTS

The main signs in rodent hosts with heavy infections of echinostomes are weakness, watery diarrhea, weight loss, anemia and unthriftiness.   Packed cell volume, hemoglobin and red blood cell counts have been reported in golden hamsters infected with *E. trivolvis* (Huffman *et al.*, 1986). These blood parameters increased due to the diarrhea present in the infected hamsters. No increase in eosinophils was noted in peripheral blood smears from the infected hamsters. In extra-intestinal infections of *E. trivolvis* in golden hamsters serum bilirubin concentrations increased (Huffman *et al.*, 1988).   In golden hamsters infected with *E. caproni*, hemoglobin and packed cell volumes decreased due to hemorrhage seen in association with the damage to the intestinal villi (Huffman *et al.*, 1988).

4.3. GROSS AND HISTOPATHOLOGICAL EFFECTS

Echinostomiasis in domestic pigs was reported by Bandyopadhyay *et al.* (1995). The pigs developed fatal diarrhea from a massive infection with *Artyfechinostomum oraoni*. The necropsy revealed hemorrhage and edema of the jejunum and duodenum extending up to the pyloric end of the stomach. *Echinostoma revolutum* has been reported to cause severe enteritis in pigeons (Hossain *et al.*, 1980). Griffiths *et al.* (1976) reported an outbreak of trematodiasis in domestic geese in Minnesota. Losses occurred in a large flock on summer pasture.   The trematodes *E. revolutum* and *Notocotylus attenuatus* were present in large numbers.   The birds were weak and severely emaciated. Lesions in the small intestine consisted of mild hyperemia together with a severe catarrhal enteritis.   Histopathological examination of a large renal abscess revealed two invading flukes in one bird.

Most reports describing the lesions caused by infection with echinostomes are from rodent models, particularly mice and hamsters. Pathological effects were reported by Bindseil and Christensen (1984) in the small intestine of conventional mice and congenitally athymic, nude mice infected with *E. caproni* (referred to in the paper as *E. revolutum*).   Pathological effects of echinostome infections in animals varies depending on the worm burden.   In the golden hamster, ballooning of the small intestine and cecum may occur (Figure 4).   Congestion of intestinal blood vessels is grossly apparent and worms may be seen through the serosal surface.   Lymphatic nodules of the small intestine are often enlarged (Figure 4).   Huffman *et al.* (1986) described the pathological lesions in experimental echinostomiasis in the golden hamster.   Destruction of villi occurred in areas of the small intestine where *E. trivolvis* adults were attached (Figure 5).   Erosion and desquamation of the intestinal epithelia occurred with an increase in the number of intestinal glands and proliferation of goblet cells.   Lymphocytic infiltration of the villi was the primary response.   A few eosinophils were also present.   Echinostomiasis has produced significant mortality in commercial duck production in Europe and Asia (Kishore and Sinha, 1982).   Mucha *et al.* (1990) reported that echinostomes from mallard ducks were attached loosely to the mucosal surface of the intestine. A slight abrasion of the mucosal surface at the site of attachment was the only damage noted. In my laboratory we have investigated the infectivity and comparative pathology of *E. caproni*, *E. revolutum*, and *E. trivolvis* in the domestic chick. Damage to the intestinal villi of all the infected hosts was observed at the site of worm attachment along with a proliferation of goblet cells. The pathology of extraintestinal infections with *E. trivolvis* in the golden hamster was reported by Huffman *et al.* (1988). The occurrence of *E. trivolvis* in extraintestinal sites results from worm crowding in heavy infections. The worms migrate up the common bile duct and can be found in the liver, gall-bladder and pancreas. The migration of worms through the liver damages blood vessels resulting in hemorrhage and mononuclear infiltration. Granulomas composed of multinucleate giant cells developed in the liver surrounding the echinostome eggs. Secondary bacterial invasion followed the migration tracts of the echinostomes, resulting in hepatic necrosis (Huffman *et al.* 1986). Granulation tissue in the liver was another response to the worm. Heavy infections also resulted in *E. trivolvis* adults migrating anteriad into the stomach. McMaster *et al.* (1995) reported that relative splenic weights decreased in golden hamsters infected with *E. trivolvis* or *E. caproni*. Work in my laboratory did not see any increase in splenic or hepatic weights in domestic chicks infected with either *E. caproni*, *E. trivolvis* or *E. revolutum*.

4.4. INFLAMMATORY CELL MECHANISMS

Huffman and Fried (1990) have suggested that a complex set of interrelating factors may govern the immune response in echinostome infections. The golden hamster shows only a limited capacity to expel primary and secondary *E. trivolvis* and *E. caproni* infections (Franco *et al.*, 1986, 1988: Christensen *et al.*, 1990; Huffman *et al.*, 1992). Fried and Huffman (1996) reviewed the possible effector mechanisms which may act against *E. caproni*. Additional information on inflammatory cell mechanisms in rodent hosts infected with echinostomes can be found in Chapter 12.

4.5. CONTROL AND TREATMENT

Echinostome infections may be controlled in captive animals by eliminating contact with the intermediate hosts. Control of echinostomiasis in wild animals would be impractical, as control of the intermediate hosts would be necessary.

Hosier *et al.* (1988) reported that infections of ICR mice with either *E. trivolvis* or *E. caproni* were eliminated with oxyclozanide (Zanil) at 340 mg/kg body weight. Mauer *et al.* (1995) used different concentrations of clorsulon, rafoxanide, mebendazole and arprinocid in ICR mice infected with *E. caproni*. Doses of 100 mg/kg of clorsulon, and 50 mg/kg of rafoxanide were 100% effective in eliminating the echinostome infections. Faust (1996) infected golden hamsters and Swiss Webster mice with *E. trivolvis* or *E. caproni*. At day 14 post infection, all animals were treated with oral doses of oxyclozanide, clorsulon, rafoxanide, paratox and piperazine. Oxyclozanide, clorsulon and rafoxanide were effective in eliminating the infections. Oxyclozanide has been used on duck farms in Poland for treatment of the trematode, *Notocotylus attenuatus*. Administered orally or via feed, a dose up to 45 mg/kg can be used without host side effects (Robertson and Courtney, 1995). This drug could be used to treat echinostomiasis. Echinostomiasis has produced significant mortality in commercial duck production in Europe and Asia (Kishore and Sinha, 1982). Clorsulon is an FDA approved drug in the treatment of *Fasciola hepatica* for beef cattle and lactating and nonlactating dairy cattle. Through oral administration, clorsulon, at 3.75 mg/kg provide complete efficacy for adult flukes. Repeated doses of clorsulon are recommended in endemic areas of liver fluke disease (Robertson and Courtney, 1995). Since the safety index of rafoxanide is marginal, this may not be the drug of choice to treat echinostome infections. At doses of 80 mg/kg for cattle, a loss of appetite and diarrhea occur. Even at lower doses of 45 mg/kg in sheep, rafoxanide can cause pathologic changes such as cataracts and optic nerve degeneration (Robertson and Courtney, 1995).

5. Effects of Experimental Echinostome Infections in Rodents, Mallard Ducks and Domestic Chicks

5.1 SUPERIMPOSED AND CONCURRENT INFECTIONS

Holmes (1990) suggested three major selective forces for niche restriction in intestinal helminths: specialization, reproductive efficiency, and competition. Behnke (1987) and Christensen *et al.* (1986) have suggested that niche restrictions are side effects of immune mechanisms, and are not related to any selective forces acting on niches. The selective factor most frequently used in analysis of the niche that an organism will occupy is competition. Intraspecific competition will extend the niche occupied by an organism with inter-specific competition a major force restricting niche overlap. Negative interactions may be indicated by reductions in establishment, growth, maturation, or reproduction; displacement of one species is also common (Dobson, 1985). Dobson (1985) noted that the extent of competition is markedly affected by intensity of the interacting species. With relatively low populations, neither intra - nor interspecific competition is likely to occur; with relatively high populations, both are likely to occur. The occurrence of concurrent infections with two or more helminth species in domestic animals and wildlife is of interest in studies of multiple parasitism. *Echinostoma caproni* has been used in concurrent studies with other helminths. Andreassen *et al.* (1990) found that superimposing the intestinal tapeworm Hymenolepis diminuta on mice with an established *E. caproni* infection resulted in destrobilization and expulsion of the tapeworm. Mechanisms associated with this interaction were not elucidated. Iorio *et al.* (1991) examined concurrent infections of *E. caproni* and *E. trivolvis* in female ICR mice. In the concurrent infections, 13 (59%) of 22 mice were infected with both species and the percentage of worm recovery was 73% for *E. caproni* and 14% for *E. trivolvis*. There was no difference in worm distribution of either species in single versus concurrent infections. In concurrent infections at 14 days post infection, there was a significant decrease in the body area of worms of both species compared to single worm species. In concurrent infections in mice, interspecific competition produced inhibitory effects on worm growth in both species. Huffman *et al.* (1992) studied intra- and inter-specific competition between *E. caproni* and E. *trivolvis* in the golden hamster and established five age classes of both species of echinostomes in the hamster. The location of worms in concurrent infections suggested that competition between the two species did not occur and worms of both species were clustered. Fujino *et al.* (1996) reported that rapid expulsion of the intestinal trematodes *E. trivolvis* and *E. caproni* from C3H/HeN mice occurred after infection with *Nippostrongylus braziliensis*. Expulsion of the trematodes occurred due to the increase of mucins by hyperplastic goblet cells associated with the primary *N. braziliensis* infection. Holben (1993) reported that in concurrent infections with E. caproni and *N. braziliensis* gross pathological changes included the increase production of mucous and an increase in the number of lymphatic nodules along the length of the intestine. The presence of an established *N. braziliensis* infection decreased significantly the percent recovery of *E. caproni*. A blood eosinophilia occurred in all infected animals.

5.2. FERTILITY AND FECUNDITY IN ANIMALS INFECTED WITH ECHINOSTOMES

Bindseil and Hau (1991) showed that infection of BALB/cBOM mice with *E. caproni* had a negative influence on pregnancy: fewer fetuses were present in infected mice than in controls. Ovulation, fertilization, and egg implantation were not affected. Pregnancy and lactation in the host do not appear to affect the course of cestode or trematode infections. Reproductive processes in the female host, however, increased susceptibility to infection with nematodes (Ogilvie and Jones, 1973). The environment into which the young are born contains a myriad of infectious organisms from which the neonate must be protected. There are two routes whereby the neonate may gain protection from infectious organisms: via the placenta before birth and via colostrum and milk after birth (Paul, 1989). Both of these routes transfer maternal antibodies to the young (Carlier and Truyens, 1995). Bindseil *et al.* (1990) reported that infection of BALB/cABom mice with *E. caproni* reduced mouse fertility, as determined on day 18 of pregnancy by counting and weighing fetuses from infected versus non-infected females. They suggested that effects on fertility were not due to worm-induced lesions in the intestine but to some undefined pathophysiological disorders. Huffman *et al.* (1998) reported that average litter size was not significantly different between hamsters infected with *E. trivolvis* and uninfected mothers. They also reported that the infectivity in the progeny from infected mothers showed decreased infectivity at ages

TABLE 7. Fecundity results for *Echinostomes* from experimentally infected hosts

Host	*Echinostome* species and appearance of first eggs in the feces (days)	Reference
Golden hamster	*E. paraensei* (14)	Meece and Nollen (1996)
	E. trivolvis (9)	Franco *et al.* (1986)
	E. caproni (8)	Issacson *et al.* (1989)
ICR mice	*E. trivolvis* (15)	Hosier and Fried (1986)
	E. caproni (8)	Hosier and Fried (1991)
NMRI mice	*E. caproni* (14)	Odaibo *et al.* (1989)
Swiss Webster	*E. trivolvis* (15)	Hosier and Fried (1986)
mice	*E. caproni* (8)	Hosier and Fried (1986)
Jird	*E. caproni* (14)	Mahler *et al.* (1995)
Chick embryo[a]	*E. caproni* (8)	Chien and Fried (1992)
Domestic chick	*E. trivolvis* (12)	Freid (1984)
	E. caproni9	Fried *et al.* (1989)
	E. revolutum[b] (12)	Humpries *et al.* (1997)
Mallard duck	*E. trivolvis* (15)	Much *et al.* (1990)
Pigeon	*E. revolutum*[c] (8)	Lie and Umathevy (1965)

[a] This species of *Echinostome* grown on the chick chorioallantoic membrane, [b] North America, [c] Europe

6, 7, and 8 wk post infection. Neonates from uninfected mothers showed a greater infectivity when compared with animals born from infected hosts.

The decreased echinostome infectivity may be the result of passive transfer of maternal antibodies. Simonsen *et al.* (1991) showed that hamsters developed a delayed positive humoral response to infection with *E. caproni*. Huffman *et al.* (1998) reported that spleens from infected mothers showed a depletion of white pulp in response to chronic infection. The adrenal cortex from infected mothers infected with *E. trivolvis* was widened and composed pre-dominately of lipid-poor reticularis type cells. The increased relative adrenal weights may reflect cortisol effects related to stress associated with infection (Cotran *et al.* 1989).

5.3. NUTRITIONAL DEFICIENCIES AND ECHINOSTOME EFFECTS

Occurrence of disease involves interactions among three entities, the disease agent, host characteristics, and the environment. Nutritional deficiencies generally reduce the capacity of the host to resist infection. An infectious disease usually makes a

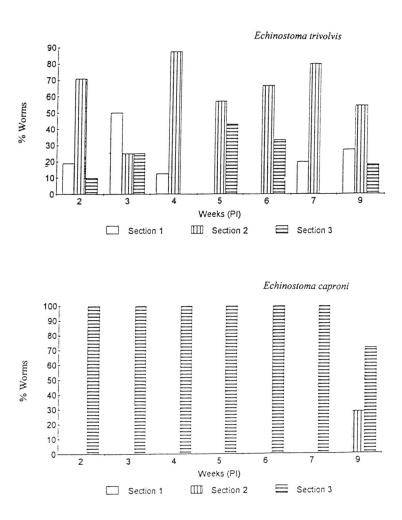

Figure 1. Worm distribution in the small intestines of golden hamsters each fed 25 cysts of either *E. trivolvis* or *E. caproni*. The small intestine was divided into three equal sections from the pyloric sphincter to the ileo-cecal valve; section 1 was posterior to the pyloric sphincter and section three was the most posteriad. No worms were found in section 1 of the small intestines of hamsters infected with *E. caproni*

co-existing malnutrition worse. The interaction between nutrition and infection is more detrimental for the host then the independent effects of the two (Eve, 1981). Synergistic effects of nutrition on infectious diseases may include the following: 1. reduction in antibody formation; 2. diminished inflammatory response and altered wound healing; 3. alteration in tissue integrity; 4. interference with production of non-specific protective substances; and 5. variations in endocrine activity (Scrimshaw et al., 1968).

Parasites are an expected component of wild animal populations and are a function of the host/environmental relationship. Some wildlife species are capable of increasing in number to the point where the population exceeds the nutritional carrying capacity. One indicator of exceeding nutritional carrying capacity is the presence of heavy helminth burdens (Eve, 1981). Sudati et al. (1996, 1997) reviewed the salient literature on the effects of host diet on helminth growth and development and noted that most work on this topic was done on cestodes and nematodes in various vertebrate hosts. Relatively little information is available on the synergistic effects of gastrointestinal trematodes in animals with nutritional deficiencies. The rodent and chick/echinostome model can be used to study this synergistic effect.

Rosario and Fried (1999) investigated the effects of a protein-free diet (PFD) on worm recovery, growth and distribution of E. caproni in ICR mice. Infected mice on the PFD exhibited hair loss, a condition known as 'wasting syndrome' (Poleschchuk et al., 1988). Along with severe weight-loss of infected mice on PFD extensive damage occurred to the gastrointestinal tract. The parasites from mice on PFD were also affected. They showed decreases in body area and dry weight. There was a posteriad shift of worms over the course of infection in hosts on the PFD.

Burkholder (1999) did not observe any synergistic effects of selenium deficiency in Swiss Webster mice infected with E. caproni. No changes in mouse body weight, basal metabolic rate or damage to the intestinal tract was found when compared to control animals. In my laboratory Swiss Webster mice on Zn and Fe deficient diets showed greater infectivity with E. caproni when compared to infected mice on control diets. The worms showed a decrease in dry weight and in the number of eggs per gram of feces when compared to controls. Baek et al. (1996) examined the effect of zinc deficiency on the response of CBA mice to infection with E. caproni. CBA mice fed a zinc deficient diet gained less weight than controls and zinc deficiency delayed worm expulsion. Zinc deficiency resulted in prolonged IgM response, a delayed IgG response and an increased IgA response. Resistance to challenge infection was slightly, but not significantly, affected by zinc deficiency.

Beaver (1937) speculated that Echinostoma trivolvis (referred to as E. revolutum in his monograph) developed poorly in pigeons that may have been depleted of vitamins A and D. Simpkins and Fried (1999) reported on the effects of a vitamin A, D, and E deficiency on the infectivity, growth and development of E. trivolvis in domestic chicks. The worms from chicks on vitamin deficient diets showed a decline in weight and body and organ size when compared to worms from chicks on normal diets. No significant weight differences in chicks on experimental diets versus control diets were

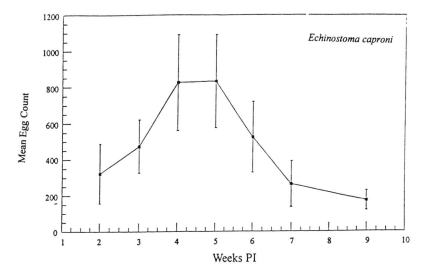

Figure 2. Mean number of eggs per gram of feces at 2, 3, 4, 5, 6, 7 and 9 weeks postinfection (PI). Results are based on feces collected from five hosts infected with *E. caproni* at each week PI

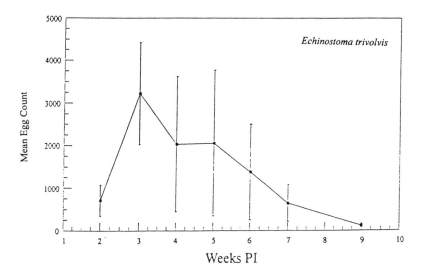

Figure 3. Mean number of eggs per gram of feces at 2, 3, 4, 5, 6, 7 and 9 weeks postinfection (PI). Results are based on feces collected from five hosts infected with *E. trivolvis* at each week PI

noted. A greater infectivity (95%) was noted in experimental animals compared to that of the controls (85%).

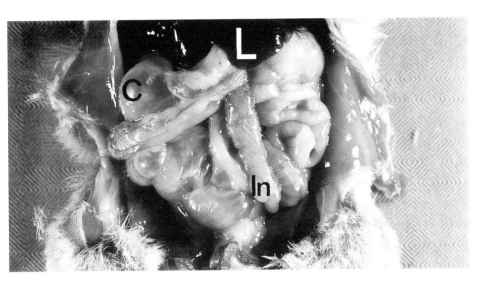

Figure 4. Ballooning of the small intestine and cecum in the golden hamster infected with *E. trivolvis*. Liver (L), Cecum (C), Lymphatic Nodule (LN)

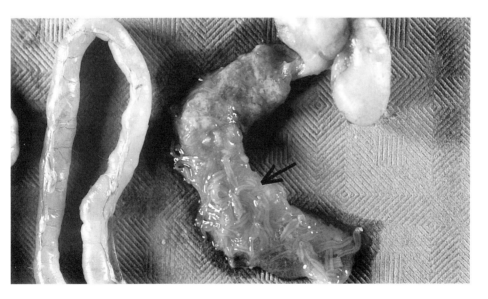

Figure 5. Severe enteritis in a golden hamster infected with *E. trivolvis*. Arrow indicates the echinostomes attached to the mucosal surface of the small intestine

6. References

Alder, K.J. (1995) Helminths of the opossum, *Didelphis virginica*, in Southern Illinois, with a compilation of all helminths reported from this host in North America. *Journal of the Helminthological Society of Washington* 62, 197-208.

Andreassen, J., Odaibo, A.B. and Christensen, N.O. (1990) Concurrent infections of the trematode *Echinostoma caproni* and the tapeworm *Hymenolepis diminuta* and *Hymenoplepis microstoma* in mice. *Journal of Parasitology* 76, 573-575.

Baek, J., Simonsen, P.E., Friis, H. and Christensen, N.O. (1996) Zinc deficiency and host response to helminth infection: *Echinostoma caproni* in CBA mice. *Journal of Helminthology* 70, 7-12.

Bandyopadhyay, A.K., Maji, A.K., Manna, B., Bera, D.K., Addy, M. and Nandya. (1995) Pathogenicity of *Artyfechinostomum oraoni* in naturally infected pigs. *Tropical Medicine and Parasitology* 46, 138-139.

Barrows, P.L. and Hayes, F.A. (1977) Studies on endoparasites of the mourning dove (*Zenaida macroura*) in the southeastern United States. *Journal of Wildlife Diseases* 13, 24-28.

Beaver, P.C. (1937) Experimental infections on *Echinostoma revolutum* (Froelich), a fluke from birds and mammals. *Illinois Biological Monographs* 15, 1-96.

Behnke, J.M. (1987) Evasion of immunity to nematode parasites causing chronic infections. *Parasitology* 26, 2-71.

Benbrook, E.A. and Sloss, M.W. (1961) Veterinary Clinical Parasitology, 3rd edition Ames Iowa State University Press.

Bindseil, E. and Christensen, N.O. (1984) Thymus-independent crypt hyperplasia and villous atrophy in the small intestine of mice infected with the trematode *Echinostoma revolutum*. *Parasitology* 88, 431-438.

Bindseil, E. and Hau, J. (1991) Negative effect on early post-implantation pregnancy and progesterone levels in mice infected with the intestinal trematode *Echinostoma caproni*. *Parasitology* 102, 387-390.

Bindseil, E., Andersen, L.L.I. and Hau, J. (1990) Negative influence of extragenital *Schistosoma mansoni* and *Echinostoma caproni* infections on fertility and maternal murine alpha fetoprotein levels in the circulation of female mice. *International Journal of Feto-Maternal Medicine* 3, 236-241.

Brindley, P.J. and C. Dobson. (1981) Genetic control of liability to infection with *Nematospiroides dubius* in mice: selection of refractory and liable populations of mice. *Parasitology* 83, 51-56.

Burkholder, J. (1999) Effects of selenium deficiency on worm recovery, growth and distribution of *Echinostoma caproni* in Swiss Webster mice. Master's Thesis, East Stroudsburg University.

Carlier, Y. and C. Truyens. (1995) Influence of maternal infection on offspring resistance towards parasites. *Parasitology Today* 11, 94-99.

Chien, W.Y. and B. Fried. (1992) Cultivation of excysted metacercariae of *Echinostoma caproni* to ovigerous adults in the allantois of the chick embryo. *Journal of Parasitology* 78, 1019-1023.

Christensen, N.O., Nydal, R., Frandsen, F. and Nansen, P. (1981) Homologous Immunotolerance and decreased resistance to *Schistosoma mansoni* in *Echinostoma revolutum* infected mice. *Journal of Parasitology* 67, 164-166.

Christensen, N.O., Knudsen, J., Fagbemi, B.O. and Nansen, P. (1985) Impairment of primary expulsion of *Echinostoma revolutum* in mice concurrently infected with *Schistosoma mansoni*. *Journal of Helminthology* 59, 333-335.

Christensen, N.O., Knudsen, J. and Andreassen, J. (1986) *Echinostoma revolutum*: resistance to secondary and superimposed infections in mice. *Experimental Parasitology* 61, 311-318.

Christensen, N.O., Simonsen, P.E., Odaibo, A.B. and Mahler, H. (1990) Establishment: Survival and fecundity in *Echinostome caproni* (Trematoda) infections in hamsters and jirds. *Journal of the Helminthological Society of Washington* 57, 104-107.

Cotran, R.S., Kumar, V. and Robbins, S.L. (1989) Robbins Pathologic Basis of Disease W.B. Saunders Company, Philadelphia, PA.

Dobson, A.P. (1985) The population dynamics of competition between parasites. *Parasitology* 91, 317-347.

Eve, J.H. (1981) Management Implications of Disease. In: *Diseases and Parasites of White-Tailed Deer,* edited be Davidon *et al.* Heritage Printers, Charlotte, NC 458pp.

Faust, S.K. (1996) The effects of anthelmintic treatment in golden hamsters *(Mesocricetus auratus)* and Swiss Webster mice infected with *Echinostoma caproni* or *Echinostoma trivolvis*. Masters Thesis, East Stroudsburg University.

Franco, J., Huffman, J.E. and Fried, B. (1986) Infectivity, growth, and development of *Echinostoma revolutum* (Digenea: Echinostomatidae) in the golden hamster, *Mesocricetus auratus*. *Journal of Parasitology* 72, 142-147.

Franco, J., Huffman, J.E. and Fried, B. (1988) The effects of crowding on adults of *Echinostoma revolutum* (Digenea: Echinostomatidae) in experimentally infected golden hamsters, *Mesocricetu auratus*. *Journal of Parasitology* 74, 240-243.

Frazer, B.A., Fried, B., Fujino, T. and Sleckman, B.P. (1999) Host-parasite relationships between *Echinostoma caproni* and RAG-2 deficient mice. *Parasitology Research* 85, 337-342.

Fried, B. (1984) Infectivity, growth and development of *Echinostoma revolutum* (Trematoda) in the domestic chick. *Journal of Helminthology* 58, 241-244.

Fried, B. and Weaver, L.J. (1969) Effects of temperature on the development and hatching of eggs of the trematode of *Echinostoma revolutum*. *Transactions of the American Microscopical Society* 88, 253-257.

Fried, B. and Butler, M.S. (1978) Infectivity, excystation, and development on the chick chorioallantois of the metacercaria of *Echinostoma revolutum* (Trematoda). *Journal of Parasitology* 64, 175-177.

Fied, B. and Nelson, D.D. (1978) Host-parasite relationships of *Zygocotyle lunata* (Trematoda) in the domestic chick. *Parasitology* 77, 49-55.

Fried, B. and Alenick, D.S. (1981) Localization, length and reproduction in single-and multiple-worm infections of *Echinostoma revolutum* (Trematoda). *Parasitology* 82, 49-53.

Fried, B. and Freeborne, N.E. (1984) Effects of *Echinostoma revolutum* (Trematoda) adults on various dimensions of the chicken intestine, and observations on worm crowding. *Proceedings of the Helminthological Society of Washington* 51, 297-300.

Fried, B. and Sousa, K.R. (1990) Single and five worm infections of *Echinostoma caproni* (Trematoda) in the ICR mouse. *International Journal for Parasitology* 20, 125-126.

Fried, B. and Huffman, J.E. (1996) The biology of the intestinal trematode *Echinostoma caproni*. *Advances in Parasitology* 38, 311-368.

Fried, B., Donovick, R.A. and Emili, S. (1988) Infectivity, growth and development of *Echinostoma liei* (Trematoda) in the domestic chick. *International Journal for Parasitology* 18, 413-414.

Fried, B., Mueller, T.J. and Frazer, B.A. (1997) Observations of *Echinostoma revolutum* and *Echinostoma trivolvis* in single and concurrent infections in domestic chicks. *International Journal for Parasitology* 27, 1319-1322.

Fujino, T. and Fried, B. (1993) Expulsion of *Echinostoma trivolvis* (Cort, 1914) Kanev, 1985 and retention of *Echinostoma caproni* Richard, 1964 (Trematoda: Echinostomatidae) in C3H mice: pathological, ultrastructural and cytochemical effects on the host intestine. *Parasitology Research* 79, 286-292.

Fujino, T., Yamada, M., Ichikawa, H., Fried, B., Arizomo, N. and Tada, I. (1996) Rapid expulsion of intestinal trematodes *Echinostoma trivolvis* and *Echinostoma caproni* from C3H/HeN mice after infection with *Nippostrongylus brasiliensis*. *Parasitology Research* 82, 577-579.

Fujino, T., Ichikawa, H., Fukuda, K. and Fried, B. (1998) The expulsion of *Echinostoma Trivolvis* caused by goblet cell hyperplasia in severe combined immunodeficient (SCID) mice. *Parasite* 5, 219-222.

Gower, W.C. (1937) Studies on the trematode parasites of ducks in Michigan with Special reference to the mallard. p. 1-94, Agricultural Experimental Station, Michigan State College.

Griffiths, H.J., Gonder, E. and Pomerory, B.S. (1976) An outbreak of trematodiasis in domestic geese. *Avian Diseases* 20, 604-606.

Hansen, K., Nielsen, J.W.W., Hinsbo, O. and Christensen, N.O. (1991) *Echinostoma caproni* in rats: worm population dynamics and host blood eosinophilia during primary infections with 6, 25 and 50 metacercariae and resistance to secondary and superimposed infection. *Parasitology Research* 77, 686-696.

Holben, D.M. (1993) Concurrent infection of *Echinostoma caproni* and *Nippostrongylus brasiliensis* in the golden hamster, *Mesocricetus auratus*. Master's Thesis East Stroudsburg University.

Holmes, J.C. (1990) Competition, contacts, and other factors restricting niches of parasitic helminths. *Annales de Parasitologie Humaine et Comparee* 65, 69-72.

Hosier, D.W. and Fried, B. (1986) Infectivity, growth, and distribution of *Echinostoma revolutum* in Swiss Webster and ICR mice. *Proceedings of the Helminthological Society of Washington* 53, 173-176.

Hosier, D.W. and Fried, B. (1991) Infectivity, growth and distribution of *Echinostoma caproni* (Trematoda) in the ICR mouse. *Journal of Parasitology* 77, 640-642.

Hosier, D.W., Fried, B. and Szewczak, J.P. (1988) Homologous and heterologous resistance of *Echinostoma revolutum* and *Echinostoma liei* in ICR mice. *Journal of Parasitology* 74, 89-92.

Hossain, M.I., Dewan, M.L., Baki, M.A. and Mondel, M.M.H. (1980) Pathology of *Echinostoma revolutum* infection in pigeons. *Bangladesh Veterinary Journal* 14, 1-3.

Huffman, J.E. and Fried, B. (1990) *Echinostoma* and Echinostomiasis. *Advances in Parasitology* 29, 215-269.

Huffman, J.E., Michos, C. and Fried, B. (1986) Clinical and pathological effects of *Echinostoma revolutum* (Digenea: Echinostomatidae) in the golden hamster, *Mesocricetus auratus*. *Parasitology* 93, 505-515.

Huffman, J.E., Iglesias, D. and Fried, B. (1988a) *Echinostoma revolutum:* Pathology of extraintestinal infection in the golden hamster. *International Journal for Parasitology* 18, 873-874.

Huffman, J.E., Alcaide, A., and Fried, B. (1988b) Single and concurrent infections of the golden hamster, *Mesocricetus auratus,* with *Echinostoma revolutum* and *Echinostoma liei* (Trematoda: Digenea). *Journal of Parasitology* 74, 604-608.

Huffman, J.E., Murphy, P.M. and Fried, B. (1992) Superimposed infections in golden hamsters infected with *Echinostoma caproni* and *Echinostoma trivolvis* (Digenea: Echinostomatidae). *Journal of the Helminthological Society of Washington* 59, 16-21.

Huffman, J.E., Pekala, R.F., Taylor, M.L., and Fried, B. (1998) The effects of *Echinostoma trivolvis* infections on the fertility and fecundity of golden hamsters *(Mesocricetus auratus)* and on the infectivity of their progeny. *Journal of the Helminthological Society of Washington* 65, 266-269.

Humphries, J.E., Reddy, A. and Fried, B. (1997) Infectivity and growth of *Echinostoma revolutum* (Froelich, 1802) in the domestic chick. *International Journal for Parasitology* 27, 129-130.

Iorio, S.L., Fried, B. and Hosier, D.W. (1991) Concurrent infections of *Echinostoma caproni* and *Echinostoma trivolvis* in ICR mice. *International Journal for Parasitology* 21, 714-718.

Jeyarasasinghan, U., Heyneman, D., Lim, H.K. and Mansour, N. (1972) Life cycle of a new echinostome from Egypt, *E. liei* sp. nov. (Trematoda: Echinostomatidae). *Parasitology* 65, 203-222.

Keymer, A. (1980) Density-dependent mechanisms in the regulation of intestinal helminth populations. *Parasitology* 84, 573-587.

Kim, S. and Fried, B. (1989) Pathological effects of *Echinostoma caproni* (Trematoda) In the domestic chick. *Journal of Helminthology* 63, 227-230.

Kinsella, J.M. and Forrester, D.J. (1972) Helminths of the Florida duck, *Anas platyrhynchos fulvigula*. *Proceedings of the Helminthological Society of Washington* 39, 173-176.

Kishore, N. and Sinha, D.P. (1982) Observations on *Echinostoma revolutum* infection in the rectum of domestic ducks *(Anas platyrhynchos domesticus)*. *Agricultural Science Digest* 2, 57-60.

Leidy, J. (1888) Trematodes of the muskrat. *Proceedings of the Academy of Natural Science*. Philadelphia 40, 126

Lie, K.J. and Umathevy, T. (1965) Studies on Echinostomatidae (Trematoda) in Malaya. VIII. The life history of *Echinostoma audyi* sp. n. *Journal of Parasitology* 51, 781-788.

Lie. K.J. and Basch, P.F. (1967) The life history of *Echinostoma paraensei* sp. n. (Trematoda: Echinostomatidae). *Journal of Parasitology* 53, 1192-1199.

Lie, K.J. and Nasemary, S. (1973) Studies on Echinostomatidae (Trematoda) in Malaysia. XV1. The life history of *Echinostoma ilocanum* (Garrison, 1908). *Proceedings of the Helminthological Society of Washington* 40, 49-65.

Mabus, J., Huffman, J.E. and Fried, B. (1988) Humoral and cellular response to infection with *Echinostoma revolutum* in the golden hamster, *Mesocricetus auratus*. *Journal of Helminthology* 62, 127-132.

Mahler, H., Christensen, N.O. and Hindsbo. (1995) Studies on the reproductive capacity of *Echinostoma caproni* (Trematoda) in hamsters and jirds. *International Journal for Parasitology* 25, 705-710.

Maurer, K., Decere, M. and Fried, B. (1995) Effects of the anthelmintics clorsulon, rafoxanide, mebendazole, and aprinocid of *Echinosotma caproni* in ICR mice. *Journal of Helminthology* 70, 95-96.

McMaster. R.P., Huffman, J.E. and Fried, B. (1995) The effects of dexamethasone on the course of *Echinostoma caproni* and *Echinostoma trivolvis* infections in the golden hamster (*Mesocricetus auratus*). *Parasitology Research* 81, 518-521.

Meece, J.K. and Nollen, P.M. (1996) A comparison of the adult and miracidial stages of *Echinostoma paraensei* and *Echinostoma caproni*. *International Journal for Parasitology* 26, 37-43.

Mohandas, A. and Nadakal, A.M. (1978) In vivo development of *Echinostoma malayanum* Leiper, 1911 with notes on effects of population density, chemical composition, and pathogenicity, and *in vitro* excystment of the metacercaria
(Trematoda: Echinostomatidae). *Zeitschrift fur Parasitenkunde* 55, 139-151.

Moravec, F., Barus, V., Rysavy, B. and Yousip, F. (1974) Observations on the development of two echinostomes, *Echinoparyphium recurvatum* and *Echinostoma revolutum*. The anatagonists of human schistosomes in Egypt. *Folia Parasitologia* 21, 107-126.

Mucha, K.H., Huffman, J.E. and Fried, B. (1990) Mallard ducklings (*Anas platyrhynchos*) experimentally infected with *Echinostoma trivolvis* (Digenea). *Journal of Parasitology* 76, 590-592.

Nasincova, V. (1986) Contribution to the distribution and life history of *Echinostoma revolutum* (Trematoda) in central Europe. *Vest cs. Spolec Zoologica* 50, 70-80.

Odaibo, A.B., Christensen, N.O. and Ukoli, F.M.A. (1988) Establishment, survival and fecundity in *Echinostoma caproni* infections in NMRI mice. *Proceedings of the Helminthological Society of Washington* 50, 265-269.

Ogilvie, B.M., Jones, V.E. (1973) Immunity in the parasitic reklationship between helminths and hosts. *Progress in Allergy* 17, 93-144.

Paul, W.E. (1989) Fundamental Immunology. Raven Press, New York.
Poleschchuk, V.P., Balayan, M.S., Frolova, M.P. and Dokin, V.P. (1988) Diseases of wild-caught moustached tamanins (*Saguinus mystax*) in captivity. *Zeitschrift fur Versuchstierkunde* 31, 69-75.

Ramalingam, S. and Samuel, W.M. (1978) Helminths in the great horned owl *Bubo virginianus* and snowy owl *Nyctea scandiaca* of Alberta, Canada. *Canadian Journal of Zoology* 56, 2454-2456.

Rankin, J.S. Jr. (1937) An ecological study of parasites of some North Carolina salamanders. *Ecological Monographs* 7, 169-267.

Richard, J. and Brygoo, E.P. (1978) Life cycle of the trematode *Echinostoma caproni* Richard, 1964 (Echinostomatoidea). *Annales de Parasitologie Humaine et Comparee (Paris)* 53, 265-275.

Rim, H.J. (1982) *CRC Handbook Series in Zoonoses* (G.V. Hillyer and C.E. Hopla eds). Pp. 53-69. CRC Press, Boca Raton, Florida.

Robertson, E. L. and Courtney, C.H. (1995) *Veterinary Pharmacology and Therapeutics* H.R. Adams (ed). Iowa State Univ. Press, Ames.

Rosario, C. and Fried, B. (1999) Effects of a protein-free diet on worm recovery, growth and distribution of *Echinostoma caproni* in ICR mice. *Journal of Helminthology* 73, 167-169.

Saito, S. (1984) Development of *Echinostoma hortense* in rats, with special reference to the genital organs. *Japanese Journal of Parasitology* 33, 51-61.

Scrimshaw, N.S., Taylor, C.E. and Gordon, J.E. (1968) *Interactions of Nutrition and Infection*. WHO Monograph Series. No. 57. World Health Organization, Geneva, Switzerland.

Senger, C.M. 1954. Notes on the growth, development and survival of two echinostome trematodes. *Experimental Parasitology* 6, 491-496.

Shaw, M.G. and Kocan, A.A. (1980) Helminth fauna of waterfowl in Central Oklahoma. *Journal of Wildlife Diseases* 16, 59-64.

Simpkins, H.L. and Fried, B. (1999) Effects of a diet deficient in vitamins A, D, and E on infectivity, growth, and development of *Echinostoma trivolvis* in domestic Chicks. *Parasitology Research* 85,

873-875.

Simonsen, P.E., Estambale, B.B. and Agger, M. (1991) Antibodies in the serum of golden hamsters experimentally infected with the intestinal trematode *Echinostoma caproni. Journal of Helminthology* 65, 239-247.

Sirag, K.I., Christensen, N.O., Frandsen, F., Monrad, J. and Nansen, P. (1994) Homologous and heterologous resistance in *Echinostoma revolutum* infections in mice. *Parasitology* 80, 479-486.

Skryabin, K.I. (1979) *Trematodes of Animals and Man,* Vol. 1. U.S. Department of Commerce, National Technical Information Service.

Smith, H.J. (1978) Parasites of red foxes in New Brunswick and Nova Scotia. *Journal of Wildlife Diseases* 14, 366-370.

Stanley, J.G. and Rabalais, F.C. (1971) Helminth parasites of the red-winged blackbird, *Agelaius phoeniceus* and common grackle, *Quiscalus quiscula* in north-western Ohio. *Ohio Journal of Science* 71, 302-303.

Sudati, J.E., Reddy, A and Fried, B. (1996) Effects of high fat diets on worm recovery, growth and distribution of *Echinostoma caproni* in ICR mice. *Journal of Helminthology* 70, 351-354.

Sudati, J.E., Rivas, F. and Fried, B. (1997) Effects of a high protein diet on worm recovery, growth and distribution of *Echinostoma caproni* in ICR mice. *Journal Helminthology* 71, 351-354.

Threlfall, W. (1980) Helminth parasites of an American flamingo from New Foundland, Canada. *Proceedings of the Helminthological Society of Washington* 48, 89-90.

Wiley, C.H. (1941) The life history and bionomics of the trematode *Zygocotyle lunata* (Paramphistomidae). *Zoologica* 26, 65-88.

Wilkinson, J.N., Canaris, A.G. and Broderson, D. (1977) Parasites of waterfowl from Southwest Texas, I: the northern cinnamon teal, *Anas cyanoptera septentrionalium. Journal of Wildlife Diseases* 13, 62-63.

Yao, G., Huffman, J.E. and Fried, B. (1991) The effects of crowding on adults of *Echinostoma caproni* in experimentally infected golden hamsters. *Journal of Helminthology* 65, 248-254.

HUMAN ECHINOSTOMIASIS: MECHANISMS OF PATHOGENESIS AND HOST RESISTANCE

M.A. HASEEB[a], AND L.K. EVELAND[b]

[a]Department of Microbiology and Immunology and Medicine, State University of New York, Health Science Center, New York, New York 11203, USA, and [b]Department of Biological Sciences, California State University, Long Beach, California 90840, USA

B. Fried and T.K. Graczyk (eds.),
Echinostomes as Experimental Models for Biological Research, 83–98.
© 2000 Kluwer Academic Publishers. Printed in the Netherlands.

1. Introduction

Human echinostomiasis refers to infection of the human intestine by digenetic trematodes of the family Echinostomatidae which primarily infect a variety of birds and mammals other than humans.   The life cycles of these worms are completed in nature without human involvement and the natural definitive hosts may also serve as reservoirs for human infection.   Human echinostomiasis, caused by about 20 species belonging to eight genera (Table 1), is thus considered a zoonosis.

Although echinostomiasis occurs worldwide, most human infections are reported from foci in east and southeast Asia.   Fortunately, human echinostomiasis is relatively rare, yet the foci of transmission remain endemic owing to the local dietary preferences.   Most of these endemic foci are localized in China, India, Indonesia, Japan, Korea, Malaysia, Philippines, Russia, Taiwan and Thailand (Fig. 1).   One or two cases have also been reported from each of the following countries: Brazil, Egypt, Italy, Mexico and Romania (see Section 3. Epidemiology).

Echinostomes are distomate, hermaphroditic flukes with a distinctive horseshoe-shaped collar on the dorsolateral aspects of the oral sucker.   The collar bears prominent spines whose number is characteristic of the species.   Based on the arrangement of collar spines, the echinostome species that infect humans can be divided into three groups: Echinostomatinae which includes species with collar spines arranged in two rows, interrupted ventrally but not dorsally (*Artyfechinostomum, Echinoparyphium, Echinostoma, Paryphostomum, Hypoderaeum*); Himasthlinae which includes species with collar spines arranged in one row, interrupted ventrally (*Himasthla*); and Echinochasminae whose members have collar spines arranged in one row, interrupted both dorsally and ventrally (*Echinochasmus, Episthmium*) (Beaver *et al.*, 1984).

This chapter will document occurrence and distribution of human echinostomiasis. It will outline the parasitological features (including life cycle and diagnostic forms) of parasites and epidemiologic characteristics that determine the transmission and prevalence of echinostomiasis in humans.   Although little is known about features of human disease and the underlying mechanisms, there is extensive information from experimental work which will form the basis for outlining mechanisms likely to be operative in pathogenesis and host resistance in human echinostomiasis.

TABLE 1. Members of the Echinostomatidae reported from humans

Species	Number of collar spines	Geographic distribution	Definitive hosts other than man	Source of human infection
Artyfechinostomum mehrai	39-42	India	Pig, rat	Snail
Echinochasmus angustitestis	Unknown	China	Unknown	Fish
Echinochasmus fujilanensis	Unknown	China	Unknown	Fish
Echinochasmus japonicus	24	China, Japan, Korea	Cat, dog, rat	Fish
Echinochasmus liliputanus	Unknown	China	Unknown	Fish
Echinochasmus perfoliatus (=Echinostoma perfoliatus)	24	China, Egypt, Japan, Italy, Romania, Russia	Cat, dog, fox, hog, rat	Fish
Echinoparyphium paraulum	37	Russia	Dove, duck, goose, swan	Unknown
Echinoparyphium recurvatum (=Euparyphium recurvatum)	45	Egypt, Indonesia, Taiwan	Fowl, rat	Frog, snail, tadpole
Echinostoma cinetorchis	37	Japan, Korea, Taiwan	Rat	Frog, tadpole
Echinostoma echinatum (=Echinostoma lindoense)	37	Brazil, Indonesia, Japan, Thailand	Duck, rat	Mussel, snail
Echinostoma hortense	27-28	Japan, Korea	Dog, rat	Fish, frog
Echinostoma ilocanum (=Euparyphium ilocanum)	51	China, India, Indonesia, Malaysia, Philippines, Thailand	Dog, rat	Snail

Echinostoma japonicum	Unknown	Japan, Korea	Unknown	Unknown
Echinostoma macrorchis	45	Indonesia, Japan, Korea, Taiwan	Rat	Snail
Echinostoma malayanum (=*Euparyphium malayanum*)	43	China, India, Indonesia, Malaysia, Philippines, Singapore, Thailand	Rat	Fish, snail, tadpole
Echinostoma melis (=*Euparyphium jassyense* =*Echinostoma jassyense*)	27	China, Romania, Taiwan	Unknown	Tadpole
*Echinostoma revolutum**	37	China, Egypt, Indonesia, Malaysia, Mexico, Taiwan, Thailand	Duck, goose, rat	Clam, snail, tadpole
Episthmium caninum	Unknown	Thailand	Dog	Fish
Himasthla muehlensi	Unknown	Colombia (?), New York (?)	Gull	Clam
Hypodereaum conoideum	47-53	Thailand	Duck, fowl	Snail, tadpole
Paryphostomum sufrartyfex	39-42	India	Dog, hog, rat	Snail

* The taxonomy of echinostomes with 37 collar spines has undergone extensive revisions in recent years. It is likely that *Echinostoma revolutum* was reported for *Echinostoma echinatum*. See Huffman and Fried (1990); Kanev (1994); Kanev *et al.* (1995); and Chapter 2. Life cycles have not been worked out for *Echinochasmus angustitestis*, *Echinochasmus fujilanensis*, *Echinochasmus liliputanus*, *Echinoparyphium paraulum*, *Episthmium caninum*, *Himasthla muehlensi* and *Paryphostomum sufrartyfex*. See Section 3. (Epidemiology) for more information on *Himasthla muehlensi*.

2. Life Cycle

The echinostomes that infect humans reside as adults in the intestine of man or their natural definitive hosts and their eggs are voided with the host feces. The operculate eggs are partially developed when laid and miracidial development takes place outside the host in about two to three weeks. In fresh water echinostome eggs hatch and the miracidia begin to search for a snail. Upon penetration of an appropriate snail host, miracidia develop into mother sporocysts which in turn give rise to mother rediae, daughter rediae, and finally cercariae in about six to seven weeks post-infection.

The cercariae have strong tails and a collar of spines similar to their respective adults. Upon release from the snail (first) intermediate host, cercariae swim to locate a second intermediate host in which they encyst to become metacercariae. The cercariae may encyst within the redia or the snail in which they developed. They may also encyst in other snails, bivalves, amphibians and fresh water fish. When the second intermediate host, with encysted metacercariae, is ingested by a definitive host, the metacercariae excyst and the developing worms usually attach to the mucosa of the small intestine where they develop into adults in about two weeks (Fig.2). Echinostome adults tend to aggregate in the host and there is evidence from studies *in vitro* for a chemical signal(s) mediating clustering of adult worms (Fried *et al.*, 1980; Haseeb and Fried, 1988). Worm aggregation in the host may be an adaptation to facilitate cross-fertilization (see Chapter 7). As with the second intermediate host, echinostomes exhibit little host specificity for a definitive host. However, host-preference for some echinostomes in experimental vertebrate hosts has been documented (Hosier and Fried, 1986; Huffman *et al.*, 1988a,b; Odaibo *et al.*, 1988, 1989). Their natural definitive hosts include a variety of birds and mammals, and the same echinostome species may develop in birds and mammals. The likely reason for widespread and occasionally distant occurrence is that they infect birds and are thus carried to these locations (see Table 1). Human infections occur most commonly in areas where it is customary to eat raw or undercooked mollusks, fresh water fish or tadpoles (Belding, 1965; Beaver *et al.*, 1984; Fried and Huffman, 1996; see Section 3. Epidemiology).

3. Epidemiology

Current incidence of echinostomiasis is difficult to determine with any accuracy because of unavailability of systematic surveys in recent years. We are largely relying on historic surveys, limited recent surveys and occasional case reports.

The prevalence of *Echinostoma echinatum*, one of the four or five echinostomes infecting humans in Indonesia, in residents of Lake Lindu Valley during the period 1937-1956 was reported to be 42.6% and in some areas it was as high as 96% (Carney *et al.*, 1980). More recent surveys have reported it to be encountered occasionally (Clarke *et al.*, 1974; Kusharyono et al., 1991) or in 1% of the survey specimens (Carney, 1991). This significant decline in the incidence of echinostomiasis in this area has been attributed to the ecological changes brought about by the introduction of a new fish, *Tilapia mossambica*, which resulted in almost complete disappearance of clams (*Corbicola* spp.) that served as second intermediate hosts and were part of the local diet (Carney *et al.*, 1980; Graczyk and Fried, 1998).

Echinostoma ilocanum and *Echinostoma malayanum* have an overall prevalence of 3% in the Philippines (Eduardo, 1991); however, in northern Luzon the prevalence

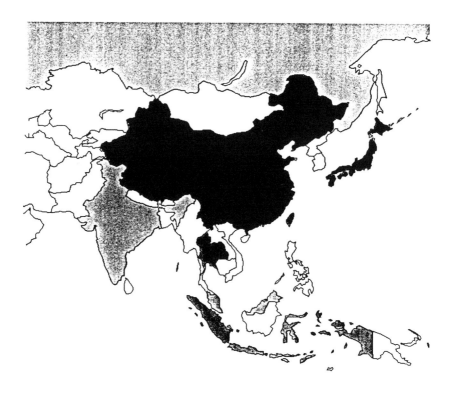

Figure 1. Occurrence of members of the Echinostomatidae causing human echinostomiasis in southeast Asia. Darkness of shading reflects the number of echinostome species reported from a particular area: countries with fewer echinostome species appear lighter and those with more species appear darker. Areas from which human echinostomiasis have not been reported are not shaded

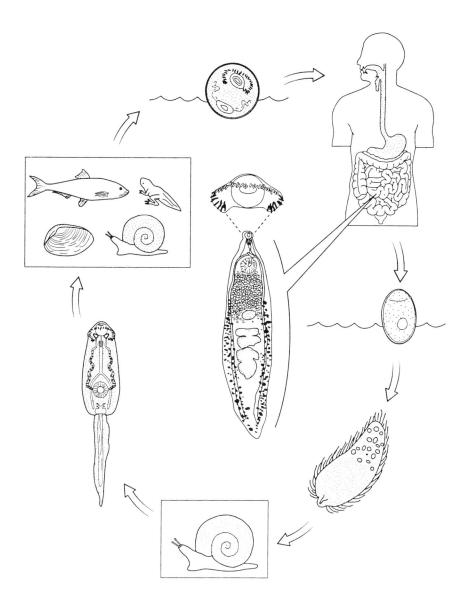

Figure 2. A generalized life cycle of echinostomes which infect humans. Its purpose is to document: 1) the developmental form (metacercaria) infective for humans; 2) residence of adult worms in the human small intestine; 3) passage of echinostome eggs in human feces; and, 4) succession of echinostome developmental stages in the two intermediate hosts and the definitive host. No attempt has been made to include all likely first and second intermediate hosts or definitive hosts

Echinostoma ilocanum averages 11%, reaching 44% in some areas (Cross and Basaca-Sevilla, 1986). The prevalence of *Echinostoma ilocanum*, *Echinostoma malayanum* and *Echinostoma revolutum* in northeast Thailand has been reported to be 8.3% in males and 2.9% in females (Radomyos *et al.*, 1994). In another area of the Philippines, Echague, the overall prevalence of *Echinostoma malayanum* has been recorded to be 20% (range 0 to 35.7%) (Tangtrongchitr and Monzon, 1991). It has also been reported in a tribal community near Calcutta, India, where pigs are suspected to be the natural reservoirs of infection (Maji *et al.*, 1993). Of the five echinostome species infecting humans in Taiwan, *Echinoparyphium recurvatum*, *Echinostoma melis* and *Echinostoma revolutum* (most likely *Echinostoma echinatum*) were found to infect 11 to 65% of the people surveyed (Lu, 1982; Carney, 1991). In Fujian and Guangdong provinces of China, the prevalence of echinostome infections has been reported to be 5% (Li, 1991). A survey conducted by the Ministry of Public Health during the period 1988-1992 revealed three new species (*Echinochasmus angustitestis, Echinochasmus fujilanensis, Echinochasmus liliputanus*) whose life cycles have not been worked out (Yu *et al.*, 1994).

A prevalence of 9 to more than 50% has been recorded for the various echinostome species from Japan and Korea (see Table 1). This information is available in a recent review (Graczyk and Fried, 1998).

As stated above, a number of isolated cases have been reported from Brazil, Egypt, Italy, Mexico and Romania. In a number of instances it is not possible to ascertain where and how the infection was acquired. One of the notable cases is that of a German patient diagnosed with *Himasthla muehlensi* in Hamburg. He had lived in Colombia for six years and gave a history of eating raw clams in New York City (Belding, 1965; Beaver *et al.*, 1984). It remains undetermined whether he acquired infection in Colombia or in New York City. The origin of clams eaten in New York City also remains unknown.

4. Pathogenesis and Host Resistance

Host-parasite relationships in human echinostomiasis have not been investigated and therefore pathogenesis remains undetermined. The clinical manifestations that have been described include headache, dizziness, nausea, vomiting, abdominal pain, fever, flatulence, diarrhea with blood and mucus, anemia and eosinophilia. Severity of symptoms has generally been attributed to "heavy infection", most often without giving the worm burden. Such infections reportedly caused inflammation and ulceration of the intestinal mucosa. However, significant morbidity in association with echinostomiasis has not been reported. Since human disease is poorly documented, we have attempted to reconstruct events from studies of echinostome infections in naturally-infected and experimental animals.

In experimental infections of vertebrate hosts, echinostome metacercarial cysts most likely excyst in the jejunum or ileum and juvenile worms may colonize locally or any other part of the lower intestine. Excystation is likely affected by the local pH, temperature and bile salt concentration. The fact that metacercarial cysts of *Echinostoma caproni* implanted into the small intestines of mice develop into adult worms suggests that excystation results in response to local conditions of the small intestine and that passage through the stomach does not necessarily play a significant role in this process (Chien *et al.*, 1993). Information on the physiology of metacercarial excystation is available in a recent review (Fried, 1994). The preponderance of worms are localized in the jejunum but exhibit lack of site specificity. As they grow relatively older, they may be found in the lower gastrointestinal tract. Some species may even explore extraintestinal sites. *Echinostoma trivolvis* has been found in the common bile duct, pancreas, gall bladder and liver of experimentally-infected golden hamsters (Huffman *et al.*, 1988). Apparently, extraintestinal migration is a species-specific characteristic since in the same vertebrate model, *Echinostoma caproni* was not found in any extraintestinal location (Huffman *et al.*, 1988). As in infections with other trematodes which normally enter the common bile duct (e.g., *Clonorchis sinensis*), it is not known whether these trematodes follow any cues; chemical cues mediating site selection in trematodes remain undefined. The growth rate, worm size, fecundity and longevity are affected by the location of worms in the host intestine and the number of worms at that location (Huffman and Fried, 1990).

The echinostome armature consisting of tegumentary spines, the two muscular suckers and the large collar spines may cause physical damage to the intestinal mucosa of the definitive host. The backwardly-pointed collar spines most likely help the worm to secure itself in the mucosa. The oral sucker has been shown to draw intestinal villi in a manner similar to that used by hookworms (Fried and Huffman, 1996). The acetabulum has also been shown to grasp the mucosa in a similar manner but the grasped host tissue lacked an inflammatory response (Simonsen *et al.*, 1989). *Echinostoma malayanum* causes mucosal destruction, edema of the lamina propria and hyperplasia of the epithelial cells in experimental rats (Mohandas and Nadakal, 1978). Golden hamsters infected with *Echinostoma trivolvis* develop watery diarrhea and weight loss and eventually die. The histopathological examination of the intestine shows erosion of villi and mononuclear cell infiltrates (Huffman *et al.*, 1986). Similar observations have been made in mice infected with *Echinostoma caproni* (Bindseil and Christensen, 1984; Simonsen *et al.*, 1989). In addition to villous atrophy and crypt hyperplasia, marked dilation of the region of the intestine occupied by worms has also been reported (Bindseil and Christensen, 1984; Weinstein and Fried, 1991). These observations indicate that the disease results primarily from the physical damage caused by worms.

The life-span of adult echinostomes in experimental infections is variable depending on the combination of echinostome species and the experimental host. Apparently, different species parasitize humans for different lengths of time. *Echinostoma echinatum* reportedly may live in humans for ten months but in mice and rats it is expelled soon after attaining maturity (Carney *et al.*, 1980; Huffman and Fried, 1990). A knowledge of the factors that determine parasite longevity in a given host may contribute to our understanding of the mechanisms of pathogenesis in echinostomiasis.

In the intestines of experimental animals, both *Echinostoma caproni* and *Echinostoma trivolvis* induce changes at the cellular level and in the expression of certain glycoconjugates. The physical damage to the intestinal mucosa may be partly responsible for the altered expression and distribution of glycoconjugates. However, the concurrent increase or decrease in the number of goblet cells in mice infected with *Echinostoma trivolvis* and *Echinostoma caproni*, respectively, point to additional factors affecting glycoconjugate expression (Fujino and Fried, 1993). Studies of experimental infections of *Echinostoma trivolvis* in mice have shown that the number of goblet cells and mast cells increase with a peak coinciding with worm expulsion only in normal mice but not in congenitally athymic mice, yet the kinetics of expulsion were similar in both kinds of mice (Fujino *et al.*, 1993). These observations suggest that the goblet cell hyperplasia and mastocytosis may be coincidental to worm expulsion and that the changes in populations of these cells are most likely T cell-dependent, but the factors that are responsible for worm expulsion are T cell-independent.

Expulsion of *Echinostoma trivolvis* was slightly delayed when mice were administered dexamethasone for 5 to 7 days postinfection. In contrast to control mice, dexamethasone-treated mice did not increase their goblet cell numbers. The worm expulsion corresponded with the cessation of dexamethasone administration and an increase in the number of goblet cells (Fujino *et al.*, 1996). Consistent with these observations, when dexamethasone administration was extended to thirty days postinfection, expulsion of worms was delayed correspondingly. In addition, mice did not develop goblet cell hyperplasia throughout the period of dexamethasone administration (Fujino *et al.*, 1997). As in normal mice, dexamethasone administration in severe combined immunodeficient (SCID[1]) mice also delayed the expulsion of *Echinostoma trivolvis*. Dexamethasone administration also affected goblet cell and mast cell numbers in a manner observed in normal mice (Fujino *et al.*, 1998). In sharp contrast to studies in congenitally athymic mice (Fujino *et al.*, 1993), these observations suggest that the goblet cell hyperplasia, mastocytosis and worm expulsion mechanisms are all T cell-independent. Administration of another immunosuppressive agent, FK-506[2], resulted in selective suppression of mast cell hyperplasia but did not affect goblet cell numbers, and kinetics of worm expulsion in untreated and treated mice were similar (Fujino *et al.*, 1998). Furthermore, rapid expulsion of challenge infections (both homologous and heterologous) took place when challenge infections were given during the period of expulsion of worms of the primary infection. The worm expulsion

---

[1]SCID (scid/scid) mice lack functional B and T cells but have normal macrophages. These mice, homozygous for scid mutation, express defective V(D)J recombinase activity, which is essential for normal immunoglobulin and T cell receptor gene rearrangement (Bosma *et al.*, 1983; Bosma and Carroll, 1991).

[2]FK-506, isolated from *Streptomyces tsukubaensis*, is a macrolide with a wide spectrum of immunosuppressive properties (Kino *et al.*, 1987).

corresponded with the increase in the number of goblet cells (Fujino et al., 1996).

Unlike Echinostoma trivolvis infections, those with Echinostoma caproni do not induce goblet cell hyperplasia in normal mice but result in marked hyperplasia in RAG-2-deficient[3] mice (Weinstein and Fried, 1991; Frazer et al., 1999). Since worms survived in RAG-2-deficient mice as long as they did in ICR mice in the presence of a marked goblet cell response, the goblet cell hyperplasia is most likely an adjunct to another mechanism(s) for worm expulsion.

Goblet cell hyperplasia and mastocytosis have been reported in a variety of animal models of gut-inhabiting nematodes, cestodes and trematodes with variable contribution to worm expulsion. Expulsion of Nippostrongylus brasiliensis from rats is associated with goblet cell hyperplasia (Tiuria et al., 1995), but that of Strongyloides venezuelensis is associated with mastocytosis although one animal model (Syrian golden hamster) uniquely develops goblet cell hyperplasia (Shi et al., 1995; Tiuria et al., 1995). Hymenolepis diminuta induces both mastocytosis and goblet cell hyperplasia in rats but only the former is associated with worm expulsion (McKay et al., 1990; Ishih, 1992). Neodiplostomum seoulense also induces both mastocytosis and goblet cell hyperplasia but neither response has been attributed the primary role of worm expulsion (Chai et al., 1998).

It would be reasonable to conclude that in various intestinal helminth infections, including echinostomiasis, goblet cell hyperplasia, and thereby hypersecretion of mucus, may play a major role in worm expulsion and that there may be accessory factors involved in this process. Studies in IL-13-deficient mice, which do not develop goblet cell hyperplasia when infected with Nippostrongylus brasiliensis and fail to expel the worms (McKenzie et al., 1998), also support this notion. Goblet cell hyperplasia and mastocytosis are the only known host responses which are thought to mediate worm expulsion. Since echinostomes are not known to cause a long-term or chronic infection in any of the definitive hosts, it is likely that both nonspecific and specific host responses contribute to termination of infection. Contribution of locally-produced immunoglobulins (sIgA and IgE) or other specific responses in expulsion of echinostomes remains unknown.

Relative resistance to homologous and heterologous challenge infections has been demonstrated in experimental vertebrate hosts but owing to the lack of clinical epidemiologic data it is not known whether humans attain resistance following a primary infection or whether there is increased resistance in relation to increasing host age (See Chapter 8 for further discussion). The fact that mice infected previously or concurrently with Trypanosoma brucei, Schistosoma mansoni (infections known to elicit polyclonal lymphocyte activation) or Echinostoma revolutum had their ability to expel a challenge Echinostoma revolutum infection impaired suggests a role for specific immune mechanisms in expulsion of echinostomes from the mammalian gut

---

[3]RAG-2-deficient mice lack a part of the recombinase activating gene. Homozygous mutants fail to produce mature B or T lymphocytes (Shinkai et al., 1992).

(Christensen *et al.*, 1984, 1985, 1986).

5. Diagnosis

The human echinostome infections may remain inapparent or become symptomatic depending on worm burden. The symptoms, even when present, are often nonspecific, making a clinical diagnosis difficult, particularly in nonendemic areas. Local physicians, with the knowledge of endemicity of these infections and greater familiarity with the presenting signs and symptoms may, however, make clinical diagnoses.

Laboratory diagnosis of echinostomiasis is based on demonstration of eggs in human feces. The yellow to yellow-brown eggs are thin-shelled with an operculum that is usually difficult to see and a slight thickening of the shell at the abopercular end. They are unembryonated when passed in feces and measure 80 - 135 μm x 55 - 80 μm, depending on the echinostome species. Determination of echinostome species based on egg morphology is usually not possible. However, worms recovered following treatment may be speciated.

Since echinostome infections are rare in humans, laboratory workers in non-endemic areas are not proficient at identifying eggs or adults. It is likely that echinostome infections in immigrants and visitors from endemic areas are either missed because of nonspecific symptoms or misdiagnosed because eggs resemble other trematode eggs detected in human feces.

Immunglobulins (IgA, IgG, IgM) have been detected by enzyme immunoassays in the sera of mice and golden hamsters infected with *Echinostoma caproni* and *Echinostoma trivolvis* (Agger *et al.*, 1993; Simonsen *et al.*, 1991; Graczyk and Fried, 1994, 1995). Antibodies were detectable in mice a few days after infection but hamsters took several weeks to elicit a comparable response. In view of these observations it is not possible to predict the kinetics of a humoral response to an echinostome infection in humans. It would be premature to speculate on the diagnostic utility of serologic assays in the absence of epidemiologic correlates that may help determine specificity of such assays and may also help develop guidelines for interpretation of their results. Furthermore, such diagnostic approaches have proven impractical in other helminth infections which remain restricted to the human gastrointestinal tract.

6. Prevention and Control

Birds and wild or domestic animals are the natural definitive hosts for echinostomes, and human infections, in virtually every instance, are acquired from eating fish, mollusks or tadpoles, either raw or improperly cooked. Therefore, prevention of echinostome infections in humans is a matter of avoiding the sources of infection (Tangtrongchitr and Monzon, 1991; Haseeb and Fried, 1997).

In many areas where there are fuel shortages, or economic circumstances that preclude the purchase of fuel for cooking food, prevention may be difficult. Also, in many endemic areas traditional methods for preparing food may include fermentation, smoking or pickling, which may not kill metacercariae. Fish and mollusks prepared in this manner may also be shipped to places where unsuspecting people could acquire infection. For this reason, controls on imported products may reduce human infections in nonendemic areas. Because birds and mammals are primary sources of echinostome eggs, sanitary disposal of human waste would probably not appreciably affect the spread of echinostomiasis. However, any of these measures are also expected to control clonorchiasis and other trematode infections which are often present in areas considered endemic for human echinostomiasis (Anonymous, 1995; Haseeb and Fried, 1997).

7. References

Anonymous (1995) *Control of Foodborne Trematode Infections*. Report of a WHO Study Group. *WHO Technical Report Series* 849, 1-157.

Agger, M.K., Simonsen, P.E. and Vennervald, B.J. (1993) The antibody response in serum, intestinal wall and intestinal lumen of NMRI mice infected with *Echinostoma caproni*. *Journal of Helminthology* 67, 169-178.

Beaver, P.C., Jung, R.C. and Cupp, E.W. (1984) *Clinical Parasitology*. Lea & Febiger, Philadelphia.

Belding, D.L. (1965) *Textbook of Parasitology*. Appleton-Century-Crofts, New York.

Bindseil, E. and Christensen, N.Ø. (1984) Thymus-independent crypt hyperplasia and villous atrophy in the small intestine of mice infected with the trematode *Echinostoma revolutum*. *Parasitology* 88, 431-438.

Bosma, G.C., Custer, R.P. and Bosma, M.J. (1983) A severe combined immunodeficiency mutation in the mouse. *Nature (London)* 301, 527-530.

Bosma, M.J. and Carroll, A.M. (1991) The SCID mouse mutant: definition, characterization, and potential uses. *Annual Review of Immunology* 9, 323-350.

Carney, W.P. (1991) Echinostomiasis - a snail-borne intestinal trematode zoonosis. *Southeast Asian Journal of Tropical Medicine and Public Health* 22, 206-211.

Carney, W.P., Sudomo, M. and Purnomo, A. (1980) Echinostomiasis: a disease that disappeared. *Tropical and Geographical Medicine* 32, 101-105.

Chai, J.Y., Kim, T.K., Cho, W.H., Seo, M., Kook, J., Guk, S.M. and Lee, S.H. (1998) Intestinal mastocytosis and goblet cell hyperplasia in BALB/c and C3H mice infected with *Neodiplostomum seoulense*. *Korean Journal of Parasitology* 36, 109-119.

Chien, W.Y., Hosier, D.W. and Fried, B. (1993) Surgical implantation of *Echinostoma caproni* metacercariae and adults into the small intestine of ICR mice. *Journal of the Helminthological Society of Washington* 60, 122-123.

Christensen, N.Ø., Fagbemi, B.O. and Nansen, P. (1984) *Trypanosoma brucei*-induced blockage of expulsion of *Echinostoma revolutum* and of homologous *E. revolutum* resistance in mice. *Journal of Parasitology* 70, 558-561.

Christensen, N.Ø., Knudsen, J. and Andreassen, J. (1986) *Echinostoma revolutum*: resistance to secondary and superimposed infections in mice. *Experimental Parasitology* 61, 311-318.

Christensen, N.Ø., Knudsen, J., Fagbemi, B. and Nansen, P. (1985). Impairment of primary expulsion of *Echinostoma revolutum* in mice concurrently infected with *Schistosoma mansoni*. *Journal of Helminthology* 59, 333-335.

Clarke, M.D., Carnet, W.P., Cross, J.H., Hadidjaja, P., Oemijati, S. and Joescef, A. (1974) Schistosomiasis and other human parasitoses of Lake Lindu in Central Suwalesi (Celebes), Indonesia. *American Journal of Tropical Medicine and Hygiene* 23, 385-392.

Cross, J.H. and Basaca-Sevilla, V. (1986) Studies on *Echinostoma ilocanum* in the Philippines. *Southeast Asian Journal of Tropical Medicine and Public Health* 17, 23-27.

Eduardo, S.L. (1991) Food-borne zoonoses in the Philippines. *Southeast Asian Journal of Tropical Medicine and Public Health* 22, 16-22.

Frazer, B.A., Fried, B., Fujino, T. and Sleckman, B.P. (1999) Host-parasite relationships between *Echinostoma caproni* and RAG-2-deficient mice. *Parasitology Research* 85, 337-342.

Fried, B. (1994) Metacercarial excystment of trematodes. *Advances in Parasitology* 33, 91-144.

Fried, B. and Huffman, J.E. (1996) The biology of the intestinal trematode *Echinostoma caproni*. *Advances in Parasitology* 38, 311-368.

Fried, B., Tancer, R.B. and Fleming, S.J. (1980) *In vitro* pairing of *Echinostoma revolutum* (Trematoda) metacercariae and adults, and characterization of worm products involved in chemoattraction. *Journal of Parasitology* 66, 1014-1018.

Fujino, T. and Fried, B. (1993) *Echinostoma caproni* and *E. trivolvis* alter the binding of glycoconjugates in intestinal mucosa of C3H mice as determined by lectin histochemistry. *Journal of Helminthology* 67, 179-188.

Fujino, T., Fried, B., Ichikawa, H. and Tada, I. (1996a) Rapid expulsion of the intestinal trematodes *Echinostoma trivolvis* and *E. caproni* from C3H mice by trapping with increased goblet cell mucins. *International Journal for Parasitology* 26, 319-324.

Fujino, T., Fried, B. and Tada, I. (1993) The expulsion of *Echinostoma trivolvis*: worm kinetics and intestinal cytopathology in conventional and congenitally athymic BALB/c mice. *Parasitology* 106, 297-304.

Fujino, T., Ichikawa, H. and Fried, B. (1998a) The immunosuppressive compound FK506 does not affect expulsion of *Echinostoma trivolvis* in C3H mice. *Parasitology Research* 84, 519-521.

Fujino, T., Ichikawa, H., Fried, B. and Fukuda, K. (1996b) The expulsion of *Echinostoma trivolvis*: suppressive effects of dexamethasone on goblet cell hyperplasia and worm rejection in C3H/HEN mice. *Parasite* 3, 283-289.

Fujino, T., Ichikawa, H., Fried, B. and Fukuda, K. (1997) The expulsion of *Echinostoma trivolvis*: worm kinetics and intestinal reactions in C3H/HeN mice treated with dexamethasone. *Journal of Helminthology* 71, 257-259.

Fujino, T., Ichikawa, H., Fukuda, J. and Fried, B. (1998b) The expulsion of *Echinostoma trivolvis* by goblet cell hyperplasia in severe combined immunodeficient (SCID) mice. *Parasite* 5, 219-222.

Graczyk, T.K. and Fried, B. (1994) ELISA method for detecting anti-*Echinostoma caproni* (Trematoda: Echinostomatidae) immunoglobulins in experimentally infected ICR mice. *Journal of Parasitology* 80, 544-549.

Graczyk, T.K. and Fried, B. (1995) An enzyme-linked immunosorbent assay for detecting anti-*Echinostoma trivolvis* (trematoda) IgG in experimentally infected ICR mice. Cross-reactivity with *E. caproni*. *Parasitology Research* 81, 710-712.

Graczyk, T.K. and Fried, B. (1998) Echinostomiasis: a common but forgotten food-borne disease. *American Journal of Tropical Medicine and Hygiene* 58, 501-504.

Haseeb, M.A. and Fried, B (1988). Chemical communication in helminths. *Advances in Parasitology* 27, 169-207.

Haseeb, M.A. and Fried, B. (1997) Modes of transmission of trematode infections and their control. *In* B. Fried and T.K. Graczyk (Eds), *Advances in Trematode Biology*. pp. 31-56. CRC Press, Boca Raton, FL.

Hosier, D.W. and Fried, B. (1986) Infectivity, growth, and distribution of *Echinostoma revolutum* in Swiss Webster and ICR mice. *Proceedings of the Helminthological Society of Washington* 53, 173-176.

Huffman, J.E. and Fried, B. (1990) *Echinostoma* and echinostomiasis. *Advances in Parasitology* 29, 215-269.

Huffman, J.E., Michos, C. and Fried, B. (1986) Clinical and pathological effects of *Echinostoma revolutum* (Digenea: Echinostomatidae) in the golden hamster, *Mesocricetus auratus*. *Parasitology* 93, 505-515.

Huffman, J.E., Alcaide, A. and Fried, B. (1988a) Single and concurrent infections of the golden hamster, *Mesocricetus auratus*, with *Echinostoma revolutum* and *E. liei* (Trematoda: Digenea). *Journal of Parasitology* 74, 604-608.

Huffman, J.E., Iglesias, D. and Fried, B. (1988b) *Echinostoma revolutum*: pathology of extraintestinal infection in the golden hamster. *International Journal for Parasitology* 18, 873-874.

Ishih, A. (1992) Mucosal mast cell response to *Hymenolepis diminuta* infection in different rat strains. *International Journal for Parasitology* 22, 1033-1035.

Kanev, I. (1994) Life cycle, delimitation and redescription of *Echinostoma revolutum* (Froelich, 1802) (Trematoda: Echinostomatidae) with a discussion of its identity. *Systematic Parasitology* 28, 125-144.

Kanev, I., Fried, B., Dimitrov, V. and Radev, V. (1995) Redescription of *Echinostoma trivolvis* (Cort, 1914) (Trematoda: Echinostomatidae) with a discussion of its identity. *Systematic Parasitology* 32, 61-70.

Kino, T., Hatanak, H., Hashimoto, M., Nishiyama, M., Goto, T., Okuhara, M., Kohsaka, M., Aoki, H. and Imanaka, H. (1987) FK-506, a novel immunosuppressant isolated from a Streptomyces. I. Fermentation, isolation, and physico-chemical and biological characteristics. *Journal of Antibiotics (Tokyo)* 40, 1249-1255.

Kusharyono, C. and Sukartinah, S. (1991) The current status of food-borne parasitic zoonoses in Indonesia. *Southeast Asian Journal of Tropical Medicine and Public Health* 22, 8-10.

Li, X. (1991) Food-borne parasitic zoonoses in the People's Republic of China. *Southeast Asian Journal of Tropical Medicine and Public Health* 22, 31-35.

Lu, S.C. (1982) Echinostomiasis in Taiwan. *International Journal of Zoonoses* 9, 33-38.

McKay, D.M., Halton, D.W., McCaigue, M.D., Johnston, C.F., Fairweather, I. and Shaw, C. (1990) *Hymenolepis diminuta*: intestinal goblet cell response to infection in male C57 mice. *Experimental Parasitology* 71: 9-20.

McKenzie, G.J., Bancroft, A., Grencis, R.K. and McKenzie, A.N.J. (1998) A distinct role for interleukin-13 in Th2-cell-mediated immune responses. *Current Biology* 12, 339-343.

Maji, A.K., Bera, D.K., Manna, B., Nandy, A., Addy, M. and Bandyopadhyay, A.K. (1993). First record of human infection with *Echinostoma malayanum* in India. *Transactions of the Royal Society of Tropical Medicine and Hygiene* 87, 673.

Mohandas, A. and Nadakal, A.M. (1978) *In vivo* development of *Echinostoma malayanum* Leiper, 1911, with notes on effects of population density, chemical composition and pathogenicity and *in vitro* excystment of the metacercaria (Trematoda: Echinostomatidae). *Zeitschrift für Parasitenkunde* 55, 139-151.

Odaibo, A.B., Christensen, N.Ø. and Ukoli, F.M.A. (1988) Establishment, survival and fecundity of *Echinostoma caproni* infections in NMRI mice. *Proceedings of the Helminthological Society of Washington* 55, 265-269.

Odaibo, A.B., Christensen, N.Ø. and Ukoli, F.M.A. (1989) Further studies on the population regulation in *Echinostoma caproni* infections in NMRI mice. *Proceedings of the Helminthological Society of Washington* 56, 192-198.

Radomyos, P., Radomyos, B. and Tungtrongchitr, A. (1994) Multi-infection with helminths in adults from northeast Thailand as determined by post-treatment fecal examination of adult worms. *Tropical Medicine and Parasitology* 45, 133-135.

Shi, B.B., Ishikawa, N., Itoh, H., Khan, A.I., Tsuchiya, K., Horii, Y. and Nawa, Y. (1995) Goblet cell hyperplasia induced by *Strongyloides venezuelensis* - infection in Syrian golden hamster, *Mesocricetus auratus*. *International Journal for Parasitology* 25, 399-402.

Shinkai, Y., Rathbun, G., Lam, K-P., Oltz, E.M., Stewart, V., Mendelsohn, M., Charron, J., Datta, M., Young, F., Stall, A.M. and Alt, F.W. (1992) RAG-2-deficient mice lack mature lymphocytes owing to inability to initiate V(D)J rearrangement. *Cell* 68, 855-867.

Simonsen, P.E., Bindseil, E. and Køie, M. (1989) *Echinostoma caproni* in mice: studies on the attachment site of an intestinal trematode. *International Journal for Parasitology* 19, 561-566.

Simonsen, P.E., Estambale, B.B. and Agger, M. (1991) Antibodies in the serum of golden hamsters experimentally infected with the intestinal trematode *Echinostoma caproni*. *Journal of Helminthology* 65, 239-247.

Tangtrongchitr, A. and Monzon, R.B. (1991) Eating habits associated with *Echinostoma malayanum* infections in the Philippines. *Southeast Asian Journal of Tropical Medicine and Public Health* 22(Suppl.), 212-216.

Tiuria, R., Horii, Y., Makimura, S., Ishikawa, N., Tsuchiya, K. and Nawa, Y. (1995) Effect of testosterone on the mucosal defence against intestinal helminths in Indian soft-furred rats, *Millardia meltada* with reference to goblet and mast cell responses. *Parasite Immunology* 17, 479-484.

Weinstein, M.S. and Fried, B. (1991) The expulsion of *Echinostoma trivolvis* and retention of *Echinostoma caproni* in the ICR mouse: pathological effects. *International Journal for Parasitology* 21, 255-258.

Yu, S., Hu, L. and Jiang, Z. (1994) Report of the first nationwide survey of the distribution of human parasites in China. 1. Regional distribution of parasite species. *Chinese Journal of Parasitology and Parasitic Diseases* 12, 241-247.

MAINTENANCE, CULTIVATION, AND EXCYSTATION OF ECHINOSTOMES

BERNARD FRIED
Department of Biology, Lafayette College
Easton, Pennsylvania 18042, USA

B. Fried and T.K. Graczyk (eds.),
Echinostomes as Experimental Models for Biological Research, 99–118.

1. Introduction and Significance

The need for a consistent and reliable source of material is important for continued research on parasitic helminths. Often, only a particular stage in the life cycle of a helminth is available and workers have exploited use of that stage. For instance, Halton and colleagues (see Smyth and Halton, 1983 for review) have examined various aspects of the biology of the monogenean *Diclidophora merlangi*, by obtaining these parasites from the gills of naturally infected adult whitings, *Merlanguis merlanguis*. My laboratory has relied on obtaining *Echinostoma trivolvis* cercariae from naturally infected *Helisoma trivolvis* snails in a farm pond in Northampton County, PA for continuing studies on the biology of echinostomes (Schmidt and Fried, 1997).

The ideal situation for studies on the biology of echinostomes is to maintain the entire life cycle in the laboratory. For at least two echinostome species, *Echinostoma paraensei* and *E. caproni*, this situation has been achieved. Thus, the life cycle of *E. paraensei* is maintained in Dr. Sam Loker's laboratory and that of *E. caproni* by Dr. N. O. Christensen at the Danish Bilharziasis Laboratory (DBL) or in Dr. B. Fried's laboratory. Material from these laboratories may be available by contacting Drs. Loker, Christensen or Fried. Mailing addresses of these scientists can be found in Fried and Huffman (1996).

Although a wide variety of echinostome material may be available from numerous sources, identification of these digeneans to the species or even generic level, is often difficult. Hopefully, with information provided by Kostadinova and Gibson (Chapter 2) the task of echinostome identification at least to the generic level may be easier. It is the intent of this chapter to make readers aware of the availability of echinostome material from the wild, from commercial dealers, and from researchers who are willing to exchange material. Moreover, the ability to maintain echinostomes in vivo, in ovo, and in vitro is explored in this chapter. Table 1 provides a list of studies with useful protocols for work with echinostomes as experimental models in biology and chemistry.

Table 1

Studies from Dr. B. Fried's Laboratory that Provide Protocols for Echinostome Work*

Studies	Protocols	References
Transfer of worms to altered sites	Surgical procedures used to alter chick intestines for studies on *Echinostoma trivolvis* adults	Fried and Vonroth (1968)
Oxygen consumption	Warburg manometry and enzyme histochemical studies on oxygen uptake in *E. trivolvis*	Taft and Fried (1968)
Histochemical glycogen studies	Histochemical studies with periodic acid Schiff (PAS) to localize glycogen in *E. trivolvis*	Fried and Kramer (1968)

Transfer of worms to the chick chorioallantis	First description of echinostome cultivation in chick embryos	Fried et al. (1968)
Maintenance of eggs and miracidia	Development of E. trivolvis eggs in vitro	Fried and Weaver (1969a)
Chick infectivity studies	Exposure of chicks to E. trivolvis metacercariae	Fried and Weaver (1969b)
Oil Red O staining of whole adults	Visualization of neutral lipids in excretory system of E. trivolvis	Fried and Morrone (1970)
Experimental ablation of adults	Wound healing of E. trivolvis adults following amputation of body parts	Fried et al. (1971)
Thin-layer chromatographic (TLC) analysis of excretory- secretory products in adults	TLC analysis of lipids in E. trivolvis	Fried and Appel (1977)
In vitro culture of excysted metacercariae	Histochemical and TLC procedures to analyze lipids in E. trivolvis cultured in vitro	Butler and Fried (1977)
TLC of amino acids	TLC procedures to analyze amino acids in E. trivolvis	Bailey and Fried (1977)
Histochemical and ultrastructural studies on metacercariae	Encysted metacercariae of E. trivolvis examined by histochemical procedures and by TEM	Gulka and Fried (1979)
In vitro and ectopic encystment	Chemical and mechanical factors in cyst induction of E. trivolvis	Fried and Bennet (1979)
TLC of phospholipids	Silica gel used to analyze phospholipids in E. trivolvis	Fried and Shapiro (1979)
Pairing and aggregation studies	In vitro pairing of metacercariae and adults of E. trivolvis and characterization of worm chemoattracants	Fried et al (1980)
Metacercarial survival in vitro	Survival of E. trivolvis cysts in Locke's solution at 4 C	Fried and Perkins (1982)

In vivo studies in chicks	Infectivity, growth, and development of *E. trivolvis* in the domestic chick	Fried (1984)
Glycogen and protein analysis	Microanalytic techniques to measure glycogen and protein in adult *E. trivolvis*	Sleckman and Fried (1984)
Histochemical enzymic studies on larvae	Whole-mounts used to detect enzymes in cercariae and excysted metacercariae of *E. trivolvis*	Fried *et al.* (1984)
Scanning electron microscopy (SEM)	SEM used to study development of *E. trivolvis* from excysted metacercariae to adults	Fried and Fujino (1984)
Quantitative analysis of sterols	Gas-liquid chromatography/mass spectroscopy (GLC-MS) for the analysis of sterols in *E. trivolvis*	Chitwood *et al.* (1985)
Cultivation from cercarial stage	Cultivation of cercariae of *Himasthla quissetensis* on the chick chorioallantois	Fried and Groman (1985)
In vivo and in vitro maintenance	Procedures to maintain *E. trivolvis* in the laboratory	Fried (1985)
Hamster studies	Infectivity, growth and development of *E. trivolvis* in golden hamsters	Franco *et al.* (1986)
Mouse studies	Infectivity, growth and development of *E. trivolvis* in mice	Hosier and Fried (1986)
Feeding studies	Feeding of *E. trivolvis* in the chick and on the chorioallantois	Wisnewski *et al.* (1986)
Cercarial preference in snails	Various second intermediate snail hosts used to show preference of *E. trivolvis* cercarial encystment	Anderson and Fried (1987)
In vivo excystment	Excystment of *E. trivolvis* in the chick intestine	Fried and Kletkewicz (1987)
In vitro excystation	Factors needed to excyst metacercarial cysts of *E. trivolvis* and *E. caproni*	Fried and Emili (1988)
Cercarial behavior	Attraction of *E. trivolvis* cercariae to the dialysate of *Biomphalaria glabrata* snails	Fried and King (1989)

Acetocarmine studies on cercariae and metacercariae	A whole-mount procedure to distinguish larval stages of *E. caproni* and *E. trivolvis*	Fried and Manger (1992)
Inoculation of metacercariae into the allantois	*E. caproni* cultivated to ovigerous adults from excysted metacercariae in chick embryos	Chien and Fried (1992)
TLC studies on natural pigments in rediae	Beta carotene and lutein analyzed in rediae of *E. trivolvis*	Fried *et al* (1993)
TLC on sugars in infected snails	*E. caproni* infection in snails depleted hemolymph sugars	Perez *et al.* (1994)
Retraction and extension of collar spines	TEM and immunogold colloidal studies on structure and function of collar spines in *E. trivolvis*	Fujino *et al.* (1994)
Infection of snails with cercariae	Infectivity of neonatal snails with *E. trivolvis* cercariae	Fried *et al.* (1995)
Polymerase chain reaction (PCR) to distinguish closely related species	*E. trivolvis* and *E. caproni* adults compared by using random amplified polymorphic DNA analysis	Fujino *et al.* (1995)
Anthelmintics	Clorsulon, rafoxonide, mebendazole, and arprinocid used to treat *E. caproni* in mice	Maurer *et al.* (1996)
Immunosuppresant to establish infection in mice	Dexamethasone suppressed goblet cell hyperplasia that allowed for *E. trivolvis* development	Fujino *et al.* (1996)
In vitro egg laying	*E. caproni* egg laying in nutritive and non-nutritive media	Reddy and Fried (1996a)
Cercarial emergence	Factors involved in cercarial emergence of *E. trivolvis* from *Helisoma trivolvis*	Schmidt and Fried (1996)
Pairing of rediae in vitro	Chemoattraction factors that regulate pairing of *E. trivolvis* rediae	Reddy and Fried (1996b)
Paraesophageal glands	Histological and histochemical studies on paraesophageal glands in *E. trivolvis* and	Humphries and Fried (1996)

E. revolutum cercariae

In vivo studies in chicks	Growth and development of *E. revolutum*	Humphries *et al.* (1997)
Prevalence studies	*E. trivolvis* cercarial prevalence study in *Helisoma trivolvis* in Pennsylvania	Schmidt and Fried (1997)
Nutritional studies	Effects of a high protein diet on *E. caproni* in mice	Sudati *et al.* (1997)
Maintenance studies	Long term laboratory maintenance of *H. trivolvis* infected with *E. trivolvis*	Schmidt *et al.* (1998)
Histochemical glycogen and lipid studies	Histochemical demonstration of glycogen and lipids in cercariae of *E. trivolvis*	Fried *et al.* (1998)
Snail conditioned water (SCW)	SCW enhances miracidial release from eggs of *E. caproni*	Fried and Reddy (1999)
RAG -2- deficient mice	Growth of *E. caproni* in immunosuppressed mice	Frazer *et al.* (1999)
Proteases in echinostomes	Electrophoretic analysis of proteases in adult *E. caproni* and *E. trivolvis*	Mueller and Fried (1999)
Ultrastructure of echinostome gastrodermis	Comparative studies on lamellar gastrodermal projections in *E. caproni. E. trivolvis* and *E. paraensei*	Fujino *et al.* (1999)

_____

* Arranged in Chronological order from 1968-1999

2.   **Obtaining Material**

Preserved echinostome material is available from commercial suppliers. For example, Wards Biological (Rochester, NY) is a good source of prepared slides of *Echinostoma* adults and larval stages. Live material is available from some suppliers who ship snails infected with larval echinostomes. Carolina Biological (Elon, NC) supplies physid snails that may be infected with cercariae and metacercariae of echinostomes, but the specific identification of these echinostomes is usually not available. Frogs (both tadpole and adult stages) obtained from suppliers may contain encysted echinostome metacercariae in the kidneys (Fried and Bradford, 1997). Identification of these metacercariae to species without growing the larvae in definitive hosts is not possible. For a discussion of frogs infected with echinostome metacercariae see Fried et al. (1997).

Several suppliers can provide marine gastropods infected with larval echinostomes. Thus, Jones Biological (Long Beach, CA) supplies *Cerithidea californica* infected with larval echinostomes in the genera *Himasthla, Echinostoma, Echinoparyphium* and *Acanthoparyphium*. Martin (1972) has provided a key to the larval trematodes of *C. californica* that is helpful in identifying the echinostomes. The Marine Biological Laboratory (MBL) in Woods Hole, MA supplies *Ilyanassa obsoletus* infected with the echinostome *Himasthla quissetensis*. The Biomedical Research Institute (Rockville, MD) is a good source of obtaining uninfected *Biomphalaria glabrata* snails. These snails serve as experimental first and second intermediate hosts for a number of echinostome species (see reviews in Huffman and Fried, 1990; and Fried and Huffman, 1996).

Echinostome adults and eggs may be available from human subjects. For example, Grover *et al.* (1998) obtained adult worms and eggs of *Echinostoma ilocanum* from an 8 yr old girl in India. The adults may be used for morphologic studies (identification of the adults) and the eggs may be isolated from the feces or dissected from worms and embryonated in conditioned (dechlorinated) tap water to obtain miracidia in an attempt to infect potential snail first intermediate hosts.

Adult echinostomes are usually available in the wild mainly from the intestines of various cold-blooded and warm-blooded vertebrate hosts. Procuring warm blooded and even some cold-blooded hosts as a source of such material may require federal, state, and local permits. Necropsy procedures on these hosts may be governed by institutional guidelines. Vertebrate hosts infected with echinostome adults may also be available from commercial suppliers, zoos, hunters, and trappers. Laboratory manuals that mention echinostome species from vertebrate hosts include Cable (1940) and Schmidt (1988).

The intramolluscan stages of echinostomes, i.e., sporocysts and rediae, are available from numerous fresh water and marine gastropods. Some freshwater snail genera of importance in harboring these stages are *Biomphalaria, Physa,* and *Lymnaea*. The aforementioned snails may also harbor encysted metacercariae of echinostomes in the pericardial/kidney region or other tissues of gastropod hosts. Freshwater bivalves in various genera, i.e., *Anodonta, Corbicula, Dreissena*, may also harbor encysted metacercariae mainly in the viscera, gonads and gills. Marine bivalves such as *Mercenaria* and *Crassostrea* are also good sources of echinostome cysts. Numerous marine gastropods serve as first intermediate hosts of echinostomes and of particular importance are the genera *Ilyanassa, Cerithidea*, and *Littorina*. The role of edible freshwater and marine molluscs as potential vectors of human echinostomiasis is underexplored and the eating of uncooked shellfish is discouraged on these grounds (Graczyk and Fried, 1998).

3.   **Maintenance of Larval and Adult Echinostomes and the Intermediate and Definitive Hosts.**

Larval echinostomes can be maintained in various invertebrate and cold blooded vertebrate hosts. The stage that emerges from the egg (the miracidium) usually develops in a specific snail host. For example, in the biology of *Echinostoma caproni*, the medically important planobid snail *Biomphalaria glabrata* is a suitable laboratory first intermediate host. The allopatric species, *E. paraensei*, also cycles through this snail. Interestingly, *Schistosoma mansoni* cycles through the *B. glabrata* snail and can be cycled concurrently with either *E. paraensei* or *E. caproni*. Once the snails are infected with miracidia they are maintained in aquaria containing pond water, artificial spring water (ASW) or conditioned tap water. Snails can be maintained at 20 to 22 C where larval patency (sporocysts with fully developed cercariae) may take 6 to 8 wk. The cycle can be accelerated by maintaining the snails at 26 to 28 C (where patency may be reached in 3 to 4 wk). Snail mortality and increased feeding may also occur at the higher temperatures. Snail maintenance for echinostome studies has been described in Fried (1985). Infected snails may be necropsied about 10 days post miracidial infection to obtain sporocysts from the heart and aorta as described in Jeyarasasingham *et al.* (1972). Beyond 2 wk post-infection snails can be necropsied and the digestive gland gonad (DGG) complex dissected to obtain rediae as described in Huffman and Fried (1990) and Fried and Huffman (1996).

Snails both naturally and experimentally infected with echinostomes may be isolated in stender dishes or finger bowls (either individually or in groups) half-filled with ASW to obtain freshly released cercariae as described in Schmidt and Fried (1996). This procedure is referred to as the snail isolation procedure.

To infect second intermediate hosts (mainly snails, clams or tadpoles in the genus *Rana*) cercariae can be placed in containers with ASW along with the experimental second intermediate host. Cercariae will enter the hosts usually via the cloacal opening of the tadpole or the excretory pore of the snail. Echinostome cercariae have an affinity for the kidneys of snails and tadpoles and usually encyst in those sites shortly after host entry. This process has been described by several authors for different species of echinostomes, (Anderson and Fried, 1987; Fried *et al.* 1997a). The snail kidney and the kidney of *Rana* are a good source to obtain encysted metacercariae of echinostomes. In snails, cercariae may also make their way into the pericardial cavity and encyst there. Other snail sites may serve for the encystment of echinostome cercariae and include the foot, mantle cavity and general body surface (Andersen and Fried, 1987).

Helisoma trivolvis snails collected in the wild and infected with *E. trivolvis* may be kept at 4 C in ASW and maintained there for about a year (Schmidt *et al.*, 1998). When brought out to room temperature and fed lettuce they began to release cercariae. This procedure of placing infected snails in dormancy for a relatively long period is helpful in maintaining echinostome material in the laboratory during severe weather conditions when collecting snails in the wild is not possible. Another paper that discusses the maintenance of *E. trivolvis* in the laboratory is that of Fried (1984).

The definitive hosts of most echinostomes are birds and mammals. Some echinostome species will only develop in birds whereas others are specific to mammals. Thus, *E. revolutum* is mainly a bird form, whereas *E. trivolvis* infects birds and various mammals including the hamster in the laboratory. This echinostome is not very infective to laboratory mice and usually does not reach sexual maturity in that host. However, if mice are immunosuppressed with the corticosteroid dexamethasome, they can be used as successful hosts that will support sexually mature populations of *E. trivolvis* adults (Fried *et al.*, 1997b). On the other hand, *E. caproni* develops well in various mammal and avian hosts and numerous studies have reported on development of this echinostome in mice, birds, hamsters, and domestic chicks (see review in Fried and Huffman, 1996).

Care of a number of vertebrate animals, particularly warm blooded species, is under reasonably strict control by federal, state, and even institutional guidelines. Thus, before studies with warm-blooded hosts are initiated, familiarity with the rules concerning their care and maintenance is needed. Irwin (1997) and Fried (2000) have provided good discussions of this topic.

4.   **In Vivo Cultivation**

Most studies on adult echinostomes involve feeding the encysted metacercariae to experimental definitive hosts, usually mice, hamsters and domestic chicks. Oral infection with metacercarial cysts is usually the prime mode of infection. Cysts are fed by pipet in a minimal amount of Locke's solution or deionized water. The inoculum used varies from single cyst infections to infections with hundreds of cysts. There may not be a correlation between the cyst inoculum and the number of adult worms recovered, particularly in studies on echinostomes in experimentally infected avian hosts (Fried *et al.*, 1997a).

Whereas per os infection of cysts is recommended for exposure of chicks and hamsters, intubation of cysts by stomach tube as described on pages 134-135 in MacInnis and Voge (1970) is recommended for studies on mice. In studies on *Echinostoma* where cysts are intubated by stomach tube the expected worm recovery is about 50% when a cyst inoculum of 25 or 50 is used (Fried and Huffman, 1996). Some studies suggest that echinostome cysts pretreated in sodium bicarbonate (Fried and Butler, 1978) may enhance subsequent worm infectivity in avian hosts. However, conclusive results using this technique with echinostomes in mammalian hosts are not available, although treatment of cysts with sodium bicarbonate does enhance the infectivity of *Zygocotyle lunata* (Paramphistomatidae) in rats (Bacha, 1964).

Necropsy procedures should follow NIH guidelines for vertebrate hosts. Hosts should be lightly anesthetized with ether prior to cervical dislocation. The intestine of the host is then removed to a petri dish or finger bowl half-filled with saline or Locke's solution, opened longitudinally, the worms removed and transferred to fresh saline. The worms may be used for numerous research purposes such as biological, physiological, biochemical, and immunological studies.

Morphological studies are usually done on live worms or those fixed in either alcohol-formalin acetic-acid (AFA) or neutral buffered formalin (NBF) and then stained with various dyes, i.e., carmine or hematoxylin or others and then prepared as whole mounts. For additional information on handling, fixing and staining digeneans, including echinostomes, see Pritchard and Kruse (1982).

Criteria for successful in vitro and in ovo cultivation rely on growing juvenile echinostomes in appropriate vertebrate hosts for periods long enough to obtain ovigerous adults. Usually worms are obtained at various time periods following the exposure of hosts to encysted metacercariae to obtain juvenile and sexually mature adults. In vivo worm growth can be assessed by examining gametogenesis, organ formation, vitellogenesis, and egg formation. A series of worms showing growth and developmental changes in vivo is important to have as a standard for comparison with worms grown in ovo or in vitro. Smyth (1990) has pioneered studies and protocols for comparing worm growth in vivo with that obtained in vitro or in ovo. Although most studies on the growth of echinostomes in vivo, in vitro, and in ovo are based on morphological characteristics and use of worm length or body area to assess growth, some studies have used criteria such as worm wet and dry weight (Hosier and Fried, 1991) or worm protein and carbohydrate content as a function of worm aging in the definitive host (Sleckman and Fried, 1984). Other criteria for assessing growth and development have not been used. Studies on the developmental biology of echinostomes in vivo, in vitro, and in ovo are needed.

5.   **In Ovo Cultivation**

Several species of echinostomes have been cultivated on the chick chorioallantois (CAM) or in the allantoic sac of fertile chick embryos and this topic has been reviewed by Fried (1989), Fried and

Stableford (1991) and Irwin (1997). The technique of cultivating echinostomes in chick embryos is useful as a basic research tool and has the potential for testing anthelmintics in an ectopic site. The use of chicken eggs as an initial anthelmintic screen rather than using laboratory animals has advantages in terms of cost and not having to use vertebrate hosts for the initial screening. Techniques for preparing fertile chick eggs for cultivation studies have been described in Fried and Stableford (1991) and in Irwin (1997). These papers should be consulted prior to attempting to cultivate either larval or adult echinostomes in ovo.

Most work on echinostome cultivation in ovo has used the North American species, *Echinostoma trivolvis*. This echinostome develops from excysted metacercariae to adults on the CAM and has been used to assess temperature tolerance; i.e., wide range of temperature tolerance on the CAM, and migratory behavior (Fried and Stableford, 1991). In spite of numerous studies with *E. trivolvis* on the CAM, development of this species to ovigerous adults on the CAM has not been successful. Better success in producing sexually mature echinostomes in chick embryos has been achieved with studies on *E. caproni* in which ovigerous worms were grown in the allantoic sac and the worms produced eggs with well-developed miracidia (Chien and Fried, 1992). Unfortunately, infection of CAM-derived miracidia in *Biomphalaria glabrata* snails was not tested. Future work should attempt to infect the *B. glabrata* snail with miracidia derived from *E. caproni* grown in chick embryos, thus completing the life cycle of this echinostome without a definitive vertebrate host.

A study with one other echinostome species grown on the CAM is that of Fried and Groman (1985) in which cercariae of the marine echinostome *Himasthla quissetensis* were grown on the CAM without the production of metacercariae. The study showed that at least for this species, an obligatory encysted metacercarial stage, was not needed. This study has implications for future work with species of *Echinochasmus* which are capable of infecting humans by direct cercarial penetration associated with contaminated drinking water. Details of the infection route of species of *Echinochasmus* from the cercarial stage (by direct penetration in man), to the adult stage are not known. The dynamics of cercarial invasiveness of species of *Echinochasmus* should be examined in chick embryos.

6. In Vitro Cultivation

Several studies have been reported on the in vitro cultivation of echinostomes from excysted metacercariae to preovigerous adults. However, the goal of producing sexually mature adults in vitro beginning with the metacercarial stage has not been achieved for any species of echinostome. Attempts to cultivate echinostomes from metacercariae to adults were pursued actively during the 1960s thru 1980s. Early studies have been reviewed in Fried (1978). As reviewed more recently by Irwin (1997) the area of in vitro cultivation of digeneans has not been pursued actively in the 1990s for a variety of reasons, i.e., inherent difficulties in such studies; lack of consistent positive results leading to problems in publishing the results; very labor intensive work; and other such reasons summarized in Irwin (1997).

Most of these studies beginning with the metacercarial stage of echinostomes have used test tube cultures containing a defined medium (often NCTC 135) supplemented with various types of natural products, ie., serum, egg yolk, host mucus. A blanket of antibiotics (typically penicillin and streptomycin) is usually used to keep the cultures oligoseptic. The antibiotics provide a bacteriostatic rather than a bacteriocidal environment at concentrations that are not lethal to the excysted metacercariae and the young adults growing in culture. The use of fungicides has not been well studied with digeneans which appear not to tolerate antimycotics very well. Studies on the use of antibiotics and antimycotics for the in vitro culture of digeneans in general and echinostomes in particular need more work. The maintenance temperature for cultures is generally in the range of 37 to 42 C suggesting that echinostomes can tolerate a wide range of temperatures in vitro from the excysted metacercaria to the adult stage. The gaseous phase is usually air or about 7% CO_2 in air. Cultures are changed frequently to prevent the build up of metabolites and to presumably renew an optimal nutritional medium. Key in vitro studies on echinostomes include that of Howell (1968) reporting limited development of *Echinoparyphium recurvatum* metacercariae in a yolk-albumen-saline medium supplemented with yeast extract; that of Butler and Fried (1977) in which *E. trivolvis* (referred to as *E. revolutum*) doubled its mean body area (from encysted metacacariae to immature adults) within 14 days in culture in a medium of NCTC 135 supplemented with 20% hen's egg yolk. Perhaps the study that achieved best growth and development of *E. trivolvis* in vitro was that of Fried and Kim (1989) in which E. trivolvis (again referred to as *E. revolutum* in the paper) metacercariae were cultured in NCTC 135 supplemented with 40% chicken serum and 10% mucosal extract from the lower ileum of the domestic chick. In this medium worms increased their body area three times by 8 days in culture and showed some genital development. Clearly factors in the mucosal extract enhanced cultivation in vitro using the *E. trivolvis* model. In vitro culture with excysted metacercariae of the allopatric species *E. caproni* has not been studied extensively. Owing to the easy availability of this species and relative success growing this organism in chick embryos, additional work on *E. caproni* in vitro is worth pursuing.

Limited work is available on culturing adults of echinostomes in vitro with either defined medium or defined medium plus natural products (see Fried, 1978 for review). Adults maintained in vitro go downhill quickly, possibly due to worm autolysis and/or microbial contamination. Attempts to develop cell lines from echinostomes as with schistosomes (see Hobbs *et al.*, 1993 for discussion) have not been made. Such studies using larval or adult stages of selected species of echinostomes, particularly *E. caproni* or *E. paraense* would be worthwhile following the model studies of Weller and Wheldon (1982) and Hobbs *et al.* (1993) on *Schistosoma mansoni.*

Little work has been done culturing the intramolluscan stages of echinostomes. By contrast, considerable work on the cultivation in vitro of schistosome sporocysts has been done. The most recent study on this topic plus a review of the earlier schistosome-related work has been presented by Cousteau *et al.* (1997) who reported development of mother sporocysts of *Schistosoma japonicum* into daughter sporocysts in vitro in the presence of a *Biomphalaria glabrata* embryonic cell line (Bge cells). Loker *et al.* (1999) used the Bge cell line to culture in vitro the rediae of *E. caproni* for up to 5 days. Rediae ingested the Bge cells in culture; they also released motile cercariae in culture and some of these cercariae produced in vitro encysted metacercariae. Rediae consumed cercariae released into culture but apparently did not attack rediae of the same species or another species placed in the culture. Because of the sparcity of studies on the in vitro culture of echinostome rediae, the study of Loker *et al.* (1999) is a good one to follow for those who may wish to initiate redial cultures with species of echinostomes other than *E. caproni*.

7. **Transplantation of Larval and Adult Echinostomes to New Sites**

Perhaps more than any other digenean, larval and adult echinostomes have been used extensively in fluke transplantation studies. Echinostomes are quite adaptable in their abilities to survive the transplant procedures and establish in new hosts and sites. The ability to transplant parasites from one host to another allows the experimenter to manipulate parasites, i.e., mark them with dyes, label them with radioisotopes, experimentally ablate or damage the worms for repair studies. Thus, transplant studies allow for enhanced experimentation with echinostomes. Moreover, in terms of transplantation with larval stages, such as redial stages, investigators can increase the number of snails infected with larval stages when infections are otherwise limited by inability to make collections in the field or to infect snails in the laboratory with the miracidial stage.

Perhaps the larval stage used most frequently in transplant studies is the echinostome redial stage. Echinostome rediae have been transplanted from infected to uninfected snails to study various aspects of larval development. Donges (1963) infected snails with individual rediae of the echinostome *Isthmiophora spiculator* to determine the exact sequence of larval stages. Chernin (1967) developed a technique for transferring echinostome rediae into the cephalopedal sinus of *B. glabrata*. Heyneman (1966) successfully established echinostome infections by introducing either miracidia or rediae into the tissues of lymnaeid snails. Perhaps the most interesting finding of echinostome redial transplant work is the fact of unlimited multiplication of the rediae following their transplantation from donor infected snail to recipient uninfected snail (Donges and Gotzelmann, 1988). There appears to be no genetically programmed cessation to redial echinostome multiplication. Further work in this area is warranted.

The other larval echinostome stage that has been used in transplantion studies is the excysted metacercaria. Since the echinostome metacercaria can be excysted easily by chemical means this stage is convenient for manipulation or radioisotope labeling prior to transplantion. A recent study that examined transplantation of both encysted and excysted metacercariae used *Echinostoma caproni*. Chien *et al.* (1993) surgically implanted metacercariae of this echinostome into the small intestines of ICR mice. Worm recovery following implantation of either encysted or excysted metacercariae into the gut was greater than 90%. The recovery of sexually mature echinostomes at 10 days post-transplant suggested that larval sojourn thru the stomach of the mouse host was not a prerequisite for normal development. The ability to surgically implant echinostome metacercariae into the small intestine of a vertebrate host may be useful for studies on wound healing and regeneration.

Several studies have transplanted echinostomes into new sites or experimentally altered sites to test worm adaptability in the new environments. Beaver (1935, 1937) and Fried and Vonroth (1968) removed adults of *E. trivolvis* from naturally or experimentally infected avian and mammalian hosts and implanted them into new avian hosts via the cloaca. *E. trivolvis* adults can also be transplanted surgically into the intestines or body cavity of avian hosts (Fried and Fink, 1968; Fried and Vonroth, 1968). These studies established the fact that echinostomes are adaptable in their feeding habits and can feed on the host peritoneal lining. Fried and Vonroth (1968) successfully transplanted adults of *E. trivolvis* into a surgically altered chick gut (well nourished by an intact blood supply, but without host ingesta). The study provided unequivocal evidence that an echinostome species can feed on host mucosal tissue in the absence of ingesta in the host gut. How echinostomes feed and process their food is not well understood and transplantation studies may be useful in this area of research. Transplantation studies using ablated echinostome adults have been used to study wound healing of echinostomes. Following ablation of adult worms and placing the severed parts back into the cloaca, wound healing was noted (Fried *et al.*, 1971). This study suggested that echinostomes are capable of limited repair of damage in the absence of regeneration of lost or missing systems. Studies on regeneration and wound healing with echinostomes are worth repeating to look for potential neoblasts or other totipotent cells. A good model for such a study is that of Allen and Nollen (1991) on the eyefluke *Philophthalmus gralli*.

Chien *et al.* (1993) implanted adults of *E. caproni* into the mouse small intestine, but experimental or manipulative studies using this echinostome implanted into the mouse were not done.

8. In Vivo and In Vitro Excystation of Metacercariae

Considerable information is available on excystation of echinostome metacercariae, particularly on in vitro excystation. Fried (1994) has provided an extensive review on metacercarial excystment of digeneans in general. This topic has also been reviewed by Irwin (1997).

Numerous studies mainly on the biology and life cycle of digeneans have described the excystation of encysted metacercariae in various parts of the host alimentary tract. These studies have mainly been incidental to the life cycle or biology of the particular digenean under study. Perhaps the most complete study on in vivo excystation is that of Fried and Kletkewicz (1987) on the metacercarial cysts of *Echinostoma trivolvis* fed to domestic chicks. They fed domestic chicks large numbers of cysts and necropsied the hosts from 15 min to 24 hr later. Excystation was first seen in the lower ileum at 30 min. They tested the viability of cysts retained in the gizzard in a standard excystation medium and found that a sojourn in the gizzard for 3 hr made the cysts nonviable. Thus, long term treatment of the cysts in an acid environment was detrimental. However, cysts removed from the gizzard within 2 hr were still capable of excystation. They found that intestinal emptying was rapid in the chick (1 to 1.5 hr) and that emptying time influenced the rate and percentage of excystation in the host.

In vivo excystation with echinostome metacercarial cysts can provide valuable information on normal mechanisms of excystation and the sites where the larvae emerge from the cysts under normal conditions. Excystation observed in vivo is important to determine if the mechanisms seen in vitro during excystation are the same as that which occurs in vivo. These experiments are sometimes difficult to do in vivo because they require a large number of experimental hosts, a large number of cysts and great patience in examining the wide expanse of intestine for the relatively small cysts or excysted metacercariae. Workers attempting in vivo excystation studies with echinostomes should consult the Fried and Kletkewicz (1987) paper.

Considerable work has been done on chemical (in vitro) excystation of metacercarial cysts of echinostomes (see review in Fried, 1994 and Irwin, 1997. Species in genera other than *Echinostoma* have been studied including members of the genera *Echinoparyhium*, *Himasthla*, and *Acanthoparyphium*. From a utilitarian standpoint, chemically encysted metacercariae of echinostomes have been used for in vitro and in ovo cultivation, for immunological studies, for light and electron microscopical studies and for behavioral studies. The excysted metacercaria (newly hatched juvenile) is essentially an immature or juvenile adult not yet invested with egg precursor material nor a developed male or female system. As such, this stage in the biology of the digenean is a nice tool to initiate studies on the developmental biology of echinostomes from the juvenile to the adult stage. The excysted metacercaria can be used as the starting point for in vivo, in ovo, and in vitro studies on digeneans.

Although the details vary in terms of the echinostome species studied, it appears that most species require the synergistic effects of an alkaline-bile salts plus trypsin medium in a basal salt solution such as Hanks' Medium or Earle's Medium. In a medium designed by Fried and Roth (1974) for the excystation of *Parorchis acanthus,* the constituents consist of 0.5% bile salts plus 0.5% trypsin (various grades of typsin are usable) in Earle's BSS adjusted to a pH of 8.0 $\pm$ 0.2 with 7.5% $NaHCO_3$. This medium can be used fresh or it can be frozen for more than a year, thawed, filtered and still be used to excyst echinostome metacercariae (for details see the review in Fried, 1994). This excystation medium allows for activation of the larva within the cyst (rotation), movement of the larva against a weak area or the mucus plug area in the inner cyst of the echinostome, eventual exit through an aperture in the inner cyst. Eventually, the larva disrupts the outer cyst and excystation is complete with the organism free in the medium. Since too long a sojourn in the excystation medium may be deleterious to the newly emerged larva, it is recommended that the larva be removed to saline. The larva is then ready for use in various biological or other studies as mentioned previously. Optimal temperature for in vitro excystation ranges from 37 to 42 C. However, temperature as a variable during excystation has not been studied extensively with the echinostome model. Detailed recipes for various excystation media used for different species of echinostomes have been published in Fried (1994) and Irwin (1997).

9. References

Allen, W.B. and Nollen, P.M. (1991) A comparative study of the regenerative process in a trematode *Philophthalmua gralli*, and a planarian *Dugesia dorotocephalia*, *International Journal for Parasitology* **21**, 441-447.

Anderson, J.W. and Fried, B. (1987) Experimental infection of *Physa heterostropha*, *Helisoma trivolvis*, and *Biomphalaria glabrata* (Gastropoda) with *Echinstoma revolutum* (Trematoda) cercariae, *Journal of Parasitology* **73**, 49-54.

Bacha, W.J. Jr. (1964) Effect of salt solutions on the establishment of infections of the trematode *Zygocotyle lunata* in white rats, *Journal of Parasitology* **50**, 546-548.

Bailey, R.S., Jr and Fried, B. (1977) Thin layer chromatographic analyses of amino acids in *Echinostoma revolutum* (Trematoda) Adults, *International Journal for Parasitology* **7**, 497-499.

Beaver, P.C. (1935) Experiments on regeneration in the trematode, *Echinostoma revolutum*,. *Journal of Parasitology* **23**, 423-424.

Beaver, P.C. (1937) Experimental studies on *Echinostoma revolutum* (Froelich) a fluke from birds and mammals, Illinois Biological Monographs **15**, 1-96.

Butler, M.S. and Fried, B. (1977) Histochemical and thin layer chromatographic analyses of netural lipids in *Echinostoma revolutum* metacercariae cultured in vitro, *Journal of Parasitology* **63**, 1041-1045.

Cable, R. (1940) An Illustrated Laboratory Manual of Parasitology, Burgess Publishing, Minneapolis.

Chernin, E. (1967) Occurrence of metacercariae within echinostome rediae transplanted into *Australorbis glabratus, Journal of Parasitology* **53**, 219.

Chien, W.Y. and Fried, B. (1992) Cultivation of excysted metacercariae of *Echinostoma caproni* to ovigerous adults in the allantois of the chick embryo, *Journal of Parasitology* **78**, 1019-1023.

Chien, W.Y., Hosier,D.W., and Fried, B. (1993) Surgical implantation of *Echinostoma caproni* metacercariae and adults into the small intestine of ICR mice, *Journal of the Helminthological Society of Washington* **60**, 122-123.

Chitwood, D.J., Lusby, W.R. and Fried, B. (1985) Sterols of *Echinostoma revolutum* (Trematoda) adults, *Journal of Parasitology* **71**, 846-847.

Cousteau, C., Ataev, G., Jourdane, J., and Yoshino, T.P. (1997) *Schistosoma japonicum*: In vitro cultivation of miracidium to daughter sporocyst using a *Biomphalaria glabrata* embryonic cell line, *Experimental Parasitology* **87**, 77-87.

Donges, J. (1963) Die experimentelle Bestimmung der Anzahl der Rediengenerationen bei Trematoden, Naturwissenschaften **50**, 103-104.

Donges, J. and Gotzelmann, M. (1988) Digenetic trematodes: Multiplication of the intramolluscan stages in some species is potentially unlimitable, *Journal of Parasitology* **74**, 884-885.

Franco, J., Huffman, J.E., and Fried, B. (1986) Infectivity, growth and development of *Echinostoma revolutum* (Digenea: Echinostomatidae) in the golden hamster, *Mesocricetus auratus, Journal of Parasitology* **72**, 142-147.

Frazer, B.A., Fried, B., Fujino, T., and Sleckman, B.P. (1999) Host-parasite relationships between *Echinostoma caproni* and Rag-2-deficient mice, *Parasitology Research* **85**, 337-342.

Fried, B. (1978) Trematoda In: *Methods of Cultivating Parasites in Vitro* (Taylor, A.E.R. & Baker, J.R. Eds.) pp 151-192, Academic Press, London.

Fried, B. (1984) Infectivity, growth and development of *Echinostoma revolutum (Trematoda)* in the domestic chick, *Journal of Helminthology* **58**, *241-244*.

Fried, B. (1985) Maintenance of *Echinostoma revolutum* (Trematoda) in the laboratory, *Proceedings of the Pennsylvania   Academy of Science* **59**, 27-28.

Fried, B. (1989) Cultivation of trematodes in chick embryos, *Parasitology Today* **5**, 3-5.

Fried, B. (1994) Metacercarial excystment of trematodes, *Advances in Parasitology* **33**, 91-144.

Fried, B. (2000) Larval and adult *Echinostoma* spp. (Trematoda) in: British Society for Parasitology : Practical Exercises in Parasitology, Halton, P.W., Behnke, J.M., and Marshall, I., (eds,) Cambridge Univiersity Press, Cambridge, in press.

Fried, B. and Appel, A.J. (1977) Excretion of lipids by *Echinostoma revolutum* (Trematoda) adults, *Journal of Parasitology* **63**, 447.

Fried, B. and Bennett, M.C. (1979) Studies on encystment of *Echinostoma revolutum* cercariae. *Journal of Parasitology* **65**, 38-40.

Fried, B. and Bradford, J.D. (1997) In vitro excystation of metacercarial cysts of *Echinostoma trivolvis* from *Rana* species tadpoles, *The Korean Journal of Parasitology* **35**, 75-77.

Fried, B. and Butler, M.S. (1978) Infectivity, excystation, and development on the chick chorioallantois of the metacercaria of *Echinostoma revolutum* (Trematoda). *Journal of Parasitology* **64**, 175-177.

Fried, B., and Emili, S. (1988) Excystation in vitro of *Echinostoma liei* and *E. revolutum* (Trematoda) metacercariae, *Journal of Parasitology* **74**, 98-102

Fried, B. and Fink, L. (1968) Transplantation of *Echinostoma revloutum* (Trematoda) into the chick coelom, *Proceedings of the Pennsylvania Academy of Science* **42**, 61-62.

Fried, B. and Fujino, T. (1984) Scanning electron microsopy of *Echinostoma revolutum* (Trematoda) during development in the chick embryo and the domestic chick, *International Journal for Parasitology* **14**, 75-81.

Fried, B. and Groman, G.M. (1985) Cultivation of the cercariae of *Himasthla quissetensis* (Trematoda) on the chick chorioallantois, *International Journal for Parasitology* **15**, 219-223.

Fried, B. and Huffman, J.E. (1996) The biology of the intestinal trematode *Echinostoma caproni, Advances in Parasitology* **38**, 311-368.

Fried, B. and Kim, S. (1989) In vitro cultivation of *Echinostoma revolutum* (Trematoda) metacercariae in an extract of chick mucosal epithelium, *Proceedings of the Helminthological Society of Washington* **56**, 168-172.

Fried, B. and King, B. (1989) Attraction of *Echinostoma revolutum* cercariae to *Biomphalaria glabrata* dialysate, *Journal of Parasitology* **75**, 55-57.

Fried, B. and Kletkewicz, K. (1987) Excystation of *Echinostoma revolutum* metacercariae in the domestic chick, *Proceedings of the Helminthological Society of Washington* **54**, 267-268.

Fried, B. and Kramer, M.D. (1968) Histochemical glycogen studies on *Echinostoma revolutum, Journal of Parasitology* **54**, 942-944.

Fried, B. and Manger, P.M. Jr. (1992) Use of an aceto-carmine procedure to examine the excysted metacerceriae of *Echinostoma caproni* and *E. trivolvis* (Trematoda), *Journal of Helminthology* **66**, 238-240.

Fried, B. and Morrone, L.J. (1970) Histochemical lipid studies on *Echinostoma revolutum, Proceedings of the Helminthological Society of Washington* **37**, 122-123.

Fried, B. and Perkins, C. (1982) Survival of metacercariae of *Echinostoma revolutum* (Trematoda) in half strength Locke's solution under refrigeration, *Proceedings of the Helminthological Society of Washington* **49**, 153-154.

Fried, B. and Reddy, A. (1999) Effects of snail conditioned water from *Biomphalaria glabrata* on hatching of *Echinostoma caproni* miracidia, *Parasitology Research* **85**, 155-157.

Fried, B. and Roth, M. (1974) In vitro excystment of the metacercaria of *Parorchis acanthus. Journal of Parasitology* **60**, 465.

Fried, B. and Shapiro, I.L. (1979) Thin-layer chromatographic analysis of phospholipids in *Echinostoma revolutum* (Trematoda) adults, *Journal of Parasitology* **65**, 243-245.

Fried, B. and Stableford, L.T. (1991) Cultivation of helminths in chick embryos, *Advances in Parasitology* **30**, 107-165.

Fried, B. and Vonroth, W., Jr. (1968) Transplantation of *Echinostoma revolutum* (Trematoda) into normal and surgically altered chick intestine., *Experimental Parasitology* **22**, 107-111.

Fried, B. and Weaver, L.J. (1969a) Effects of temperature on the development and hatching of eggs of the trematode *Echinostoma revolutum. Transactions of the American Microscopical Society* **88**, 253-257.

Fried, B. and Weaver, L.J. (1969b) Exposure of chicks to the metacercaria of *Echinostoma revolutum* (Trematoda), *Proceedings of the Helminthological Society of Washington* **36**, 153-155.

Fried, B., Weaver L.J. and Kramer, M.D. (1968) Cultivation of *Echinostoma revolutum* (Trematoda) on the chick chorioallantois, *Journal of Parasitology* **54**, 939-941.

Fried, B., Austin, R.M. and Gaines, J.L. (1971) Survival and wound healing of adult *Echinostoma revolutum* following amputation of body parts. *Proceedings of the Helminthological Society of Washington* **38**, 128.

Fried, B., Tancer, R.B., and Fleming, S.J. (1980) In vitro pairing of *Echinostoma revolutum* (Trematoda) Metacercariae and adults, and characterization of worm products involved in chemoattraction. *Journal of Parasitology.* **66**, 1014-1018.

Fried, B., LeFlore, W.B. and Bass, H.S. (1984) Histochemical localization of hydrolytic enzymes in the cercariae and excysted metacercariae of *Echinostoma revolutum* (Trematoda), *Proceedings of the Helminthological Society Of Washington* **51**, 140-143.

Fried, B. Beers, K., and Sherma, J. (1993) Thin-layer chromatographic analysis of beta-carotene and lutein in *Echinostoma trivolvis* (Trematoda) rediae., *Journal of Parasitology* **79**, 113-114.

Fried, B., Idris, N., and Ohsawa, T. (1995) Experimental infection of juvenile *Biomphalaria glabrata* with cercariae of *Echinostoma trivolvis. Journal of Parasitology* **81**, 308-310.

Fried, B., Pane, P.L. and Reddy, A. (1997a) Experimental infection of *Rana pipiens* tadpoles with *Echinostoma trivolvis* cercariae. *Parasitology Research* **83**, 666-669.

Fried, B., Nanni, T.J., Reddy, A. and Fujino, T. (1997b) Maintenance of the life cycle of *Echinostoma trivolvis* (Trematoda) in Dexamethasone-treated ICR mice and laboratory-raised *Helisoma trivolvis* (Gastropoda). *Parasitology Research* **83**, 16-19.

Fried, B., Eyster, L.S., and Pechenik, J.A. (1998).Histochemical glycogen and neutral lipid in *Echinostoma trivolvis* cercariae and effects of exogenous glucose on cercarial longevity, *Journal of Helminthology* **72**, 83-85.

Fujino, T., Fried, B., and Hosier, D.W. (1994) The expulsion of *Echinostoma trivolvis* from ICR mice: extension/retraction mechanisms and ultrastructure of the collar spines. *Parasitology Research* **80**, 581-587.

Fujino, T., Takahashi, Y., and Fried, B. (1995) A comparison of *Echinostoma trivolvis* and *E. caproni* (Trematoda: Echinostomatidae) using random amplified polymorphic DNA analysis. *Journal of Helthminthology* **69**, 263-264.

Fujino, T., Ichikawa, H., Fried, B., and Fukuda, K. (1996) the expulsion of *Echinostoma trivolvis*: suppressive effects of dexamethasone on parasite cell hyperplasia and worm rejection in C3H/HeN mica, *Parasite* **3**, 283-289.

Fujino, T., Washioka, H., Sasaki, K., Tonosaki, A., and Fried, B. (1999) Comparative ultrastructure of lamellar gastrodermal projections in *Echinostoma paraensei, E. caproni*, and *E. trivolvis* (Trematoda: Echinostomatidae). *Parasitology Research* **85**, 655-660.

Grover, M., Dutta, R., Kumar, R., Aneja, S., and Mehta, G. (1998) *Echinostoma ilocanium* infection, *Indian Pediatrics* **35**, 549-552.

Graczyk, T.K. and Fried, B. (1998) Echinostomiasis: a common but forgotten food borne disease. *American Journal of Tropical Medicine and Hygiene* **58**, 501-504.

Gulka, G.J. and Fried, B. (1979) Histochemical and ultrastructural studies on the metacercarial cyst of *Echinostoma revolutum.* (Trematoda), *International Journal for Parasitology* **9**, 57-59.

Heyneman, D. (1966) Successful infection with larval echinostomes surgically implanted into the body cavity of the normal snail host. *Experimental Parasitology* **18**, 220-223.

.Hobbs, P.J., Fryer, S.E., Duimstra, J.R., Brodie, A.E., Collodi, P.A., Menino, J.S., Bayne, C.J. and Bavines, D.W. (1993) Culture of cells from juvenile worms *of Schistosoma mansoni, Journal of Parasitology* **79**, 913-927.

Hosier, D.W. and Fried, B. (1986) Infectivity, growth, and distribution of *Echinostoma revolutum* in Swiss Webster and ICR mice., *Proceedings of the Helminthological Society of Washington* **53**, 173-176.

Hosier,D.W. and Fried, B. (1991) Infectivity, growth, and distribution of *Echinostoma caproni* (Trematoda) in the ICR mouse, *Journal of Parasitology* **77**, 640-642.

Howell, M.J. (1968) Excystment and in vitro cultivation of *Echinoparyphium serratum*, *Parasitology* **58**, 583-597.

Huffman, J.E. and Fried, B. (1990) *Echinostoma* and echinostomiasis, *Advances in Parasitology* **29**, 215-269.

Humphries, J.E. and Fried, B. (1996) Histological and histochemical studies on the paraoesophageal glands in cercariae and metacercariae of *Echinostoma revolutum* and *E. trivolvis*, *Journal of Helminthology* **70**, 299-301.

Humphries, J.E., Reddy, A., and Fried, B. (1997) Infectivity and growth of *Echinostoma revolutum* (Froehlich, 1802) in the domestic chick, *International Journal for Parasitology* **27**, 129-130.

Irwin, S.W.B., (1997) Excystation and cultivation of trematodes, in B. Fried and T.K. Graczyk, (eds.), *Advances in Trematode Biology*, CRC Press, Boca Raton pp. 57-86.

Jeyarasasingham, U. Heyneman, P, Lim, H-K., and Mansour, N. (1972) Life cycle of a new echinostome from Egypt, *Echinostoma liei* sp. nov. (Trematoda: Echinostomatidae) *Parasitology* **65**, 203-222.

Loker, E.S., Cousteau, C., Ataev, G.L., and Jourdane, J (1999) In vitro culture of rediae of *Echinostoma caproni*, Parasite **6**, 169-174.

MacInnis, A.J. and Voge, M. (1970) Experiments and Techniques in Parasitology, W.H. Freeman, San Francisco, CA.

Martin, W.E. (1972) An annotated key to the cercariae that develop in the snail *Cerithidea californica*, *Bulletin of the Southern California Academy of Sciences* **71**, 39-43.

Maurer, K., Decere, M., and Fried, B. (1996) Effects of the anthelmintics clorsulon, rafoxanide, mebendazole and arprinocid on *Echinostoma caproni* in ICR mice, *Journal of Helminthology* **70**, 95-96.

Mueller, T.J., and Fried, B. (1999) Electrophoretic analysis of proteases in *Echinostoma caproni* and *E. trivolvis*, *Journal of Parasitology* **85**, 174-180.

Perez, M.K., Fried, B., and Sherma, J. (1994) High performance thin-layer chromatographic analysis of sugars in *Biomphalaria glabrata* (Gastropoda) infected with *Echinostoma caproni* (Trematoda), *Journal of Parasitology* **80**, 336-338.

Pritchard, M.H. and Kruse, S.D.W. (1982) The Collection and Preservation of Animal Parasites, University of Nebraska Press, Lincoln.

Reddy, A. and Fried, B. (1996a) Egg-laying in vitro of *Echinostoma caproni* (Trematoda) in nutritive and non-nutritive media, *Parasitology Research* **82**, 475-476.

Reddy, A. and Fried, B. (1996b) In Vitro studies on intraspecific and interspecific chemical attraction in daughter rediae of *Echinostoma trivolvis* and *E. caproni*. *International Journal for Parasitology* **26**, 1081-1085.

Schmidt, G. (1988) *Essentials of Parasitology*, 4th ed., Win C. Brown, Dubrigue.

Schimdt, K.A. and Fried, B. (1996) Emergence of cercariae of *Echinostoma trivolvis* from *Helisoma trivolvis* under different Conditions, *Journal of Parasitology* **82**, 674-676.

Schmidt, K.A. and Fried, B. (1997) Prevalence of larval trematodes in *Helisoma trivolvis* (Gastropoda) from a farm pond in Northampton County, PA with special emphasis on *Echinstoma trivolvis* (Trematoda) cercariae, *Journal of the Helmithological Society of Washington* **64**, 157-159.

Schimidt, K.A., Fried, B. and Reddy, A. (1998) Maintenance of *Helisoma trivolvis* naturally infected with *Echinostoma trivolvis* in spring water at 40°C for 300 days, *Journal of Helminthology* **72**, 91-92.

Smyth, J.D. (1990) *In Vitro Cultivation of Parasitic Helminths*. CRC Press, Boca Raton.

Smyth, J.D. and Halton, D.W. (1983) The Physiology of Trematodes, 2ND edition Cambridge University Press, Cambridge.

Sleckman, B.P. and Fried, B. (1984) Glycogen and protein content in adult *Echinstoma revolutum* (Trematoda), *Proceedings of the Helminthological Society of Washington* **51**, 355-356.

Sudati, J.E., Rivas, F., and Fried, B. (1997) Effects of a high protein diet on worm recovery, growth and distribution of *Echinostoma caproni* in ICR mice. *Journal of Helminthology* **71**, 351-354.

Taft, J. and Fried, B. (1968) Oxygen consumption in adult *Echinostoma revolutum* (Trematoda), *Experimental Parasitology* **23**, 183-186.

Weller, T.H. and Wheldon, S.K. (1982) The cultivation in vitro of cells derived from *Schistosoma masoni* I. Methodology; criteria for evaluation of cultures, and development of media, *American Journal of Tropical Medicine and Hygiene* **31**, 335-348.

Wisnewski, N., Fried, B., and Halton, D.W. (1986) Growth and feeding of *Echinostoma revolutum* on the chick chorriollantois and in the domestic chick, *Journal of Parasitology* **72**, 684-689.

ULTRASTRUCTURAL STUDIES ON ECHINOSTOMES

T. FUJINO[a] AND H. ICHIKAWA[b]

[a]Department of Biology, Faculty of Science, Yamagata University, Yamagata 990-8560, Japan, and [b]Department of Medical Zoology, Kanazawa Medical University, Ishikawa 920-0293, Japan

B. Fried and T.K. Graczyk (eds.),
Echinostomes as Experimental Models for Biological Research, 119–136.

1. Introduction

This chapter includes ultrastructural features of eggs and the other stages including miracidia, cercariae, sporocysts, rediae, metacercariae and adults of echinostomes. Most of the ultrastructural studies reported to date are on the genera *Echinostoma* Rudolphi, 1809 and *Echinoparyphium* Dietz, 1909, but include some other genera such as *Echinochasmus* Dietz, 1909, *Isthmiophora* Luhe, 1909 and Himasthla Dietz, 1909. The ultrastructure of echinostomes from naturally or experimentally infected mammalian or avian hosts can be seen in fixed specimens prepared for scanning (SEM) or transmission electron microscopy (TEM). In some reports, cytochemical, immunocytochemical or freeze fracture methods were used to localize enzymes, antigens or other materials in worm tissues or cells.

Surface ultrastructure of eggs of *Echinostoma trivolvis* and *E. caproni* was compared and miracidia of *E. caproni* were observed by SEM. The daughter rediae of *E. trivolvis* and *E. caproni* were observed by SEM, with special attention to the tegumentary papillae. The cercaria of *E. caproni* was compared with that of *E. trivolvis* and a specific difference between them was ultrastructurally observed in their tail finfolds. Argentophilic and SEM observations were made on the tegumentary papillae of some echinostome species. TEM studies revealed the ultrastructure of metacercarial cysts of *E. trivolvis* and Himasthla quissetensis. The excystment in *Himasthla leptosoma* was observed by TEM and SEM. The tegumental ultrastructure of *Echinostoma trivolvis*, *E. hortense*, *E. cinetorchis*, *Echinochasmus japonicus*, *Echinoparyphium recurvatum* and *Isthmiophora melis* adults was observed. Cytochrome c oxidase in the mitochondria of the tegument in *E. trivolvis* was cytochemically observed. The localization of tegumental surface antigens in *E. caproni* and the binding of mouse antibodies to the surface antigens of the young stages in *E. caproni* were observed by TEM. Freeze fracture technique revealed the ultrastructural differences in the lamellar gastrodermal projections in *E. paraensei*, *E. caproni* and *E. trivolvis*.

2. Eggs and miracidia

Eggs of *Echinostoma trivolvis* and *E. caproni* were compared by light microscopy (LM) and SEM (Krejci and Fried, 1994). The topography of the eggshell was smooth and similar in both *E. caproni* and *E. trivolvis*. Eggs were oval in shape and operculate, with a knob at the abopercular end. The knob of *E. caproni* and *E. trivolvis* had shallow and deep infoldings in their eggshells, respectively.

Ataev *et al.* (1998) showed by SEM that newly hatched miracidia of *E. caproni* transformed into mother sporocysts after 1 or 2 days of culture. After 10 minutes of sonication, cilia on the miracidial body surface were lost (Figures 1-4).

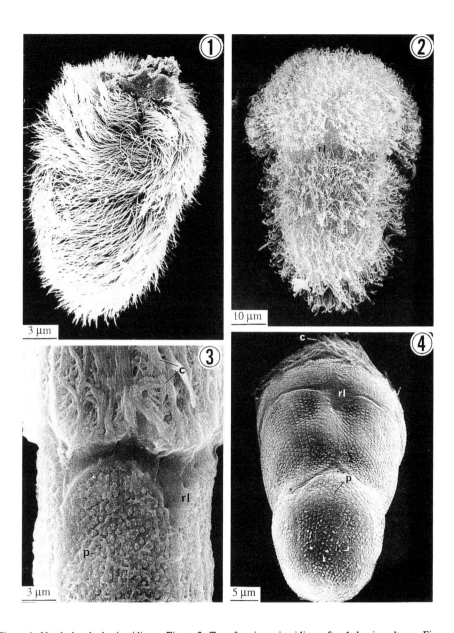

Figure 1. Newly hatched miracidium. Figure 2. Transforming miracidium after 1 day in culture. Figure 3. Transforming miracidium after 2 days in culture, showing epithelial plates with and without cilia. Figure 4. Newly hatched miracidium after 10 min of sonication. Note the absence of cilia on epithelial layer. c, cilia; p, papilla; rl, ridge layer between epithelial plates. (From Ataev *et al.*, *J. Parasitol.*, 84, 232, 1998, With permission)

3. Sporocysts, rediae and cercariae

Fried and Awatramani (1992) observed by SEM the daughter rediae of *Echinostoma trivolvis*. The rediae had a cylindrical body, two ventral ambulatory buds, a posterior papilliform process and a mouth at the anterior tip. Numerous ciliated papillae and spherical bodies surrounded the mouth region. Posterior to the mouth, the tegument was folded into transverse rings. Just posterior to the oral collar was the dorsal birth papilla with a pore in its center. The ambulatory buds had transverse tegumentary folds at their bases. On the remainder of the buds, the folds were highly ridged. Posterior to the ambulatory buds, the longitudinal ridges were oriented perpendicular to the tegumentary folds. The redial bodies are covered with microvilli, structures which probably increase the absorptive surface area; the transverse tegumentary folds and longitudinal ridges also probably increase the surface area. The spherical bodies noted on the tegumentary surface in some other species of rediae might be involved in the expulsion of waste material (Figures 5-10).

Ultrastructural differences observed by SEM were noted in the redial mouth and papilliform processes in *E. trivolvis* and *E. caproni*. The redial mouth of *E. trivolvis* had numerous uniciliate papillae and spherical bodies or knobs. The redia of *E. caproni* had unidentified knob-like structures. The papilliform processes of *E. trivolvis* had irregular tegumental foldings with abundant knob-like microvilli, whereas that of *E. caproni* had symmetrical, brick-like tegumental infoldings (Krejci and Fried, 1994).

LM and SEM were used to determine interspecific differences between various larval stages of the allopatric species *Echinostoma caproni* and *E. trivolvis*. The cercaria of *E. caproni* had one ventral tail fin-fold, whereas that of *E. trivolvis* had two (see Fried and Fujino, 1987); both species had two dorsal tail fin-folds. The middorsal tegument of the cercarial body of *E. trivolvis* had abundant spherical bodies and no uniciliate papillae, whereas *E. caproni* had relatively few spherical bodies. The acetabular lip of *E. trivolvis* was surrounded by irregular tegumental foldings and numerous uniciliate papillae, whereas the lip of *E. caproni* was surrounded by symmetrical tegumental foldings and no uniciliate papillae (Krejci and Fried, 1994).

Argentophilic and SEM observations were made on the tegumentary papillae of *Echinostoma trivolvis* (referred to as *E. revolutum* in that study) cercariae. SEM revealed that the most abundant papillae were uniciliate and arranged bilaterally on the body and the tail of the cercaria; there was a total of 18 multiciliate papillae, of which 16 were on the anterior aspect and 2 in the middle region of the body (Figures 11-16) (Fried and Fujino, 1987). The distribution and fine structure of the tegumentary papillae of the cercaria of *Himasthla secunda* were examined by LM, SEM and TEM. The localization of the sensory endings on the surfaces of the cercarial body was observed by SEM. The localization of silver within the sensory endings that was stained with silver nitrate was also examined by TEM. The ultrastructure indicated that the nerve endings showed uniciliate papillae typical of those seen in other trematode species (Chapman and Wilson, 1970).

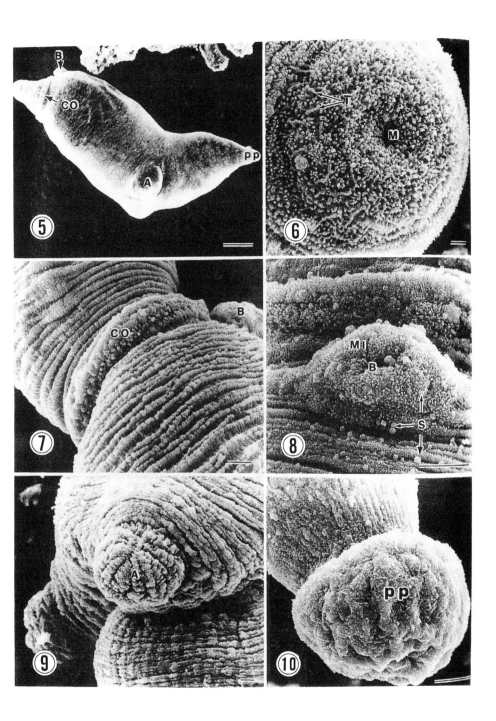

Figure 5. Lateral view of whole redial body. Bar = 100 μm. Figure 6. Mouth region of redia. Bar =
5 μm. Figure 7. Collar of redia. Note the transverse folds in the tegument both anterior and posterior to
the collar and the birth papilla posterior to the collar. Figure 8. Birth papilla covered with spherical
bodies and microvilli-like structures. Note the spherical bodies on the tegument adjacent to the birth
papilla. Figure 9. Ambulatory buds showing ridged transverse tegumentary folds. Figure 10. Fully
protruded papilliform process of the redia. Note the folded tegument. A, ambulatory buds; B, birth
papilla; CO, collar; M, mouth; MI, microvilli-like structures; pp, posterior papilliform process; S,
spherical bodies; T, tegumentary ciliated papillae. (From Fried and Awatramani, *Parasitol. Res.*, 78,
258, 1992, With permission)

4. Metacercariae

SEM observations showed that encysted metacercariae of *Echinostoma trivolvis* and *E.
caproni* are spherical or ovoidal, but *E. caproni* has a smoother outer cyst wall than
does *E. trivolvis* (Krejci and Fried, 1994).

TEM study of the metacercarial cyst of *E. trivolvis* (referred to as *E. revolutum* in
the study) revealed three cyst walls, the outer, middle and inner. The outer wall was
more electron-dense than the middle, and contained coarser granules than those in the
middle layer. The inner wall was lamellated and contained membranous whorls. The
outer and middle cyst walls stained identically with all histochemical procedure used.
These walls contained acid mucopolysaccharides and glycoprotein, whereas the inner
cyst wall contained glycoprotein (Gulka and Fried, 1979). The excysted metacercariae
of *Himasthla leptosoma*, with a reniform collar of 29 cephalic spines, were observed
by SEM and TEM. These collar spines were sometimes withdrawn into pits,
presumably by the action of a muscle complex in the collar region. Sensory papillae
were distributed in a bilaterally symmetrical arrangement around the oral sucker and
none were visible on the surface of the ventral sucker. Tegumental spines were found
only from a point some distance behind the oral collar up to the region of the ventral
sucker. The most anterior tegumentary spines were simple and peg-like, whereas those
more posteriorly located were complex in a palmate form (Figures 17-22) (Irwin *et al.*,
1984).

Laurie (1974) described the ultrastructure of the metacercarial cyst of *Himasthla
quissetensis* that was induced to encyst in various chemical solutions. The ellipsoidal
cyst was composed of an outer homogeneous layer formed by granules released from
the tegument and an inner laminated layer formed by scrolled rods originating in
subtegumental cell bodies. Stein and Basch (1977) described the ultrastructure of the
metacercarial cyst of *Echinostoma paraensei* as being composed of two well-defined
regions. The outer cyst wall is fibrous in nature, formed from secretory granules
released from the cercarial tegument. Membranous scrolls or rodlets secreted from the
subtegumental gland cells were added to form the inner cyst wall (Figures 23-25).

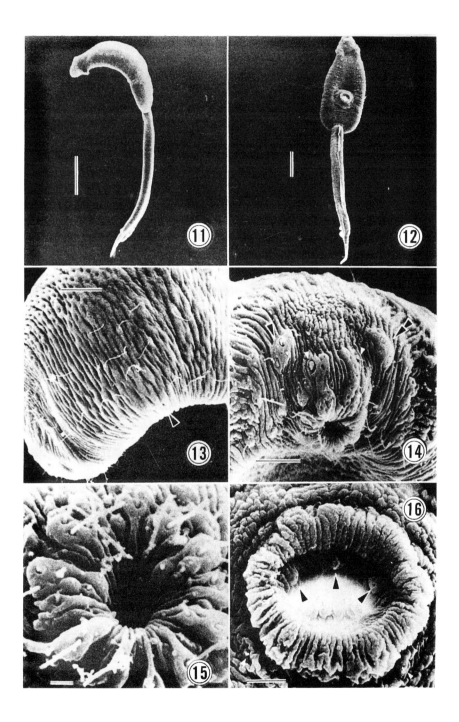

Figure 11. Body and tail in lateral view. Note dorsal and ventral tail finfolds. Bar = 50 μm. Figure 12. Body and tail in ventral view. Note the beginning of the ventral tail finfold at the body-tail junction. Bar = 50 μm. Figure 13. Lateral view of the body. Many uniciliate papillae are seen; the arrowhead indicates a multiciliate papilla. Bar= 15 μm. Figure 14. Anterior end of the body showing papillae along the collar spines. Large and small arrowheads indicate pairs of grouped multiciliate papillae. Figure 15. Oral sucker showing papillae with long and short cilia around the mouth opening. Figure 16. Acetabulum. Large arrowheads indicate 3 papillae inside the lip of the acetabulum; small arrowheads indicate a pair of papillae located close to the posterior edge of the acetabulum. Bar = 5 μm. (From Fried and Fujino, *J. Parasitol.*, 73, 1171, 1987, With permission)

5. Adults

SEM observations were made to compare the topography of *E. trivolvis* and *E. caproni* adults grown in hamsters for 15 days. The differences between the two species are in sucker morphology, collar and tegumental spines, sensory papillae on the acetabular lip and a folded protuberance of unknown function just posterior to the genital pore (Fried *et al.*, 1990). SEM was used to examine the collar spines of *E. trivolvis* from ICR mice or hamsters or chick. The collar spines of the worms were extended at 14 day p.i. in the hamster, but were retracted in worms obtained from ICR mice. It is possible that echinostomes exhibiting retracted collar spines were more likely to be expelled from hosts than were worms displaying extended spines (Kruse *et al.*, 1992). SEM of the oral collar of a 14-day-old *E. trivolvis* grown in the chick was given (Huffman and Fried, 1990). SEM of the oral collar of 15-day-old adults of *Echinoparyphium recurvatum* with 45 collar spines were studied; the middle 37 spines were arranged around the collar in a double tier, the spines of the oral tier being slightly smaller than those of the aboral tier (Figure 26) (McCarthy, 1990).

Ultrastructural observations of collar spines and the surrounding tissues associated with extension/retraction mechanisms of spines were made on *Echinostoma trivolvis* recovered from ICR mice or golden hamsters. Contraction of the muscles around the spines caused an invagination of the tegument surrounding a spine, resulting in spinal retraction. Relaxation of the muscles resulted in spinal retraction. An immunocytochemical technique using colloidal gold confirmed the presence of actin in the collar spines. Cultivation of the worms in a defined medium supplemented with fresh hamster or mouse serum showed that certain factors in mouse serum were involved in spinal retraction (Figure 27) (Fujino *et al.*, 1994).

SEM studies were made on the tegumental surfaces of *Echinostoma trivolvis* (referred to as *E. revolutum* in the study) and *Isthmiophora melis*. The tegument of both species had a cobble stone-like appearance with interspersed pits. The taxonomic position of *I. melis* and the significance of cirrus morphology as a useful taxonomic character were discussed. The distribution and possible function of the four types of sensory papillae seen in either species, ciliated, domed, button and bilobed, were discussed (Smales and Blankespoor, 1984).

Tegumental ultrastructure of *Echinostoma hortense* adults was studied by SEM. The tegument was covered with cobble stone-like cytoplasmic processes. The oral sucker

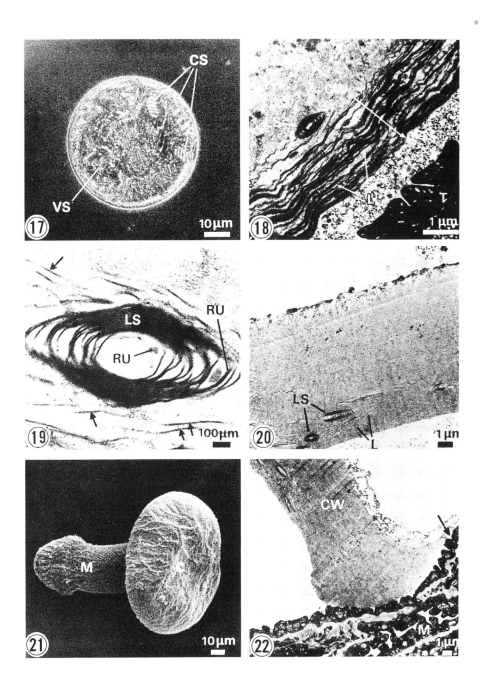

Figure 17. Light micrograph of whole cyst. Note cephalic spines and ventral sucker inside the cyst. Figure 18. TEM of a section through a metacercarial cyst wall. The outer layer is electron-lucid, containing some fine reticulate material, whereas the inner is similar but contains many lamellae, some of which are in the form of scrolls. The tegument of an encysted metacercaria is visible. Figure 19. Higher magnification of a lamellar scroll. Figure 20. The point in a cyst wall where the lamellar layer is discontinued, leaving an area devoid of lamella. A few lamellae and some lamellar scrolls are present mainly at the left of the micrograph. Figure 21. SEM of a metacercaria emerging from a cyst. Figure 22. Non-lamellar portion of a cyst wall which is breached allowing a metacercaria to escape. The outer limit of the breached cyst wall is drawn outwards (arrowed) by the body of the escaping metacercaria. C, cyst; CS, cephalic spines; CW, cyst wall; L, lamellae, LS, scrolls; M, metacercaria; RU, repeating units; T, tegument; VS, ventral sucker. (From Irwin *et al.*, *Intl. J. Parasitol.* 14, 417, 1984, With permission)

had roundly swollen sensory papillae (type II) on the ventral half of its lip and uni-ciliated knob-like papillae (type I) arranged in 2-3 rows on the dorsal outer surface. Scale-like spines each with a broad base and a round tip were distributed densely on the tegument anterior to the ventral sucker, but they became more sparse in the posterior half of the ventral surface. A few type I papillae were observed on the ventral surface. These results suggested that the tegument of this species was similar to that of *E. trivolvis* and that this species differed from the others by the number and arrangement of collar spines, and/or the type and distribution of sensory papillae (Lee *et al.*, 1986).

Tegumental ultrastructure of *Echinochasmus japonicus* adults was observed by SEM (Lee *et al.*, 1987). The worms recovered experimentally from albino rats were used. The tegument of the body was covered with cobble-stone-like spines each with a broad base and a pointed tip. They were compact in the ventrolateral tegument or dorsal surface of the anterior body. No spines were present between the two suckers and the dorsal surface of the posterior body. Two types of sensory papillae, uni-ciliated and roundly swollen sensory papillae, were identified; the former was mainly distributed on the ventral surface of the tegument and the latter on the lips of the suckers. The head crown was armed with 24 collar spines, of which the second and fourth ones were more outstretched than the others.

Comparative SEM studies were made by Kanev *et al.* (1999) on the collar region of *Echinostoma revolutum* and an intestinal digenean of fish, *Deropristis inflata* of *Deropristidae*. The collar of *D. inflata* was open dorsally and ventrally, whereas that of *E. revolutum* was closed dorsally and ventrally, forming a kidney-like ring. The collar of *D. inflata* was located posterior to the oral sucker, whereas that of *E. revolutum* was located terminally around the oral sucker. The collar of *D. inflata* had numerous collar spines which varied in size, shape and position. That of *E. revolutum* had 37 collar spines, mainly homogenous in their general morphology and a typical arrangement pattern for 37-collar-spined echinostomes. This study raises doubts about the present systematic position of the adults of *D. inflata* which possess an oral collar and crown of collar spines, but are not included in the family *Echinostomatidae*. Skrjabin (1958) placed this species in the family *Deropristidae*, but Yamaguti (1971) placed this in the family *Acanthocolpidae* instead of *Echinostomatidae*. Further studies

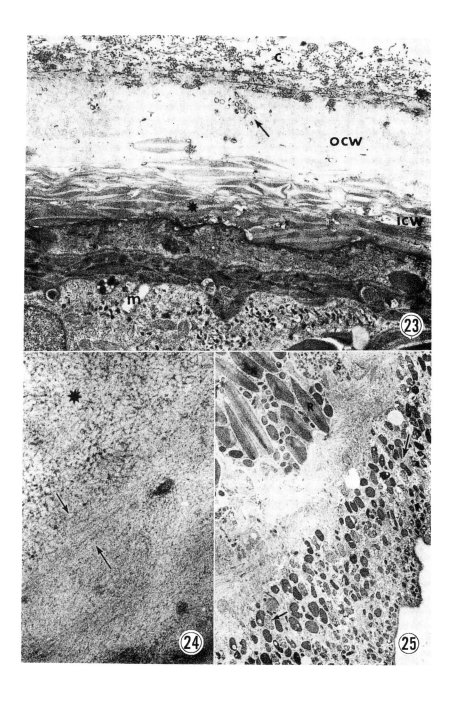

Figure 23. Cyst consisting of outer cyst wall, inner cyst wall, metacercaria and collagen fibers derived from *Biomphalaria glabrata* cells (not shown).   Figure 24. High magnification of outer cyst wall.   Note orientation of fibers (enclosed by arrows) and fibers in region denoted by asterisk.   Figure 25. Cercarial stage prior to encystment. C, collagen fibers; ICW, inner cyst wall; OCW, outer cyst wall; M, metacercaria. (From Stein and Basch, *J. Parasitol.*, 63, 1034, 1997, With permission)

Figure 26. SEM of the anterior aspect of the spiny collar of a 15-day-old adult *Echinoparyphium recurvatum*. (From McCarthy, *Parasitology*, 101, 36, 1990, With permission)

are necessary to clarify the present taxonomic situation.

The tegumental ultrastructure of juvenile and adult *Echinostoma cinetorchis* was observed by SEM (Lee *et al.*, 1992). The oral collar of 3-day-old juveniles contained 37-38 collar spines. Tongue or spade-shaped spines were distributed anterior to the ventral sucker, whereas peg-like spines were distributed posteriorly and became sparser toward the posterior end of the body. The tegument of the adults had scale-like spines, with broad and rounded tips. The spines were densely distributed anterior to the ventral sucker but became more sparse posterior to the sucker. On the dorsal surface, the tegumental spines were located mainly in the anterior part of the body. Ursone and Fried (1995) described LM and SEM of *E. caproni* during maturation in ICR mice. The morphological changes occurred during maturation in the acetabulum, collar spines, tegumental papillae, and tegumental spines.

Cytochrome c oxidase (CCO) in the mitochondria of the tegument and the tegumental and parenchymal cells was examined cytochemically in *Echinostoma trivolvis*. Mitochondria were located mainly in the basal region of the tegument. Most of them had two or three cristae that extended deep into the matrix; the cristae were mostly parallel to the long axis of the mitochondria and reacted heavily for CCO. Mitochondria in the parenchymal cells were relatively large, round or oval with undeveloped cristae in a uniformly granular matrix. These mitochondria reacted weekly for CCO (Fujino *et al.* 1995).

The surface antigens of *Echinostoma caproni* were examined. It was demonstrated that antigens released from the surface of the juveniles and 4-week-old worms during in vitro culture. SDS-PAGE and Western blot analysis indicated that four major antigens released from the surface of adult parasites had molecular masses of approximately 26, 66, 75 and 88 kDa. In vitro turn-over rate of the surface antigens was very high, with a half-life of 8-15 min in both juveniles and adults. An attempt to immunize mice with detergent-solubilized adult surface antigens failed to induce resistance of infection with metacercariae of this species (Andresen *et al.*, 1989).

The binding of mouse antibodies to the surface antigens of juvenile and 7 and 28-day-old *Echinostoma caproni* was examined by TEM of thin sections of worms treated with antibodies in a double sandwich technique with ferritin-conjugated antibody. The surface of freshly recovered sexually mature worms was covered with an irregular but often rather intensive mouse antibody containing matrix, which probably represents a layer of mouse antibody/parasite antigen complexes. It may be that the antigens were present as isolated excretions along the surface of the worms. Worm surface antigens were present in the tegument in vesicles which fused with the outer membrane of the parasite whereby their contents are released to the exterior (Simonsen *et al.*, 1990). Immune reactions to the surface of the intestinal trematode, *E. revolutum* (= *E. trivolvis*) in the SVS mouse was examined, with emphasis on the production of specific antibodies to the fluke surface. Two-week-old mice after infection with the echinostomes sera became positive for antibodies (IgG, IgM and IgA) to the surface of in vitro excysted juveniles. Antibody-or C3-covered juveniles were attacked by cells in in vitro peritoneal-cell-adherence test, but antibodies, C3 or cells

Figure 27. Longitudinal section of an extended collar spine of *Echinostoma trivolvis* adult from a hamster. Note that the tegument covering the distal region of the collar spine is thinner than that covering the proximal region (arrows). Muscle bundles are contacted. Cs, collar spine; Im interstitial material; Mu, muscle bundle; Ol outer layer; T, tegument. Bar = 1 μm. (From Fujino *et al.*, *Parasitol. Res.*, 80, 583, 1994, With permission)

had no effect on the survival of juveniles in vitro after 24 hr. It is suggested that shedding of surface-bound antibodies may be an adaptation of these flukes to withstand the hosts immune attack (Simmonsen and Andersen, 1986).

A comparative study using TEM, SEM and freeze-fracture replication was made on the gastrodermal lamellar microvilli in adults of *E. paraensei*, *E. caproni* and *E. trivolvis*. Lamellar projections by SEM were mainly planiform and rhomboidal or paddle-shaped, and they were commonly seen in the three species. The distal margins of these lamellae were mainly smooth in *E. paraensei*, whereas filiform or digitiform extensions in the margins occurred occasionally in E. caproni and frequently in *E. trivolvis*. Freeze-fracture TEM elucidated significant interspecific differences in the number of intramembranous particles (IMP)/μm^2 among the species. The ultrastructural differences in the gastrodermal lamellae of the adult echinostomes provided a taxonomically useful criterion for distinguishing *Echinostoma* species (Figures 28-30) (Fujino *et al.*, 1999).

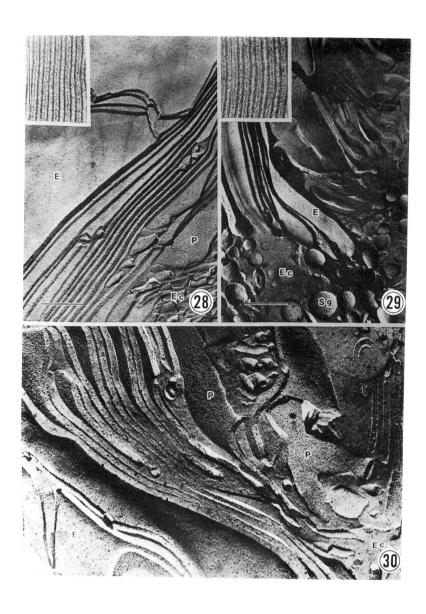

Figure 28. *Echinostoma paraensei*, Inset, TEM of the crossections of lamellar projections. Bar = 0.5 μm. Figure 29. *E. caproni*, Inset, TEM of the crossections of lamellar projections. Bar = 0.5 μm. Figure 30. *E. trivolvis*, Bar = 0.5 μm. E, ectoplasmic face; Ec, epithelial cells; P, protoplasmic face. (From Fujino *et al.*, *Parasitol. Res.*, 85, 655, 1999, With permission)

6. References

Andersen, K., Simonsen, P. E., Andersen, B. J. and Birch-Andersen, A. 1989. *Echinostoma caproni* in mice: shedding of antigens from the surface of an intestinal trematode. *International Journal for Parasitology* 19, 111-118.

Ataev, G. L., Fournier, A. and Coustau, C. 1998. Comparison of *Echinostoma caproni* mother sporocyst development in vivo and in vitro using *Biomphalaria glabrata* snails and a *B. glabrata* embryonic cell line. *Journal of Parasitology* 84, 227-235.

Chapman, H. D. and Wilson R. A. 1970. The distribution and fine structure of the integumentary papillae of the cercaria of *Himasthla secunda* (Nicoll). *Parasitology* 61, 219-227.

Fried, B. and Awatramani, R. 1992. Light and scanning electron microscopical observations of the daughter rediae of *Echinostoma trivolvis* (Trematoda). *Parasitology Research* 78, 257-259.

Fried, B. and Fujino, T. 1987. Argentophilic and scanning electron microscopic observations of the tegumentary papillae of *Echinostoma revolutum* (Trematoda) cercariae. *Journal of Parasitology* 73, 1169-1174.

Fried, B., Irwin, S. W. B. and Lowry, S. F. 1990. Scanning electron microscopy of *Echinostoma trivolvis* and *E. caproni* (Trematoda) adults from experimental infections in the golden hamster. *Journal of Natural History* 24, 433-440.

Fujino, T., Fried, B. and Hosier, D. W. 1994. The expulsion of *Echinostoma trivolvis* (Trematoda) from ICR mice: extension/retraction mechanisms and ultrastructure of the collar spines. *Parasitology Research* 80, 581-587.

Fujino, T., Fried, B. and Takamiya, S. 1995. Cytochemical localization of cytochrome c oxidase activity in mitochondria in the tegument and tegumental and parenchymal cells of the trematodes *Echinostoma trivolvis*, *Zygocotyle lunata*, *Schistosoma mansoni*, *Fasciola gigantica* and *Paragonimus ohirai*. *Journal of Helminthology* 69, 195-201.

Fujino, T., Washioka, H., Sasaki, K., Tonosaki, A. and Fried, B. 1999. Comparative ultrastructure study of lamellar gastrodermal projections in *Echinostoma paraensei*, *E. caproni* and *E. trivolvis* (Trematoda: Echinostomatidae). *Parasitology Research* 85, 655-660.

Gulka, G, J. and Fried, B., 1979. Histochemical and ultrastructural studies on the metacercarial cyst of *Echinostoma revolutum* (Trematoda). *International Journal for Parasitology* 9, 57-59.

Huffman, J. E. and Fried, B. 1990. *Echinostoma* and echinostomiasis. *Advances in Parasitology* 29, 215-269.

Irwin, S. W. B., McKerr, G., Judge, B. C. and Moran, I. 1984. Studies on metacercarial excystment in *Himasthla leptosoma* (Trematoda: Echinostomatidae) and newly emerged metacercariae. *International Journal for Parasitology* 14, 415-421.

Kanev, I., Dezfuli, B. S., Nestorov, M. and Fried, B. 1999. Scanning electron microscopy of the collar region of *Deropristis inflata* and *Echinostoma revolutum*. *Journal of Helminthology* 73, 51-57.

Krejci, K. G. and Fried, B. 1994. Light and scanning electron microscopic observations of the eggs, daughter rediae, cercariae, and encysted metacercariae of *Echinostoma trivolvis* and *E. caproni*. *Parasitology Research* 80, 42-47.

Kruse, D. M., Hosier, D. W. and Fried, B. 1992. The expulsion of *Echinostoma trivolvis* (Trematoda) from ICR mice: scanning electron microscopy of the worms. *Parasitology Research* 78, 74-77.

Laurie, J. S. 1974. *Himasthla quissetensis*: induced in vitro encystment of cercaria and ultrastructure of the cyst. *Experimental Parasitology* 35, 350-362.

Lee, S.-H., Hong, S. J. Chai, J.-Y., Hong. S.-T and Seo, B.-S. 1986. Tegumental ultrastructure of *Echinostoma hortense* observed by scanning electron microscopy. *Korean Journal of Parasitology* 24, 63-70.

Lee, S.-H., Jun, H.-S., Sohn, W.-M. and Chai, J.-Y. 1992. Tegumental ultrastructure of juvenile and adult *Echinostoma cinetorchis*. *Korean Journal of Parasitology* 30, 65-74.

Lee, S.-H., Sohn, W.-M. and Hong, S.-T. 1987. Scanning electron microscopical findings of *Echinochasmus japonicus* tegument. *Korean Journal of Parasitology* 25, 51- 58.

McCarthy, A. M. 1990. Speciation of echinostomes: evidence for the existence of two sympatric sibling species in the complex *Echinoparyphium recurvatum* (von Linstow 1873) (Digenea:

Echinostomatidae). *Parasitology* 101, 35-42.

Simonsen, P. E. and Andersen, B. J. 1986. *Echinostoma revolutum* in mice; dynamics of the antibody attack to the surface of an intestinal trematode. *International Journal for Parasitology* 16, 475-482.

Simonsen, P. E., Vennervald, B. J. and Birch-Andersen, A. 1990. *Echinostoma caproni* in mice: ultrastructural studies on the formation of immune complexes on the surface of an intestinal trematode. *International Journal for Parasitology* 20, 935-941.

Skrjabin, K. I. 1958. Trematodes of animals and man. Essentials of trematodology vol. 15, Izdatelstvo Akademia Nauk SSSR, pp. 33-37.

Smales, L. R. and Blankespoor, H. D. 1984. *Echinostoma revolutum* (Froelich, 1802) Looss, 1899 and *Isthmiophora melis* (Schrank, 1788) Luhe, 1909 (Echinostomatinae, Digenea): scanning electron microscopy of the tegumental surfaces. *Journal of Helminthology* 58, 187-195.

Stein, P. C. and Basch, P. F. 1977. Metacercarial cyst formation in vitro of *Echinostoma paraensei*. *Journal of Parasitology* 63, 1031-1040.

Ursone, R. L. and Fried, B. 1995. Light and scanning electron microscopy of *Echinostoma caproni* (Trematoda) during maturation in ICR mice. *Parasitology Research* 81, 45-51.

Yamaguti, S., 1971. *Synopsis of Digenetic Trematodes of Vertebrates*. Keigaku Publishing Company, Vol. 1. Tokyo, Japan.

REPRODUCTIVE PHYSIOLOGY AND BEHAVIOR OF ECHINOSTOMES

PAUL M. NOLLEN

Department of Biological Sciences, Western Illinois University, Macomb, Illinois, USA

B. Fried and T.K. Graczyk (eds.),
Echinostomes as Experimental Models for Biological Research, 137–148.

1. Introduction

As in most digenetic trematodes, the organs of the reproductive system of echinostomes are conspicuous and very important for continuing the life cycle of these species. The production of a multitude of eggs by this system in adults is crucial for the survival of echinostomes because of their complex life cycles. The more reproductive stages, whether they be eggs, miracidia, rediae, or cercariae, they produce, the better are their chances of survival in hostile environments where hosts may be scarce.

The importance of this system in echinostomes is reflected in the amount of body space allocated to reproductive organs. In most adults, the entire middle is taken up by the egg-filled uterus, ovary, and testes and the sides and posterior by the proliferation of vitelline glands. These glands are positioned near the gut and tegument to directly absorb nutrients. Food materials are then stored in vitelline cells which will eventually be incorporated into the egg to provide sustenance for the developing miracidia.

Many aspects of the egg-making process in echinostomes are poorly understood and need more study. Only recently have we learned anything about how adult echinostomes are attracted to each other and their mating behavior. However, studies using echinostome species rival those with schistosomes in revealing species relationships, mating patterns, and possible hybridization. Due to their availability, comparative ease in laboratory maintenance, and experimental maneuverability, echinostome species will continue to be favorite subjects for reproductive studies of digenetic trematodes.

2. Structure of the Male and Female Reproductive Systems

The basic structure of the reproductive system in echinostomes is similar to most other digenetic trematodes. The female system produces the egg, which is made up of four basic components: oocyte, vitelline cells, sperm, and eggshell. Each of these is provided by a different part of the female system and assembled into the egg in the ootype. Primary oocytes are formed by division of stem cells at the periphery of the ovary. They fill the ovary and travel to the ootype via the oviduct. Vitelline cells are produced from stem cells at the periphery of the vitelline glands, which are located laterally in the adult worm. Sperm are stored in an enlargement of the proximal uterus called the uterine seminal receptacle (USR) after being deposited in the metraterm and traveling against the flow of eggs in the uterus. From the USR, sperm are supplied to the ootype for egg-making and eventual fertilization of the oocyte. The eggshell is formed from tyrosine-rich protein granules in the vitelline cells through a quinone tanning and hardening process. Secretions of the Mehlis gland, surrounding the ootype, presumably help initiate the eggshell formation process. A Laurer's canal is found in echinostomes connecting the oviduct to the surface of the worm, but no evidence has been found that it participates in sperm transfer. In the male system, sperm are formed by the well known developmental process common to digenetic trematodes. Here stem cells at the periphery of the testes divide into primary spermatogonia. After 3 more mitotic divisions, secondary spermatocytes divide by meiosis to form bundles of 32 sperm. These sperm then travel in the sperm ducts to the anterior portion of the adult worms where they are stored in the seminal vesicle (SV). Transfer of sperm takes place via the cirrus by which they are deposited in the metraterm of the uterus either of the same worm for self-insemination or a different worm for cross-insemination.

3. Reproductive Cell Development and Movement

3.1. OOGENESIS

The ovary of echinostomes is a single organ located near the middle of the adult worm. It is spherical in structure and contains stem cells as a peripheral layer. These divide by mitosis and produce primary oocytes, which fill the ovary. Primary oocytes are large, angular cells with a prominent nucleus. Treatment of *Echinostoma caproni* adults with 3H-thymidine showed only cells at the periphery of the ovary take up the label. This indicated that only one mitotic division is involved in the production of primary oocytes. These labeled cells were followed in a timed study and filled the ovary in 132 hr (Nollen, 1990). Thus from formation of primary oocytes to delivery at the ootype takes approximately 6 days in *E. caproni*. Saito (1984) infected rats with metacercariae of *Echinostoma hortense* and found primary oocytes in the ovary of newly developed adults by 8 days and by 10 days these cells were found in the ootype.

The oviduct of *Echinostoma revolutum* has a valve-like ovicapt near the oviducal sac, which regulates the flow of primary oocytes to the ootype (Madhavi and Rao, 1972). In this species, Laurer's canal opens at the surface of the adult worm and connects to the oviduct between the oviducal sac and the ootype. The ootype is long and cylindrical and is surrounded by the cells of the Mehlis gland. A fertilization chamber was described in the anterior oviduct of *E. hortense* by Saito (1984). This may correspond to the oviducal sac described by Madhavi and Rao (1972) in *E. revolutum*, but no fertilization process was reported in the sac for this species.

3.2. VITELLOGENESIS

The vitelline glands are very prominent in echinostomes and fill the lateral portions of the adult worm from ventral sucker to the posterior end. They are located near the tegument and gut for easy transfer of nutrient material to the developing vitelline cells. The glands are connected by a vitelline duct on each side of the worm. These ducts come together and form the vitelline reservoir, which empties into the oviduct near its opening to the ootype (Madhavi and Rao, 1972; Saito, 1984). The cells at the periphery of the vitelline glands of *E. caproni* are actively undergoing mitosis and readily take up 3H-thymidine. By 60 hr, these labeled vitelline cells were found in the vitelline reservoir and newly formed eggs in the ootype indicating vitellogenesis is a rapid process (Nollen, 1990).

In a transmission electron microscopy (TEM) study of *Echinostoma miyagawai*, Chen *et al.* (1996) described the vitelline follicles as being surrounded by a basal lamina. Immature follicles had very little cytoplasm and many free ribosomes. As the cells developed, they contained more cytoplasm, more rough endoplasmic reticulum (ER), Golgi complexes, and vitelline globules (shell protein). In mature cells, the ER had concentrated around the nucleus and the edges of the cell, the vitelline globules were well formed and glycogen deposits were identified. In a similar TEM study on the vitelline cells of *E. caproni*, Schmidt (1998) identified four stages of development, earlier described for schistosomes by Erasmus *et al.* (1982). Stage 1 cells are dividing stem cells while stage 2 cells show increased activity of the ER. Stage 3 cells show shell precursor material being produced in quantity by ER and Golgi, while stage 4 represents the mature cells released into the vitelline duct with fully synthesized shell protein granules and stored glycogen. Chen *et al.* (1996) essentially described the same stages in *E. miyagawai* without identifying them.

Schmidt (1998) identified glycan vesicles in the vitelline cells of *E. caproni*. These first appear during developmental stage 4, are membrane bound, and accumulate near the nucleus and the cell periphery. These vesicles are made up of flat membrane sacs of staggered lamellar or concentric layers. Schmidt (1998) speculated that these vesicles eventually form the fluid-filled vacuoles of the mature *E. caproni* eggs and aid the hatching process of the miracidia.

3.3. SPERMATOGENESIS

The process of sperm formation is similar to that found in other digenetic trematodes. Stem cells at the periphery of the testes divide to form primary spermatogonia, which divide mitotically three times to form first secondary spermatogonia (2 cells), then tertiary spermatogonia (4 cells), and in the third division primary spermatocytes (8 cells). The cells in all of these stages adhere to each other, and, after the 2-cell stage, are found floating in the center of the testes. The 32-celled spermatid stage is formed by 2 meiotic divisions of the primary spermatocyte. After undergoing the process of spermiogenesis, bundles of 32 sperm are formed. These break up and move through the sperm ducts to be stored in the SV ready for insemination.

In a developmental study of *E. hortense*, Saito (1984) found the most mature stage of spermatogenesis in the testes 5 days after infection was the primary spermatogonia. Spermatids were detected at 7 days, sperm bundles at 8 days, and sperm in the SV at 9 days. Saito (1984) detected sperm in the USR after 10 days of development in this species. A similar study with *E. caproni* found a slightly slower developmental progression than *E. hortense* with spermatids first present at 7 days after infection and mature sperm at 10 days (Fairweather *et al.*, 1992).

When *E. caproni* adults were exposed to 3-H-thymidine and followed in a timed experiment, label found originally in the peripheral cells of the testes was observed in primary spermatocytes by 48 hr.   By 60 hr label first appeared in secondary spermatocytes and by 72 hr in spermatids.   Labeled sperm were first detected in the SV by 120 hrs.   The speed of spermatogenesis in *E. caproni* was faster than other species of blood flukes and eyeflukes previously studied (Nollen, 1990).

3.3.1. *Spermiogenesis*

The process of transforming the 32-cell spermatids, still in syncytial contact with each other, into mature sperm has been extensively studied in several species of echinostomes.   A study using *E. caproni* by Iomini and Justine (1997) showed spermiogenesis in this species was identical to that found in most platyhelminths and consisted of a proximo-distal fusion of 3 processes followed by elongation of the spermatid.   The processes fused are a single median cytoplasmic structure with longitudinal microtubules and 2 flagella with 9 + '1' structure.   The nucleus migrates from the syncytial mass to the distal part of the mature sperm.

Tubular glycylation was studied in the process of spermiogenesis in E. caproni by Iomini *et al.* (1998).   They found this process to take place in the final stages of spermiogenesis and that 2 distinct functional classes of microtubules are made up of different tubular isoforms.   Two types of axonemes were shown to have different levels of glycylation.

3.3.2. *Sperm Structure*

The sperm of echinostomes is similar to that of other digenetic trematodes in being very long and filiform.   It is impossible to tell which part of the sperm is the head or the tail.   Since the nucleus migrates to the posterior end of the mature sperm during spermiogenesis, it is the last part of the sperm to enter the oocyte at fertilization (Justine, 1995).

The sperm of *E. caproni* is 255 μm in length and has a cylindrical anterior extremity and a broad posterior extremity.   This sperm also exhibits a bilaterally asymmetrical region in cross section, which causes it to coil or form an angle after fixation unlike sperm studied in other species of digenetic trematodes (Iomini and Justine, 1997).

4. Egg Formation

Assembly of the echinostome egg takes place in the ootype. There the primary oocyte, a sperm, and vitelline cells enter the oocyte chamber and an eggshell forms around them.   Schmidt (1998) observed that from 25-31 vitelline cells were incorporated into each *E. caproni* egg.   The shell is formed by quinone tanning and a hardening system common to many species of digenetic trematodes. Gupta and Puri (1986) found proteins, phenolase, and phenols in the eggshell-forming system of *Echinostoma antigonus* indicating sclerotization through quinone bonds.

The shell proteins of the vitelline cells of *E. caproni* readily incorporate 3-H-tyrosine. This tyrosine-containing protein was found in the shells of eggs in the proximal uterus 4 days after adult worms were labeled. By 6 days these eggs were found in the proximal uterus indicating egg production is rapid in *E. caproni*. Newly formed eggs require only 2 days to be transported through the uterus (Nollen, 1990). The presence of large amounts of tyrosine in the eggshell is further evidence for quinone bonding of the protein material.

In a comparative study of the reproductive capacity of *E. caproni* in hamsters and birds, Mahler *et al.* (1995) estimated the total number of eggs in the uterus of adult worms. They found the total number of eggs increased from around 600 at day 15 after infection to over 1,700 by day 50. Little difference in egg production was seen in infections of hosts with 6 or 25 metacercariae. Thus, if it takes 2 days to fill the uterus (Nollen, 1990), peak egg production would be approximately 850/day. At that rate, an egg would be produced approximately every 2 min by an *E. caproni* adult.

Kusaura (1966) followed fertilization and formation of the egg in *E. hortense*. He described the sperm penetrating the oocyte in the fertilization chamber, an enlargement of the oviduct. Following that, the oocyte was surrounded by vitelline cells in the ootype and the eggshell formed around the entire mass. The oocyte then started meiosis and extruded 2 polar bodies. Fusion of the sperm and oocyte nuclei followed with the subsequent development of the miracidial embryo. Kasaura (1966), Terasaki *et al.* (1982) and Saito (1984) found the 2n number of 20 chromosomes for *E. hortense*.

5. Chemoattraction of Adults

In some species of echinostomes, adult worms are confined to a small area of the host's intestine, but for other echinostome species the adults may be spread out. Three well-studied species of *Echinostoma* illustrate this phenomenon. *Echinostoma paraensei* and *E. caproni* are very site specific in uncrowded infections, in the duodenum and ileum of the small intestine, respectively. On the other hand, *E. trivolvis* is usually found spread throughout the small intestine. In those echinostomes that do not cluster, communication for the purposes of mating may be important for maintaining a viable species. Thus *E. trivolvis* adults may depend on chemical cues to find each other for cross-insemination.

In vitro studies of echinostomes have demonstrated chemoattraction between adults. Fried *et al.* (1980) used agar-containing petri dish cultures to investigate worm pairing in *E. trivolvis* (referred to in that paper as *E. revolutum*). They observed significant pairing between immature and mature adults. Further investigation with excretory-secretory products found the free sterol fraction of the lipophilic extract significantly attractive for single worms. Worm pairing for *E. trivolvis* was also demonstrated on the chick chorioallantois by Fried and Diaz (1987). Using the same petri dish system, Fried and Haseeb (1990) investigated intra- and interspecies chemoattraction between *E. caproni* and *E. trivolvis* adults. *Echinostoma trivolvis* exhibited significantly greater intraspecific attraction than *E. caproni*. This attraction was also significantly greater than the interspecific attraction between both species. These studies indicate lipid fractions emitted by adults serve as an attractant in some species of echinostomes.

Trouve and Coustau (1998) investigated the differences in excretory-secretory products of three strains of *E. caproni* by electrophoresis. They found that although the three strains shared most polypeptide bands, a few were strain specific. This difference could have led to evolutionary processes such as selective mating and local host adaptation to produce strain differences.

6. Mating Behavior

The determination of mating patterns of adult digenetic trematodes has been a subject of interest since most are hermaphroditic and could either self-inseminate or cross-inseminate depending on the situation. Echinostomes, being hermaphroditic, have been the subject of many observations and studies concerning mating. Saito (1984) noted that most fixed specimens of *E. hortense* he observed had their cirrus inserted into their own metraterm indicating self-insemination. Single worm infections of *E. trivolvis* (referred to as *E. revolutum* in these studies) and *E. caproni* self-fertilized and produced viable eggs (Beaver, 1937; Fried *et al.*, 1990). Thus, evidence is available that echinostomes are capable of self-insemination with resulting viable eggs. To determine the mating behavior of echinostome adults in both single and multiple infections would take more sophisticated experimentation.

6.1. INTRASPECIES MATING

It is apparent that echinostomes are capable of self-insemination, but whether they self- and cross-inseminate in multiple infections is another question. The mating patterns of echinostomes in multiple infections have been the subject of several investigations. To determine these patterns, sperm were labeled in vitro with 3-H-tyrosine in a donor worm which was then transplanted singly with nonlabeled recipient worms. After a period of time to allow for mating activities, usually 4 to 5 days, the worms were recovered and processed for autoradiography. If labeled sperm were found in the USR of a donor worm it was evidence for self-insemination. If labeled sperm were found in the USR of unlabeled recipient worms it was evidence for cross-insemination (Nollen, 1990).

Using these techniques, the mating pattern of three closely related 37-collar-spined echinostome species, *E. caproni*, *E. trivolvis*, and *E. paraensei*, were investigated in separate studies (Nollen, 1990; 1993; 1996a). The results showed that all have an unrestricted mating pattern, where the donor worm will self-inseminate and also cross-inseminate with other worms present in the infection. This pattern is similar to that found in two species of amphistomes, but unlike that found in eyeflukes where donor worms never self-inseminated in multiple infections (Nollen, 1996a). The percentage of self- and cross-insemination varied widely within the three echinostome species. These data would be more valuable if larger numbers of mating situations were available in these studies. Among the three species, *E. paraensei* consistently showed the least amount of sperm in the USR. This may be a reflection of a smaller amount of sperm transferred during copulation or the large amount of sperm used to produce eggs in these comparatively large adults.

These studies indicate that echinostome adults are unrestricted in their mating habits, being capable of self-insemination by themselves and both self- and cross-insemination in groups. This would allow for maximum egg production in both single and multiple infections and help insure the survival of the species in spite of low infection rates.

Self- and cross-fertilization was studied in strains of *E. caproni* in an elegant series of experiments by Trouve *et al.* (1996). They used three geographic isolates of *E. caproni* from Africa including those from Madagascar, Egypt, and Cameroon. These studies confirmed that *E. caproni* follows a nonrestrictive mating behavior previously noted by Nollen (1990). Since the previous studies using radiolabeled sperm only followed sperm transfer, this study proved that the transferred sperm were used to fertilize the oocyte and give viable offspring. In infections in mice of two adults of the same strain and one adult of a different strain, they found a marked mating between individuals of the same isolate. In infections of 3 different strains, multiple fertilizations were observed. Thus, one strain remains receptive to others after an initial insemination and may be involved in cross-fertilization as a sperm donor or a sperm recipient. In one case a strain, although receiving sperm from two other isolates, preferentially produced offspring from its own sperm.

6.2. INTERSPECIES MATING

It is well known that some species of echinostomes are closely related and can be confused taxonomically. This is true of at least 3 species of 37-collar-spined echinostomes used in interspecies mating studies. At one time *E. caproni* and *E. paraensei* were combined as *E. caproni* on the basis of their having the same snail host, *Biomphalaria glabrata* (Christensen *et al.*, 1990). Later life cycle and biochemical studies showed they were distinct species (Sloss *et al.*, 1995; Morgan and Blair, 1995; Meece and Nollen, 1996; Petrie *et al.*, 1996). Studies with electrophoresis (Sloss *et al.*, 1995) and DNA analysis (Morgan and Blair, 1995) have found all three of these to be valid species. The possibility exists that, in spite of this genetic distance, these species might exhibit weak mating barriers, trade sperm, and hybridize.

All three of these species can be grown in hamsters, but, as mentioned previously, only *E. trivolvis* is not site specific in laboratory infections. In experimental studies it was noticed that some scattering of worms is found when worms are transplanted as adults and not all *E. caproni* and *E. paraensei* find their normal habitats in the ileum and duodenum, respectively. With these factors taken into consideration, a series of studies was carried out to investigate the possibility of interspecies mating in the 37-collar-spined echinostomes.

The first of these studies looked at mating behavior between *E. caproni* and *E. paraensei* in concurrent infections in mice (Nollen, 1996b). Here no interspecies mating was found when either *E. caproni* or *E. paraensei* served as the sperm donor, but self-insemination was observed in both species. However, when a donor worm was given a choice of both species of recipient worms, it carried out the unrestricted mating patterns observed in previous studies (Nollen, 1990; 1996a). It would self- and cross-inseminate with the same species but not cross-inseminate with the opposite species. Thus, mating barriers seem to be strong between these species.

A second study investigated mating behavior between *E. caproni* and *E. trivolvis* in concurrent infections in hamsters. In this case interspecies mating was detected when *E. caproni* was the sperm donor and *E. trivolvis* the recipient, but not the converse (Nollen, 1997a). Interspecies mating was very low (13%; 5 of 39 worms) when compared to the normal rate for *E. caproni* intraspecies mating (52%; 15/29) (Nollen, 1990). As with the previous study on interspecies mating, when single donor adults of either *E. caproni* or *E. trivolvis* had a choice of both recipient species, no interspecies mating took place. Thus, in concurrent infections of *E. caproni* and *E. trivolvis* mating barriers are weaker than between *E. caproni* and *E. paraensei*. The possibility of producing hybrids between *E. caproni* and *E. trivolvis* has not yet been pursued.

The final interspecies mating combination between the 37-collar-spined echinostomes was carried out with concurrent infections of *E. paraensei* and *E. trivolvis* in hamsters (Nollen, 1999). Results of this study show that the *E. trivolvis* sperm donor cross-inseminated with only 1 of 72 possible recipient *E. paraensei* worms. As in a previous interspecies mating study utilizing *E. paraensei* as a sperm donor (Nollen, 1996b) it failed to mate with all 59 *E. trivolvis* recipient worms. Again when both *E. trivolvis* and *E. paraensei* serve as sperm donors with both species of recipient worms present, no interspecies mating took place, but the normal unrestricted mating pattern was observed with the same species. This study gives us an intermediate picture between the other two studies, with a very low rate of interspecies mating between *E. trivolvis* and *E. paraensei*.

The strength of mating barriers should indicate species relationships if species are defined as interbreeding populations. In these studies utilizing concurrent infections of echinostome species (Nollen, 1996b; 1997a; 1999), *E. paraensei* never mated with *E. caproni* or *E. trivolvis*. On the other hand, *E. caproni* would mate with *E. trivolvis* but not *E. paraensei*, while *E. trivolvis* would mate with *E. paraensei* but not *E. caproni*. According to these mating barrier studies, *E. paraensei* was more distant from both *E. caproni* and *E. trivolvis* than the latter species were to each other. A similar genetic relationship was found by electrophoresis of isozymes (Sloss *et al.*, 1995) between these three echinostome species, with *E. paraensei* being more distant from *E. caproni* and *E. trivolvis* than they were to each other.

A fourth 37-collar-spined echinostome species, *Echinostoma revolutum*, was described from Europe (Kanev, 1994; Kanev *et al.*, 1995). This species was recently found in the USA (Sorenson *et al.*, 1997). Further interspecies mating studies with *E. revolutum* and the three species already used in these types of studies might elucidate the genetic relationship between these four closely related species.

7. Abnormalities of the Reproductive System

Abnormal development of the reproductive organs is well known in digenetic trematodes (Bakke, 1988; Nollen, 1997b). Both nutritional and developmental problems may induce changes in the reproductive system. Echinostomes also show these irregularities.

Pande (1980) working with *Echinostoma malayanum* noted small worms with arrested development of the reproductive system. He found mere genital rudiments for testes and ovaries in worms that should have had fully developed organs. In a study on mating behavior of *E. caproni*, Nollen (1990) looked at hundreds of adults and noted many anomalies especially in transplanted worms. These included missing and degenerate testes and ovaries, but the most common observation was accumulation of vitelline cells in the vitelline system. In many cases these cells accumulated in the space between the ovary and the anterior testes and obliterated all tissue in this area. Egg production ceased in these worms in most cases. In a report on several digenetic trematode species including *E. revolutum*, *Echinoparyphium recurvatum*, and *Echinoparyphium rubrum*, Kanev *et al.* (1994) described several types of abnormalities in the male reproductive system. Among these were missing anterior and posterior testes and different testicular forms.

It is known that in most digenetic trematodes species, the reproductive organs and especially the testes are sensitive to external conditions. Thus, care should be taken when encountering these abnormal worms to compare them to large numbers of normal adults in the same infection so as not to describe them as a new species or include them in experimental results.

Acknowledgement
Many thanks to Jeanne Stierman, Western Illinois University Reference Librarian, for her help with the data base searches for this chapter.

8. References

Bakke, T.A. (1988) Abnormalities in adult Digenea, with special reference to *Phyllodistomum umblae* (Fabricius) (Platyhelminthes: Gorgoderidae). *Zoologica Scripta* **17**, 123-134.

Beaver, P.C. (1937) Experimental studies on *Echinostoma revolutum* (Froelich) a fluke from birds and mammals. *Illinois Biological Monographs* **15**, 7-96.

Chen, K., Chen, L., Liu, Y., Xu, J. and Ji, L. (1996) Vitellogenesis of *Echinostoma miyagawai* and *Hypodereaum conoideum* (Digenea: Echinostomatidae). *Journal of the Shanghai Agricultural College* **14**, 79-85.

Christensen, N.O., Fried, B. and Kanev, I. (1990) Taxonomy of 37-collar spined *Echinostoma* (Trematoda: Echinostomatidae) in studies on the population regulation in experimental rodent hosts. *Angewandte Parasitologie* **31**, 127-130.

Erasmus, D.A., Popiel, I. and Shaw, J.R. (1982) A comparative study of vitelline cells in *Schistosoma mansoni, S. haematobium, S. japonicum*, and *S. matthei. Parasitology* **84**, 283-287.

Fairweather, I., Hayes, R.G.J., Stitt, A.W., Halton, D.W. and Fried. B. (1992) Spermatogenesis and spermiogenesis in the gut trematode, *Echinostoma caproni. British Society for Parasitology, Spring Meeting, 1992 (Abstract).*

Fried, B. and Diaz, V. (1987) Site-finding and pairing of *Echinostoma revolutum* (Trematoda) on the chick chorioallantois. *Journal of Parasitology* **73**, 546-548.

Fried, B. and Haseeb, M.A. (1990) Intra- and interspecific chemoattraction in *Echinostoma caproni* and *E. trivolvis*, adults in vitro. *Journal of the Helminthological Society of Washington* **57**, 72-73.

Fried, B., Tancer, R.B. and Fleming S.J. (1980) In vitro pairing of *Echinostoma revolutum* (Trematoda) metacercariae and adults, and characterization of worm products involved in chemoattraction. *Journal of Parasitology* **66**, 1014-1018.

Fried, B., Huffman, J.E. and Weiss, P.M. (1990) Single and multiple worm infections of *Echinostoma caproni* (Trematoda) in the golden hamster. *Journal of Helminthology* **64**, 75-78.

Gupta, V. and Puri, M. (1986) Studies on the egg shell formation in trematode parasites. *Indian Journal of Helminthology* **38**, 44-48.

Iomini, C. and Justine, J-L. (1997) Spermiogensis and spermatozoon of *Echinostoma caproni* (Platyhelminthes: Digenea): Transmission and scanning electron microscopy, and tubulin immunocytochemistry. *Tissue and Cell* **29**, 107-118.

Iomini, C., Bre, M-H., Levilliers, N. and Justine, J-L. (1998) Tubulin polyglycylation in Platyhelminthes: Diversity among stable microtubule networks and very late occurrence during spermiogenesis. *Cell Motility and the Cytoskeleton* **39**, 318-330.

Justine, J-H. (1995) Spermatozoal ultrastructure and phylogeny in the parasitic Platyhelminthes, in Jamieson, B.G.M., Ansio, J. and Justine, J-L. (eds.) Advances in Spermatozoal Phylogeny and Taxonomy. *Memoires du Museum National d'Histoire Naturelle* **166**, 55-86.

Kanev, I. (1994) Life-cycle delimitation and redescription of *Echinostoma revolutum* (Froelich, 1802) (Trematoda: Echinostomatidae). *Systematic Parasitology* **28**, 125-144.

Kanev, I., McCarthy, A.M., Radev, V. and Dimitrov, V. (1994) Dimorphism and abnormality in the male reproductive system of four digenean species (Trematoda). *Acta Parasitologica* **39**, 107-109.

Kanev, I., Fried, B., Dimitrov, V. and Radev, V. (1995) Redescription of *Echinostoma trivolvis* (Cort, 1914) (Trematoda: Echinostomatidae) with a discussion on its identity. *Systematic Parasitology* **32**, 61-70.

Kasura, T. (1966) Studies on chromosomes of reproductive cells and fertilization of *Echinostoma hortense* Asada (1926) (Trematoda, Echinostomatidae). *Okayama Igakkai Zasshi* **78** 929-942. (English Abstract).

Madhavi, R. and Rao, K.H. (1972) Anatomy of the female reproductive system in digenetic trematodes: Part 1. Echinostomatoidea. *Rivista di Parassitologia* **33**, 173-182.

Mahler, H., Christensen, N.O. and Hindsbo, O. (1995) Studies on the reproductive capacity of *Echinostoma caproni* (Trematoda) in hamsters and jirds. *International Journal for Parasitology* **25**, 705-710.

Meece. J.K. and Nollen, P.M. (1996) A comparison of the adult and miracidial stages of *Echinostoma paraensei* and *E. caproni. International Journal for Parasitology* **26**, 37-43.

Morgan, J.A.T. and Blair, D. (1995) Nuclear rDNA ITS sequence variation in the trematode genus *Echinostoma*: an aid to establishing relationships within the 37-collar-spine group. *Parasitology* **111**, 609-615.

Nollen, P.M. (1990) *Echinostoma caproni*: mating behavior and the timing of development and movement of reproductive cells. *Journal of Parasitology* **76**, 784-789.

Nollen, P.M. (1993) *Echinostoma trivolvis*: mating behavior of adults raised in hamsters. *Parasitology Research* **79**, 130-132.

Nollen, P.M. (1996a) The mating behaviour of *Echinostoma paraensei* grown in mice. *Journal of*

Helminthology **70**, 43-45.

Nollen, P.M. (1996b) Mating behaviour of *Echinostoma caproni* and *E. paraensei* in concurrent infections in mice. *Journal of Helminthology* **70**, 133-136.

Nollen, P.M. (1997a) Mating behaviour of *Echinostoma caproni* and *E. trivolvis* in concurrent infections in hamsters. *International Journal for Parasitology* **27**, 71-75.

Nollen, P.M. (1997b) Reproductive physiology and behavior of digenetic trematodes, in B. Fried and T.K. Graczyk (eds.) *Advances in Trematode Biology*, CRC Press, pp 117-147.

Nollen, P.M. (1999) Mating behaviour of *Echinostoma trivolvis* and *E. paraensei* in concurrent infections in hamsters. *Journal of Helminthology* **73**, in press.

Pande, V. (1980) Erratic development of *E. malayanum* in albino rats and its significance. *Indian Journal of Animal Science* **50**, 1117-1121.

Petrie, J.L., Burg, E.F. III and Cain, G.D. (1996) Molecular characterization of *Echinostoma caproni* and *E. paraensei* by random amplification of polymorphic DNA (RAPD) analysis. *Journal of Parasitology* **82**, 360-362.

Saito, S. (1984) Development of *Echinostoma hortense* in rats, with special reference to the genital organs. *Japanese Journal of Parasitology* **33**, 51-61.

Schmidt, J. (1998) Glycan vesicle formation in vitellocytes and hatching vacuoles in eggs of *Echinostoma caproni* and *Fasciola hepatica* (Digenea). *Tissue and Cell* **30**, 416-426.

Sloss, B., Meece, J., Romano, M. and Nollen, P. (1995) The genetic relationships between *Echinostoma caproni*, *E. paraensei*, and *E. trivolvis* as determined by electrophoresis. *Journal of Helminthology* **69**, 243-246.

Sorenson, R.E., Kanev, I., Fried, B. and Minchella, D. (1997) The occurrence and identification of *Echinostoma revolutum* from North American *Lymnaea elodes*. *Journal of Parasitology* **83**, 169-170.

Terasaki, K., Moriyama, N., Tani, S. and Ishida, K. (1982) Comparative studies on the karyotypes of *Echinostoma cinetorchis* and *Echinostoma hortense* (Echinostomatidae: Trematoda). *Japanese Journal of Parasitology* **31**, 569-574.

Trouve, S., Renaud, F., Durand, P. and Jourdane, J. (1996) Selfing and outcrossing in a parasitic hermaphrodite helminth (Trematoda: Echinostomatidae). *Heredity* **77**, 1-8.

Trouve, S. and Coustau, C. (1998) Difference in adult excretory-secretory products between geographical isolates of *Echinostoma caproni*. *Journal of Parasitology* **84**, 1062-1065.

IMMUNOBIOLOGY OF THE RELATIONSHIP OF ECHINOSTOMES WITH SNAIL INTERMEDIATE HOSTS

COEN M. ADEMA, KELLI K. SAPP*, LYNN A. HERTEL, AND ERIC S. LOKER

*Biology Department, University of New Mexico, Albuquerque, NM 87131, USA. *Current Address: Biology Department, High Point University, University Station, High Point, NC 27262.*

B. Fried and T.K. Graczyk (eds.),
Echinostomes as Experimental Models for Biological Research, 149–173.
© 2000 *Kluwer Academic Publishers. Printed in the Netherlands.*

1. Introduction

Over the last three to four decades, echinostomes have proven to be excellent model organisms to study the immunobiological interactions between snails and digenean parasites. In this chapter the use of the term "echinostome" will refer to the 37 collar spine species within the genus *Echinostoma* that have received attention, rather than to all members of the digenean order Echinostomatiformes, the superfamily Echinostomatoidea or the family Echinostomatidae. This chapter will mostly focus on two echinostome species, *Echinostoma paraensei* and *Echinostoma caproni*, that have been studied with respect to their interactions with the intermediate host, the planorbid snail *Biomphalaria glabrata*. Other echinostome species are mentioned when they are relevant to the discussion. Although some of the nomenclature for these species may be disputed, they will be designated by the names used in the original research papers.

The molluscan first intermediate host is crucial for the echinostome life-cycle: extensive asexual parasite reproduction occurs in the circulatory system and tissues of the snail host. Snails are not passive hosts. They possess potent internal defenses that can recognize and destroy multicellular pathogens. The abilities of a given echinostome species to survive a host defense response may restrict the range of suitable first intermediate hosts to a particular snail species or strain. Host-parasite specificity may be imposed here more than at any other point in the echinostome life cycle.

An understanding of echinostome survival strategies in the snail host may be applied towards the control of other digeneans. Information gained will help shape our understanding of parasite virulence and parasite-host specificity. Insights into the functioning of the immunobiology of invertebrates, such as snails, will shed light on the evolution of non-self recognition and internal defense systems. Several relevant review papers are available for an overview of this topic (see e.g van der Knaap and Loker, 1990; Yoshino and Vasta, 1996; Adema and Loker, 1997; Lardans and Dissous, 1998). This chapter aims to present recent developments in the field of echinostome-snail immunobiology.

2. Echinostome Life Cycle

One of the remarkable features of the biology of echinostomes is the complexity of their life cycles. It should be noted that some exceptions apply to the following description of the life cycle of *E. caproni* (see e.g., Schell, 1970; Yamaguti, 1970). Figure 1 shows the life cycle stages of *E. caproni* that relate to the intermediate host. Hermaphroditic adult echinostomes reproduce sexually in the small intestine of a vertebrate definitive host. Following passage from the host's intestine into surface water, a miracidium embryonates in an echinostome egg. Once fully developed, the miracidium hatches. Covered with ciliated plates, the miracidium swims to, and penetrates a snail first intermediate host.

In general, descriptions of digenean life cycles are based on dissections or histological sections of snails at different time points post infection. Thus they are based on a series of discrete "snap shots" rather than continuous observation. Consequently, some aspects of our current understanding of the intramolluscan stages of digeneans may need revision. The development of intramolluscan stages of *E. caproni* and *E.*

paraensei has been described in several studies (Ataev *et al.*, 1997; 1998, Lie and Basch, 1967; Sapp *et al.*, 1998). The miracidium transforms into a sporocyst, as part of this process, the parasite sheds the ciliated plates during penetration. The parasite migrates via the venous blood system through the body of the snail and arrives in the ventricle of the heart at 2-4 days. The sporocyst attaches with the posterior papiliform process to the junction between the ventricle and the aorta of the snail. It is unclear why this is the preferred site of development. Ataev *et al.*, (1998) have suggested that an increased oxygen concentration in the heart, as compared to elsewhere in the snail may be beneficial to the parasite. Although digeneans generally adapt their energy metabolism to a more anaerobic environment after entry into a snail, echinostomes may retain facultative aerobic capabilities as has been described for other digeneans such as *Schistosoma mansoni* (Tielens *et al.*, 1992). Also, the heart may be a location in which the blood flow of the snail is most constant, ensuring a continuous supply of nutrients.

It has been shown that fractions of echinostome products stimulate the activity of the amebocyte producing organ (APO) or hemopoietic organ (Noda, 1992) This organ, situated next to the heart, is where circulating phagocytic blood cells (hemocytes, also termed amebocytes) are generated (Jeong *et al.*, 1983). Echinostomes may affect host hemocyte functions to their benefit (see section 5 below). A location in the heart, in close proximity of the APO would facilitate any influence that the parasites mediate towards hemocytes while the latter are being produced.

Shortly after penetration by the parasite into the snail host, the primary germinal cells within sporocysts start to develop into mother rediae, the next life stage. The development continues while a sporocyst is in the heart. By day 8-11, the first mother redia releases itself by tearing through a thin spot or "dimple" (Ataev *et al.*, 1997) that develops in the wall of the sporocyst. About 5-7 rediae may develop from the primary germinal cells within a single sporocyst (Ataev *et al.*, 1997). Approximately two days post infection new, secondary germinal cells within the sporocyst differentiate from originally small, nondescript cells. Subsequently, additional redial embryos develop from these secondary germinal cells. However, only 1-3 additional rediae may mature and be released before the sporocyst degenerates about 20 days post infection (Ataev *et al.*, 1997). One or two *E. caproni* mother rediae develop and are released ahead of others. Initially rediae stay within the ventricle and aorta of the snail. When these sites become filled, rediae begin to colonize the ovotestis of the snail (Ataev *et al.*, 1997).

Rediae are highly active larval stages that have a mouth, pharynx and a gut. Rediae can consume host hemocytes and other tissues within a snail (Lim and Heyneman, 1972). Motile, young rediae in particular have been observed to attack and feed upon intramolluscan larvae of the same or a different trematode species (Basch and Diconza, 1975; Lie *et al.*, 1975; Lim and Heyneman, 1979; Sousa, 1992). In fact, echinostomes may exert a natural regulatory effect on the intramolluscan stages of schistosomes, digenean parasites of humans. While some field studies to assess the magnitude of this effect have been performed (Lim and Heyneman, 1972; Rysavy *et al.*, 1973; Lie, 1973), echinostomes have not been applied on a broad scale to reduce schistosome transmission in the field.

The redial development of *E. paraensei* differs from that of *E. caproni* in the development of a so-called precocious mother redia (PMR; Sapp *et al.*, 1998). This

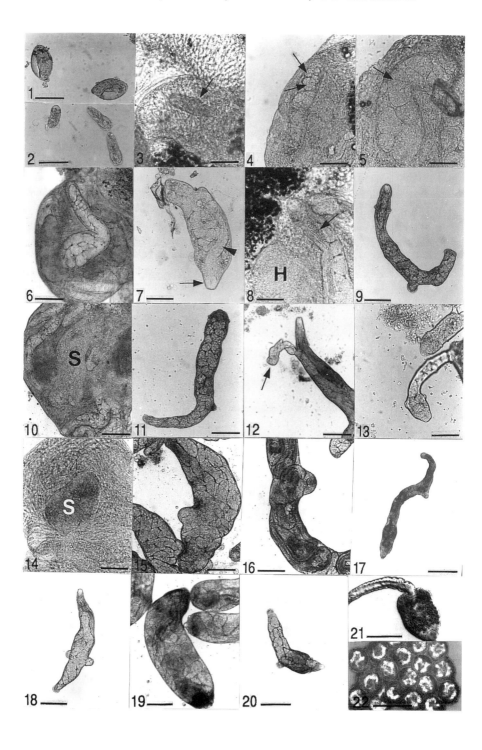

Figure 1. (opposite page) Photomicrographs of living life cycle stages of *Echinostoma caproni*, most freshly-dissected from *Biomphalaria glabrata* (Brazil). Scale bar = 100µm unless otherwise indicated. **1.** Embryonated eggs, each containing a developed miracidium. **2.** Hatched miracidia. **3.** Intraventricular sporocyst (arrow), 4 days post-infection (dpi). **4.** Intraventricular sporocyst, 6 dpi. Note germinal balls (arrows) developing within sporocyst. **5.** Intraventricular sporocyst, with developing embryos (arrow) of mother rediae 8 dpi. **6.** Intraventricular sporocyst, with elongate embryos of mother rediae present within the sporocyst, 10 dpi. Scale bar = 200µm. **7.** Newly-emerged mother redia, 11 dpi. Note pharynx (arrow) and gut that is about half the length of the body (arrowhead). **8.** Mother redia, with cells (probably hemocytes) in the gut (arrow), near a large mass of hemocytes (H), 11 dpi. **9.** Large mother redia, with germinal balls within, 13 dpi. Scale bar = 200µm. **10.** Sporocyst (S) in the ventricle, surrounded by mother rediae, 15 dpi. Scale bar = 200µm. **11.** Mother redia, with prominent germinal balls but no embryos identifiable as daughter rediae yet evident, 15 dpi. Scale bar = 200µm. **12.** Mother redia with the birth of a daughter redia (arrow), 21 dpi. Scale bar = 200µm. **13.** Detail of birth of daughter redia, 21 dpi. **14.** Shrunken sporocyst (S) in the ventricle, 22 dpi. Scale bar = 200µm. **15.** Large mother redia packed with developing daughter rediae. Some cercarial embryos also seen within, 23 dpi. Scale bar = 200µm. **16.** Large redia, probably a mother redia, with cercarial embryos within. 30 dpi, Scale bar = 200µm. **17.** Very large redia, probably a mother redia, with cercarial embryos within. 30 dpi, Scale bar = 571µm. **18.** Small redia, probably a daughter emerged from mother redia. 30 dpi, Scale bar = 200µm. **19.** Large rediae with peculiar blunted shape, containing cercarial embryos, 60 dpi. Scale bar = 200µm. **20.** Small redia, possibly a grand-daughter redia, indicating continued production of rediae, 60 dpi. Scale bar = 200µm. **21.** Emerged cercaria, Scale bar = 235µm. **22.** Cluster of metacercariae from pericardium of *B. glabrata* as a second intermediate host. Note the metacercariae are stuck together by surrounding hemocytes. Scale bar = 300µm.

PMR is a mother redia that develops in, and is released from, the mother sporocyst before others. The PMR develops to about twice the length of sibling mother rediae and has a significantly larger pharynx. The PMR attaches to the ventricle wall and remains with the mother sporocyst for 31 days or more. Experiments showed that the presence of a PMR reduced the success rate of subsequent *E. paraensei* sporocysts in establishing successful intraventricular infections in the snail host. Clearly, the PMR is specialized in both morphology and behavior. Its actions can be viewed as providing protection for the sporocyst. This developmental stage has not been observed from other echinostome species (*E. caproni*, *Echinostoma liei*, *Echinostoma trivolvis*), nor was it reported in the original description of the life history of *E. paraensei* (Lie and Basch, 1967). One possibility, as noted by Sapp *et al.* (1998), is that the PMR may have resulted from the long-term laboratory maintenance of the *E. paraensei* life cycle. Laboratory infections are repeatedly performed with high numbers of parasites. Such conditions may favor parasites that develop increased capabilities of intraspecific competition.

Mother rediae reproduce asexually, initially by producing daughter rediae and later by generating cercariae. As an exception, the above mentioned PMR has only been observed to produce daughter rediae. In turn, daughter rediae also produce new generations of daughter rediae and cercariae. While rediae of the related *Fasciola gigantica* were reported to degenerate and die after one or perhaps more rounds of reproduction (Dinnik and Dinnik, 1956), the fate of individual echinostome rediae remains unclear in the absence of specific studies.

During their development, rediae display different morphologies. Newly released daughter rediae are slender, clear and very motile while mature rediae that produce progeny are large and less active. These morphological and maturation differences seem to be associated with different functions within the host. The PMR provides a clear example of specialized behavior. As stated above, young daughter rediae of *E. paraensei* seem more prone to predatory behavior towards other intramolluscan parasites than are mature rediae. Also, young rediae affect snail hemocyte function and large, more developed rediae were not observed to mediate such an effect (section 6 of this chapter). Similar observations of asynchronous development of rediae by the related fasciolid parasites (Rakondravao *et al.*, 1992; Rondelaud, 1994) are also suggestive of a degree of specialization. Strikingly, the presence of specialized life stages and apparent division of labor (defense, reproduction and alteration of the snail host physiology) observed for echinostomes and related digeneans is reminiscent of the intra-host biology of polyembryonic insect parasitoids. The latter display morphological specialization and division of labor among genetically identical progeny during their parasitic life phase (Strand and Grbic, 1997).

The first cercariae leave the first intermediate host at 5-6 weeks post-infection. The production and release of cercariae continues throughout the life of the snail host. A swimming, free-living cercaria penetrates and encysts as metacercaria inside a second intermediate host which can be a mollusc, amphibian or fish. As a side note, metacercariae of the genus *Ribeiroia* (Echinostomatoidea) have recently been in the public spot light because they have been implicated in causing deformities in amphibians (Johnson *et al.*, 1999). Relative to other life stages of echinostomes, metacercariae and their interactions with snail intermediate hosts have been less studied. Within *B. glabrata*, metacercariae of *E. paraensei* reside primarily in the pericardium. Metacercariae are interesting from an immunobiological standpoint in that they are "recognized" as invaders and encapsulated by hemocytes, yet their cyst wall protects them from damage. On the whole, hemocyte-encapsulated metacercariae are unharmed and infective for the final host (Fried and Huffman, 1996). At experimentally elevated numbers, *E. liei* metacercariae caused mortality of especially young *B. glabrata* (Kuris and Warren, 1980). Following ingestion by the suitable definitive host, metacercaria develop into adult worms, thus completing the life cycle.

In conclusion, the first intermediate host snail harbors several different echinostome stages: sporocysts, rediae and cercariae. Intramolluscan development of echinostomes emphasizes the production of numerous rediae, in multiple generations. During different stages of ontogeny, rediae may be specialized for different reproductive or physiological functions. These metabolically active, growing and reproducing stages represent the most complex part of the echinostome life cycle. As is evident from the life cycle description, the life stage transitions and associated activities of the parasite take place over a prolonged period of time at a physically intimate interface with the first intermediate host, namely the blood and the tissues of a snail.

3. Snail Defenses

A successful echinostome infection has grave consequences for a snail. Echinostomes reduce host fitness to zero by bringing about cessation of host reproduction. This phenomenon, termed parasitic castration is assumed to benefit the parasite by freeing up nutrients and space within the snail. Additionally, parasitic castration may afford stability of the parasite's habitat by preventing physiological changes associated with reproduction of the host (Sorensen and Minchella, 1998). Given that digeneans are estimated to have originated 200-400 million years ago (Gibson, 1987), their associations with molluscs are ancient. As a result of their pronounced effects, abundance and long-standing associations with molluscs, snails should be expected to have developed defense mechanisms to counter infection by digeneans such as echinostomes.

The first indication that snails are not passive hosts for echinostomes is provided by the fact that echinostomes are restricted in their snail host range. *Echinostoma caproni* can complete its development in *Biomphalaria pfeifferi* and *B. glabrata* (Ataev *et al.*, 1997; 1998). Although *E. paraensei* can develop in the planorbid *B. glabrata* and in the physid snail *Physa rivalis,* experimental infections of *Biomphalaria straminia* were unsuccessful (Lie and Basch, 1967). Additionally, Sullivan (1988) lists *Biomphalaria obstructa*, *Helisoma trivolvis* and *Physa virgata* as unsuitable snail hosts for *E. paraensei*. Some reports regarding the successful development of a particular echinostome species in different species or even genera of snails as observed in the field may erroneously result from the notoriously difficult taxonomic identification of echinostomes based on morphological features (e.g., Kanev *et al.*, 1995).

What factors determine whether a given echinostome can or cannot utilize a certain snail as first intermediate host? A particular snail species may be **non-susceptible** or **refractory** for an echinostome by providing an unsuitable physiological environment with regards to pH of the blood fluid or concentration of nutrients. However, even if a snail offers the right environment for an echinostome parasite, snails also have potent internal defense systems that can efficiently manage and eliminate pathogens (for reviews see e.g., van der Knaap and Loker, 1990; Lardans and Dissous, 1998). Therefore, snails can also be immunobiologically incompatible or **resistant** for certain echinostomes. If a snail's internal defenses fail to eliminate an invading echinostome, that snail is immunobiologically compatible or **susceptible**. Between these extreme conditions, a stalemate is possible in which an invading parasite does not develop normally or even ceases development but where the snail defenses can not kill and eliminate the parasite (Richards *et al.*, 1992).

The internal defense of invertebrates, including that of snails, differs from that of vertebrates by the absence of a lymphocytic system. Thus, snails do not have B-cells, T-cells, antibodies or T-cell receptors (Schluter *et al.*, 1997). Regardless, the internal defense system (IDS) of snails is highly potent. Snails are considered to employ lectins as non-self recognition factors (Richards and Renwrantz, 1991). Lectins are non-enzymatic, multivalent proteins that can bind specific carbohydrates. Pathogens are recognized as non-self by so-called pattern recognition (Janeway, 1989); lectins bind to repetitive carbohydrate moieties present on the surface of pathogens, including digenean sporocysts (e.g., Uchikawa and Loker, 1991; Horak and van der Knaap, 1997). This event triggers an internal defense response, possibly via a signaling site on the lectin that is exposed through a conformational change resulting from binding to a specific ligand (Hosoi, 1998). The nature and structure of snail lectins is largely unknown. The defense response comprises humoral and cellular components that work together to kill pathogens. Various humoral factors mediate opsonic, agglutinating or lytic activities. Professional phagocytes are present in the blood (circulating hemocytes) and in the tissues. The cells phagocytose or encapsulate pathogens and then exert cell-mediated cytotoxicity, comprising mechanical aspects, lysosomal enzymes and generation of toxic

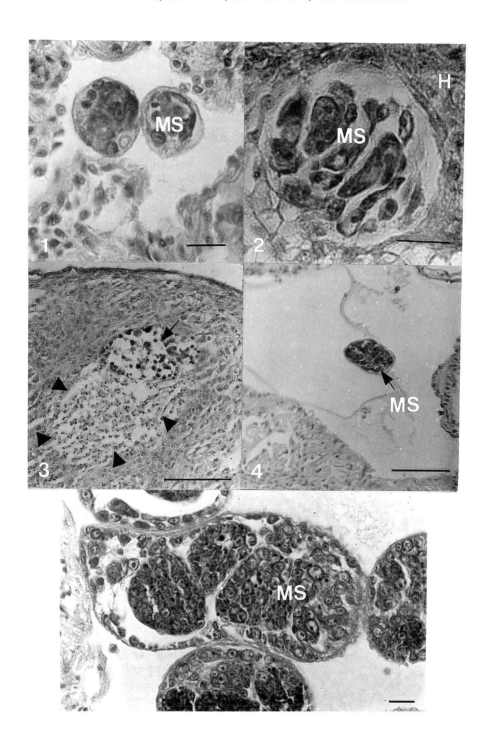

Figure 2. (opposite page) Histological sections showing the fate of *Echinostoma caproni* **mother sporocysts in** *Biomphalaria glabrata* **snails from a resistant strain (1-4) or a susceptible strain (5).**
1. Mother sporocysts (MS) at two days post exposure in the aorta. **2.** Mother sporocysts (MS) encapsulated by hemocytes (H) in the ventricle, at three days post infection. **3.** A hemocyte capsule, showing mother sporocyst remains (arrow) at the center, surrounded by a zone of dead hemocytes (arrow heads) at seven days post infection. **4.** Unencapsulated mother sporocyst (MS) in the pericardial cavity at five days post infection. The mother sporocyst has not developed and contains densely packed cells. **5.** Normally developed mother sporocyst (MS) in the ventricle of a susceptible *Biomphalaria glabrata* snail. Scale bars represent 20μm in 1,2,4 and 100 μm in 3 and 5. This Figure was supplied courtesy of Dr G. Ataev (Zoology, Faculty of Biology, Russian Pedagog University, St Petersburg, Russia) and Dr C. Coustau (Centre de Biologie et d'Ecologie Tropicale et Méditerranéenne, Université de Perpignan, France).

reactive oxygen intermediates (for reviews see e.g van der Knaap and Loker, 1990; Yoshino and Vasta, 1996; Adema and Loker, 1997; Lardans and Dissous, 1998).

Normally, these internal defense capabilities allow snails to eliminate digenean infections. Resistance in snails to certain digeneans occurs naturally in the field (Paraense and Correa, 1963). Extensive studies have shown a genetic basis for immunobiological susceptibility or resistance for schistosomes in laboratory maintained strains of *B. glabrata* (Richards *et al.*, 1992). Recently, susceptibility and non-susceptibility for an echinostome, *E. caproni*, were also shown to be heritable in specifically selected strains of *B. glabrata* (Langand and Morand, 1998; Langand *et al.*, 1998). Ataev and Coustau (1999) demonstrated that the non-susceptible *B. glabrata* strain was in fact immunobiologically resistant to *E. caproni*. As shown in Figure 2, the infection with *E. caproni* evoked a rapid, effective internal defense response from *B. glabrata*. Intraventricular sporocysts were encapsulated and then eliminated by hemocytes. In these experiments, some sporocysts had settled abnormally in the pericardial cavity. Although they were not encapsulated by hemocytes, these sporocysts still showed signs of degradation and failed to develop. This intriguing observation provides evidence that soluble *B. glabrata* defense factors directly contribute to elimination of incompatible echinostomes. This may be a widespread and under-appreciated feature of the molluscan internal defenses. Sapp (1999) found that cell-free plasma from lymnaeid snails was highly toxic for intramolluscan larvae of echinostome species that employed planorbid snails as natural intermediate hosts. Similarly, plasma from planorbid snails killed intramolluscan stages of schistosome parasites that usually develop in lymnaeid snails. Identification and characterization of the factor(s) involved are in progress.

Much of the above knowledge was obtained from study of the planorbid snail *B. glabrata*. This particular snail attracted research attention because it is one of several species from the genus *Biomphalaria* that transmits a human blood fluke, *S. mansoni*. Increased understanding of the internal defense of *B. glabrata* could provide a means towards control of the transmission of this human pathogen. Analysis of the immunobiological interactions between *B. glabrata* and additional digeneans such as echinostomes has provided new information that complements the results obtained from the *B. glabrata-S. mansoni* studies.

4. Snail Lectins: Enlightenment from Echinostomes

4.1. ECHINOSTOME INFECTION EVOKES PRODUCTION OF SNAIL PLASMA POLYPEPTIDES

As stated above, little is generally known about identity and structure of invertebrate lectins that serve as mediators of non-self recognition. However, some of the most extensively studied snail lectins were discovered using the *E. paraensei/B. glabrata* model, including . some of the first invertebrate lectins to be characterized at the molecular level.

Wounding or experimental infection with schistosomes (eukaryote) or bacteria (either Gram + or Gram -) did not result in consistent, obvious changes in the composition of polypeptides present in the cell-free hemolymph (plasma) of *B. glabrata*, as visualized by SDS-polyacrylamide gel electrophoresis (SDS-PAGE; Zelck *et al.*, 1995; Adema *et al.*, 1999). Although the SDS-PAGE methods used do not resolve proteins smaller than 10-20kDa, the negative observations may indicate a lack of response by snails to certain pathogens. In contrast, infection with *E. paraensei* induced changes in plasma of the 10R2 strain, and more prominently so in the M-line of *B. glabrata* (Loker and Hertel, 1987). Protein concentrations in plasma increased from 2 days post infection (dpi) and remained elevated for up to 60 dpi. Starting at 4 dpi, silver-stained SDS-PAGE profiles revealed two groups of previously undetected plasma polypeptides as broad, fuzzy bands (suggestive of heterogeneity) centering at 200kDa and 80-120kDa. Some weaker bands were also present between 45 and 66kDa.

4.2. SNAIL LECTINS THAT REACT WITH PARASITE PRODUCTS

The 200kDa (termed group 1 molecules or G1M) and 80-120kDa (group 2 molecules; G2M) polypeptides were also present in a particulate matter (PM) that occurred from 1 dpi in the hemolymph of infected snails but not in control snails. Later studies demonstrated that G1M and G2M, present at 8 dpi in M-line *B. glabrata* that habor *E. paraensei* infections, are plasma lectins that mediate non-self recognition. G1M and G2M were identified as the factors in *B. glabrata* plasma that agglutinated several types of vertebrate erythrocytes in a fashion sensitive to monosaccharides, including L-fucose (Couch *et al.*, 1990). Plasma from M-line *B. glabrata* infected with *E. paraensei*, contains several polypeptides that bind latex beads. G1M and G2M were most consistently associated with an opsonic effect resulting in higher adherence rates of hemocytes to the latex beads *in vitro* (Uchikawa and Loker, 1992). A detailed monosaccharide affinity study demonstrated that G1M and G2M also bound several other monosaccharides in addition to L-fucose (Monroy *et al.*, 1992). This broad specificity likely resulted from the heterogeneous nature of the G1M and G2M polypeptides that was disclosed by two-dimensional gel electrophoresis (Monroy and Loker, 1993). G1M derived from individual snails, displayed a smear along a broad spectrum of isoelectric values, G2M resolved into 5-6 spots. This confirmed that the broad fuzzy banding observed on SDS-PAGE gels indicated that G1M and G2M consisted of heterogeneous groups of polypeptides. Finally, western blotting experiments employing antibodies raised in rabbits against purified and deglycosylated

G1M and G2M, showed that G1M and G2M also bind to Gram + and Gram - bacteria and to sporocysts and rediae of *E. paraensei* in a lectin-like fashion (sensitive to monosaccharide inhibition; Hertel *et al.*, 1994).

To allow molecular characterization, partial amino acid sequences have been obtained from both G1M and G2M following gel purification. PCR experiments using primers designed from the amino acid information have thus far failed to yield consistent results. The PCR reactions may be hampered by the inherent heterogeneity of the polypeptides. Possibly, the amino acid sequences were each derived from different representative members of either G1M or G2M. As a result, the primer targets on the genomic DNA are located too far away from each other to easily facilitate regular PCR amplification.

The polypeptides G1M and G2M were initially identified in the particulate matter (PM) that occurs *in vivo* from 1 dpi in M-line *B. glabrata* infected with *E. paraensei*. The significance of PM remains unclear. Remarkably, *in vivo* PM occurs in such large amounts that one would expect it to hinder the circulation of hemolymph within infected snails. Although parasite secreted/excreted products (SEP) are a likely target to be bound and precipitated by G1M/G2M, several efforts have failed to demonstrate unequivocally parasite-derived components in PM. Perhaps parasite SEP occurs only in very low concentrations in the circulation of the snail host. Under those conditions PM can form but the small amounts of SEP involved escape detection by SDS-PAGE electrophoresis, including probing with anti-SEP antibodies (Loker *et al.*, 1992). The antibodies may simply not recognize the relevant *in vivo* SEP components.

However, PM formation could be replicated *in vitro* by a precipitation reaction resulting from mixing SEP obtained *in vitro* from *E. paraensei* sporocyst cultures with plasma from M-line *B. glabrata* infected with *E. paraensei*. SDS-PAGE analysis showed that this *in vitro* PM did contain SEP components. Moreover, a different, third group of snail-derived plasma polypeptides was also present in *in vitro* PM obtained with plasma collected at 4 dpi from *E. paraensei*-infected *B. glabrata* (Adema *et al.*, 1997a). These polypeptides centered as a broad and fuzzy, prominent band at 65kDa on SDS-PAGE gels, suggesting that these polypeptides were of a heterogeneous composition. Precipitate formation was not enzymatically driven; it was insensitive to several proteases and it was not initiated by several proteases. The precipitate reaction was sensitive to the inhibiting monosaccharide L-fucose, indicating that the 65kDa polypeptides bound parasite products in a lectin-like fashion, just like G1M and G2M did. The 65kDa polypeptides react with non-self as demonstrated by binding to *E. paraensei* SEP, to cilia from miracidia (Adema *et al.*, 1997a), and sporocysts of *E. paraensei* (Adema, unpublished data).

The 65kDa polypeptides differ from G1M/G2M in ways other than by molecular weights alone. G1M/G2M have higher binding affinities. The precipitation of 65kDa polypeptides required increased amounts of SEP. The concentration-dependent fashion in which 65kDa polypeptides bound and precipitated SEP is reminiscent of precipitin-reactions between antigens and antibodies (Adema *et al.*, 1997b). The 65kDa polypeptides are present at 2-6 dpi in the plasma of *E. paraensei* infected *B. glabrata*. G1M/G2M occur between 4-16 dpi. The massive amounts of the 65kDa snail plasma polypeptides in *B. glabrata* induced by *E. paraensei* represented a novel observation. Previous work had emphasized the time point of 8 dpi, when G1M/G2M are consistently present. At that time, the plasma concentrations of the 65kDa polypeptides in *B. glabrata* have already diminished considerably.

4.3. CHARACTERIZATION OF FIBRINOGEN-RELATED PROTEINS (FREPS)

The molecular characterization of the 65kDa polypeptides was described by Adema *et al.* (1997b). The amino acid sequences from internal peptide fragments (Genbank accession numbers P80742 - P80745) obtained from the 65kDa polypeptides displayed similarities to fibrinogen β/γ (FBG). This was surprising since molluscs do not display blood clotting (Millar and Ratcliffe, 1994). Nucleotide primers were designed for PCR experiments. The sequence of a PCR amplicon (354 bp) displayed further similarities to FBG, hence the *B. glabrata* M-line 65kDa polypeptide-derived sequence was named fibrinogen-related protein 1 (BgMFREP1, GenBank accession number U82471). A cDNA library that represented mRNA from *B.* glabrata snails harboring 4 day old *E. paraensei* infections yielded several clones that hybridized with the 354 bp BgMFREP1 sequence. The cDNA inserts were sequenced. Computer-aided analysis disclosed that, in addition to BgMFREP1, three related yet very different FREP-encoding sequences had been obtained from M-line *B. glabrata* (GenBank accession numbers; BgMFREP2: U82479; BgMFREP3: U82480; BgMFREP4: U82478).

FREPs have a unique domain structure. They comprise an N-terminal variable (V)-type immunoglobulin superfamily (IgSF) domain separated by an interceding region (ICR, variable in length between FREPs and without similarity to known sequences) from a C-terminal FBG domain (see Figure 3). While each of the respective domains of different FREPs displays conserved residues that identify it as an Ig domain (Kurosawa and Hashimoto, 1996) or an FBG domain (Doolittle, 1992) the sequence of these domains differs considerably between different FREPs (Adema *et al.*, 1997b). Ongoing analysis suggests that additional different FREP-encoding sequences occur in the genome of M-line *B. glabrata*. Clearly, the FREPs form a gene family. FREPs are the only sequences known to combine FBG and IgSF domains. Very few invertebrate

Figure 3. Basic anatomy of a *Biomphalaria glabrata* FREP. This schematic domain structure (at deduced amino-acid level) of a fibrinogen-related protein (FREP) is based on sequences obtained from 4 different 65kDa subunit-based FREPs. An N-terminal, variable-type immunoglobulin superfamily (V-type Ig) domain is separated by an interceding region (ICR) from a fibrinogen β/γ (FBG) domain at the C-terminus. The Ig-domains contain several conserved residues specific for the Ig superfamily. Two cysteine residues (about 80 residues apart) that allow formation of the typical intra-chain Ig-loop are indicated as examples. The Ig sequences differ between FREPs. The ICRs show no similarities to known sequences, and vary considerably in length and in amino acid composition among FREPs. Canonical residues demonstrate that the C-terminal domain sequences are fibrinogen β/γ–related. While the FBG sequences are the most conserved among the FREPs, they are also distinct between FREPs. Pending additional sequence information, brackets indicate the putative nature of the N-terminus observed only in BgMFREP2 (adapted from Adema *et al.*, 1997b).

sequences display V-type Ig domains (Kurosawa and Hashimoto, 1996), of those only FREPs function in an internal defense-related role. RT-PCR experiments confirmed that FREP expression is upregulated following infection of *B. glabrata* with E. *paraensei* (Adema *et al.*, 1997b). The same technique showed that at least some FREPs are produced by circulating hemocytes, the phagocytic defense cells of *B. glabrata*. The scientific significance of these FREP sequences, uniquely resulting from the study of echinostome-host immunobiology, is underlined by the conclusion of DuPasquier and Flajnik (1999) that available molluscan IgSF sequences now *"permit speculation about relationships with other invertebrate immune systems and (…) with vertebrates"*.

The observed FBG domain likely conveys lectin-like properties to the FREPs. In fact, non-self recognition may have been an ancestral function of FBG domains (Adema *et al.*, 1997b). Several other lectins have been found to employ FBG domains for binding carbohydrates such as a sialic acid specific lectin from the slug *Limax flavus* (Kurachi *et al.*, 1998), P35, a human opsonic serum lectin (Matsushita *et al.*, 1996), and techylectins from *Tachypleus tridentatus* (horseshoe crab; Gokudan *et al.*, 1999). A functional role for the Ig-domain of FREPs remains to be elucidated.

4.4. SNAIL LECTINS AND ANTI-ECHINOSTOME REPONSES

The lectins described also occur in M-line *B. glabrata* harboring echinostomes destined not to develop. Several lines of evidence show that G1M, G2M and FREPs are part of rapid and complex anti-echinostome response mounted by M-line *B. glabrata*. *Echinostoma paraensei* miracidia hatched from deliberately aged eggs penetrate but are unable to develop in snails. Also, the infection rates in large adult *B. glabrata* are lower. In both these cases, G1M/G2M and FREPs were présent in the plasma of exposed M-line snails (Adema *et al.*, 1997a). The incompatible *Echinostoma trivolvis* also induced G1M/G2M and FREP production in M-line *B. glabrata* (Adema *et al.*, 1999). Precipitation reactions disclosed that both E. *paraensei* (compatible) and *E. trivolvis* (incompatible) induced similar plasma polypeptides in M-line *B. glabrata*. The plasma polypeptides cross-reacted with SEP derived from sporocysts or rediae of either *E. paraensei* or *E. trivolvis*. The use of redial SEP to evoke precipitation reactions from *B. glabrata* plasma disclosed yet another group of echinostome-induced plasma polypeptides, visible as a broad diffuse band centered at 53kDa on SDS-PAGE gels (Adema *et al.*, 1999). The planorbid snail *Helisoma trivolvis* displays a reverse compatibility to *E. paraensei* and *E. trivolvis* relative to *B. glabrata*. In response to either echinostome species, *H. trivolvis* also produced a complex profile of plasma polypeptides that were cross-reactive with SEP from different echinostome species and stages. The *H. trivolvis* plasma polypeptides formed broad bands on SDS-PAGE gels, centering at molecular weights of 80-120kDa, 65kDa and 53kDa, similar to G2M, FREPs and the 53kDa of *B. glabrata*. A G1M-like band was not observed from *H. trivolvis* (Adema *et al.*, 1999).

Except for the 65kDa subunit-based FREPs of *B. glabrata*, all of these snail plasma polypeptides remain uncharacterized. However, all the snail polypeptides present in the precipitates were bound by anti-fibrinogen monoclonal antibodies (Adema *et al.*, 1999). This suggests that perhaps all these parasite product-reactive plasma polypeptides contain FBG domains. In that case, it would be too restrictive to designate only the 65kDa plasma polypeptides as FREPs. Currently, the presence of an FBG domain seems most likely for the 65kDa plasma polypeptides from *H. trivolvis* which have the same molecular weight as the *B. glabrata* FREPs. Although this protein band can not be associated with specific sequence information, PCR experiments, employing degenerate primers specific from BgMFREP1 and genomic DNA from *H. trivolvis* as template, disclosed a partial FREP-encoding sequence (HtrFREP1, GenBank accession U82475). Moreover, similar sequences were obtained from three other different strains of *B. glabrata*, and other planorbid snails *Biomphalaria alexandrina* and *Bulinus truncatus* (Adema *et al.*, 1997b; GenBank accession numbers U82472 through U82477).

Taken together, these observations suggest that some planorbid snails respond to compatible and incompatible echinostomes (and not to bacteria or wounding) by increased production of several heterogeneous plasma polypeptides. At least some of these polypeptides are produced by circulating hemocytes, snail defense cells. The plasma polypeptides bind with lectin-like properties to a wide range of carbohydrate moieties on erythrocytes or immobilized on carrier matrices, bacteria, parasites and parasite products. The binding then causes agglutination, opsonization or precipitation. Based on these observations, the plasma polypeptides are expected to serve as non-self recognition factors in the snail internal defense system. The heterogeneity of each of these groups of plasma polypeptides may enable recognition of a broad range of pathogens.

Several questions remain regarding the structure and functioning of the plasma lectins described. The lectins are identified by the molecular weights of their subunits as they occur on SDS-PAGE gels. For instance, non-denaturing gel separations suggest that FREP subunits of 65kDa combine into covalently-associated hexamers of about 400kDa. In turn these hexamers associate to form larger units of 1600kDa (Adema *et al.*, 1997a). It remains to be determined whether such large lectin complexes are homopolymers composed of many subunits of one particular FREP protein or whether they comprise diverse FREP polypeptides to form heteropolymers. Also, the high amounts of FREPs induced by *E. paraensei*-infection in *B. glabrata*, remain puzzling since FREPs are not observed in *in vivo* PM. Perhaps FREPs react only with increased concentrations of carbohydrate moieties that may occur close to, or at the surface of, the intramolluscan parasite. In the circulation, the higher affinities of G1M and G2M for parasite SEP may effectively inhibit FREPs from participating in precipitate formation.

5. Echinostome Interference

5.1. SURVIVAL IN A SNAIL HOST

Following the description of the snail internal defense system, it may seem remarkable the echinostome parasites survive to complete their lifecycle in specific snail hosts. *In vitro* studies have shown that digenean parasites are vulnerable to the cytotoxic effector substances elaborated by snail hosts, such as reactive oxygen intermediates (Adema *et al.*, 1994a) and lytic plasma components (Sapp, 1999). For survival, an echinostome parasite must prevent an effective defense response from the snail host. Two strategies have been proposed by which digenean parasites may escape their host defenses.

Some digeneans may prevent recognition as non-self within the host by either absorbing snail host factors onto their surface (**molecular masking**) or by **molecular mimicry** through expressing gene products that mimic host-inherent factors (Yoshino and Boswell, 1986; Dissous and Capron, 1995). In this way the parasite does not trigger a host response and does not have to contend with the snail internal defenses system.

A second survival strategy of digeneans was highlighted by the work of Lie, Heyneman and others. They observed that hemocytes of *B. glabrata* snails that harbored echinostome infections lost their ability to adhere to non-self objects. Furthermore the hemocytes accumulated in irrelevant locations in the snail (Lie and Heyneman, 1976). Infection with echinostomes caused *B. glabrata* that were selected for resistance against schistosomes to become susceptible for *S. mansoni* (Lie *et al.*, 1977). Immuno-modulatory effects were weaker and slower to develop with *S. mansoni* (Lie and Heyneman, 1977a). The inhibitory effects upon snail defenses increased with parasite numbers (Lie and Heyneman, 1977b). Normally, *S. mansoni* sporocysts were encapsulated by hemocytes obtained from *B. glabrata* strains that were resistant to schistosomes. However, in schistosome-resistant *B. glabrata* with experimental double infections, *S. mansoni* sporocysts seemed to be protected from encapsulation in proximity of *E. paraensei* sporocysts (Lie, 1982). These *in vivo* observations led to the **interference** theory (Lie, 1982) which proposed that digenean larvae, echinostomes in particular, achieve immunological compatibility with snails by affecting functional aspects of the host internal defenses, mostly by targeting snail hemocytes.

5.2. *IN VITRO* STUDY OF INTERFERENCE

Efforts to elucidate the mechanisms of interference have relied mostly on *in vitro* approaches. The *in vitro* killing of *S. mansoni* sporocysts by hemocytes of uninfected, resistant *B. glabrata* was significantly reduced when *E. paraensei* larvae were included in the assay (Loker *et al.*, 1986). Plasma from echinostome-infected snails had no inhibitory effect on hemocyte-mediated killing of schistosomes. This confirmed that hemocytes were directly affected by echinostome larvae. Moreover, this showed that interference could be studied *in vitro*. Infection with *E. paraensei* leads to significantly increased numbers of circulating hemocytes in *B. glabrata*. A large proportion of these hemocytes displays a rounded morphology and has a reduced ability to adhere to plastic slides *in vitro* (Noda and Loker, 1989a) and to phagocytose fixed sheep red blood cells (SRBC) *in vitro* (Noda and Loker, 1989b). Hemocytes of compatible *B. glabrata* were observed to bind but not to damage sporocysts of *S. mansoni*. Hemocytes were significantly less likely to bind sporocysts or rediae of *E. paraensei*. Hemocytes did adhere to echinostome larvae that were killed by glutaraldehyde fixation (Loker *et al.*, 1989). The apparent requirement of metabolic activity indicates that *E. paraensei* actively maintains a hemocyte-repellent surface and/or mediates interference through release of soluble factors.

To facilitate the study of parasite-released soluble factors, miracidia of *E. paraensei* were transformed *in vitro* to sporocysts. Parasite secreted/excreted products (SEP) were obtained from the culture medium and analyzed for composition and effect on *B. glabrata* hemocytes (Loker *et al.*, 1992). The SEP consisted of a complex protein mixture. Culture of *E. paraensei* in the presence of ^{14}C labeled amino acids yielded few radio-labeled SEP proteins on the first day, indicating that the parasites initially released mostly preformed products. At day two, the parasite actively synthesized proteins as evidenced by the incorporation of the radio-isotope in many SEP components. SEP collected on day 1 inhibited the *in vitro* phagocytosis of SRBC by hemocytes of *B. glabrata*. SEP collected on later days also reduced this phagocytic activity. Additionally, hemocytes in the presence of SEP rounded up and failed to adhere to plastic slides. The exclusion of vital dye by affected cells indicated that SEP did not kill hemocytes. The activity of SEP was heat- and protease-sensitive. Following fractionation based on native molecular weight, SEP components of more than 100kDa were found to mediate the rounding effect. Finally, a pretreatment of hemocytes from schistosome-resistant 13-16R1 *B. glabrata* with *E. paraensei* SEP inhibited the killing of *S. mansoni* sporocysts by these hemocytes. Clearly, *in vitro* obtained SEP from *E. paraensei* "*interfered with hemocyte functions in ways inferred from earlier classic in vivo studies of trematode-snail interactions*" (Loker *et al.*, 1992).

Previously, echinostome-mediated interference *in vivo* was interpreted to be selective (Lie *et al.*, 1981), affecting hemocytes such that they could not attack *E. paraensei* sporocysts, yet remained able to encapsulate injected latex beads, miracidial plates derived from *S. mansoni* and nematode larvae. In contrast, the above *in vitro* approaches showed that *E. paraensei* SEP is able to inhibit both phagocytosis and killing of *S. mansoni* by hemocytes (Loker *et al.*, 1992). This suggests a more indiscriminate inhibition of hemocyte defense functions by SEP. The apparent selective nature of interference *in vivo* may be due to localized effects where interference effects are strongest close to *E. paraensei* intramolluscan larvae and much less pronounced further removed from this parasite (Loker, 1994).

The process of interference mediated by *E. paraensei* towards snail hemocytes was studied by time-lapse video microscopy (Adema *et al.*, 1994). Interactions between glass-adherent hemocytes of *B. glabrata* and either a single sporocyst, small redia or metacercaria from *E. paraensei* were recorded and then viewed under time compression. In this fashion, slow movements of snail hemocytes could be interpreted as a continuous process over time. Hemocytes moved away from sporocysts and small rediae of *E. paraensei*, but not from metacercariae. In the presence of sporocysts, hemocytes rounded up while rediae caused hemocytes to assume a stringy, beaded morphology. These changes did not require physical contact between hemocytes and the parasite. This is consistent with the notion that *E. paraensei* mediates interference through soluble factors. Over two hours these effects combined to form a cell-free "halo" around the parasite larvae. Metacercariae exerted no obvious changes in hemocyte behavior of morphology. The hemocytes of the non-host gastropod *Helix aspersa* were not affected by the presence of any *E. paraensei* larvae. This suggested that *E. paraensei*-mediated interference towards hemocytes exhibits a degree of specificity.

However, Sapp (1999) showed that interference mediated by sporocysts of *E. paraensei* was rather indiscriminate. *Echinostoma paraensei* caused rounding of hemocytes of the non-host snails *Lymnaea stagnalis* and *Stagnicola elodes*. A remarkable, novel observation was that small rediae of *E. paraensei* caused rounding of *B. glabrata* hemocytes, whereas large rediae did not affect hemocyte morphology. This may provide an indication for division of labor between echinostome intramolluscan larvae (see section 2 of this chapter). Another striking finding was that not all echinostome species affect snail hemocytes similarly. Sporocysts of *E. trivolvis*, a close relative of *E. paraensei* (Morgan and Blair, 1995), did not cause rounding of hemocytes obtained from either *B. glabrata*, *L. stagnalis*, *S. elodes* or *H. aspersa* (Sapp, 1999). It should be noted that the effect of *E. trivolvis* towards the morphology of hemocytes of its natural snail host remains unclear. The hemocytes of *H. trivolvis* do not spread well on glass or plastic microscopic slides and therefore could not be used in this assay system. However, the adherence capabilities to sporocysts of hemocytes from *H. trivolvis* and the other four snails species listed in this paragraph could be tested. Snail hemocytes readily adhered to *E. trivolvis*. Concordant with previous work (Loker *et al.*, 1989), hemocyte adherence to *E. paraensei* sporocysts was minimal in the same experiments.

The original *in vivo* studies performed on digenean-mediated interference had demonstrated that among other echinostome species, *E. paraensei* in particular was the stronger mediator of interference (Lie, 1982). While the schistosome *S. mansoni* relies more heavily on preventing recognition as non-self for survival in the snail host compared to *E. paraensei* (see Loker and Adema, 1995), this parasite also mediates interference with host defenses *in vivo* (Lie and Heyneman, 1977a). SEP from *S. mansoni* does not cause hemocyte rounding but this SEP was found to interfere *in vitro* with protein synthesis and release (Lodes *et al.*, 1991), motility (Lodes and Yoshino, 1990), and defense capabilities (reduced phagocytic activity and production of superoxide; Connors and Yoshino, 1990; Connors *et al.*, 1991) of *B. glabrata* hemocytes. It seems that digeneans in general employ some combination of both interference and immunomasking/mimicry strategies to evade snail host defenses. In light of this discussion, the distinction between schistosomes as utilizing mostly molecular masking and mimicry as a survival technique and echinostomes as relying more on interference (Loker and Adema, 1995) becomes somewhat blurred.

5.3. INTERFERENCE, HOW IS IT MEDIATED?

Due to the pronounced rounding effects towards snail hemocytes, *E. paraensei* is an excellent model organism to study aspects of echinostome-mediated interference with regards to both characterization of the factors involved and the cellular processes in snail hemocytes that are the target for interference. A convenient assay that consists of combining SEP with hemocytes followed by checking for the occurrence of cell rounding, is available to aid in the search for SEP factors that mediate interference. Degaffé and Loker (1997) showed that this assay provides relevant data to investigate *in vivo* interference. The potency of a particular SEP sample (derived from *in vitro* transformed sporocysts) to cause hemocyte rounding *in vitro* was correlated to the infection success of *E. paraensei* miracidia from the same batch of eggs. Additionally, SEP had a greater rounding effect on hemocytes from hemolymph samples obtained from juvenile snails compared to those of hemolymph from adult *B. glabrata*. These effects resulted from quantitative rather than qualitative differences between hemolymph samples from juvenile and adult snails. The SEP-mediated rounding percentages became similar when hemocyte concentrations and hemolymph volumes were adjusted to be similar for hemolymph samples from juvenile and adult *B. glabrata*. This provides another indication of how adult snails may become more resistant to digenean infection with age due to maturation of their internal defense system (Richards, 1984; Loker *et al.*, 1987). This maturation includes increased numbers of circulating hemocytes (see also Dikkeboom *et al.*, 1985).

Interference may rely on more than one component of SEP. Fractionation based on native molecular weight indicated that SEP components in excess of 100kDa (Loker *et al.*, 1992) inhibited the spreading of *B. glabrata* hemocytes. Other biological activities are also present in SEP. In a separate study, components of *E.* paraensei SEP in the range of 10-30 and 30-100kDa (native molecular weights) inhibited motility of *B. glabrata* hemocytes as determined in chemotaxis chambers (Noda, 1993). The molecular characterization of several candidate interference factors is in progress. A polypeptide of 49kDa (denatured MW) was selected because it is the most prominent component of *E. paraensei* SEP, and it is precipitated by G1M, G2M and FREPs *in vitro* (Adema *et al.*, 1997a). Another SEP component of 51kDa was uniquely present in size-exclusion fractionated SEP samples that displayed a high hemocyte-rounding activity. Internal amino acid sequences were obtained from both polypeptides. Initial PCR experiments have yielded bands awaiting further analysis. Although interference-mediating may be toxins, ionophores, proteases, cytokines, neuroendocrine-like factors, the nature of these factors could also be completely novel. Functional assays are needed to confirm that sequences obtained actually encode functional interference factors.

The random sequencing of expressed sequence tags (EST) was applied to document additional factors relevant to the intramolluscan biology of *E. paraensei*. A total of 151 unique ESTs was obtained from a cDNA library that was constructed with mRNA obtained from developing miracidia within eggs, hatched miracidia, and sporocysts of *E. paraensei* (Adema *et al.*, 2000). No obvious candidate interference factors were found among the 64 ESTs that had significant similarities with known sequences (GenBank). The remaining 87 EST sequences showed no similarities to known sequences. Future work may reveal whether some of these unknown ESTs represent factors with which *E. paraensei* mediates interference. The EST-sequences were submitted to the GenBank/dbEST databases as reference information for both phylogenetic analysis and study of general trematode biology.

The search for interference-mediating factors may become more focused if the interference mechanism is known. One possibility is that SEP components bind to hemocyte receptors and activate signal transduction pathways that affect the cellular cytoskeleton to cause rounding of hemocytes. Signal transduction often depends on changes in intracellular calcium ion concentrations (for review, see Clapham, 1995). Confocal microscopy, using fluorescent markers to monitor calcium levels was applied to study the effect of SEP from *E. paraensei* on hemocytes (Hertel *et al.*, 1999). When exposed to *E. paraensei* sporocysts, sporocyst SEP, or daugther rediae, hemocytes were more likely to round up and exhibit a calcium transient. Some hemocytes showed one response or the other but not both. Hemocytes did not respond significantly to large rediae, to sporocysts of *S. mansoni*, or to bacterial lipopolysaccharides. The results show that hemocytes exhibit a complex spectrum of responses to different stimuli, and they confirm that sporocysts and small rediae of *E. paraensei* have more striking effects on hemocytes than the other parasite stages tested. Future applications of this technique may help to further define the mode of *E. paraensei*-mediated interference at the level of signal transduction within the target hemocytes.

6. New Developments

Many recent research developments have created excellent conditions to further pursue a more complete understanding of interference capabilities of *E. paraensei* and how these interact with *B. glabrata*. The development of the *E. caproni*-resistant *B. glabrata* strain (Langand and Morand, 1998; Langand *et al.*, 1998; Ataev and Coustau, 1999) provides a new model for comparative study of digenean-snail immunobiology in addition to the currently studied models such as *B. glabrata* and *E. paraensei* or *S. mansoni*.

Remarkably, an *in vitro* culture system has been developed in which cultured Bge cells (a cell line derived from embryonic tissues of *B. glabrata*; Hansen, 1976) support the *in vitro* growth, development and reproduction of most or even all intramolluscan stages of *S. mansoni* (Yoshino and Laursen, 1995; Ivanchenko *et al.*, 1999), *S. japonicum* (Coustau *et al.*, 1997), *E. caproni* (Ataev *et al.*, 1998; Loker *et al.*, 1999) and *Fascioloides magna* (deer fluke; Laursen and Yoshino, 1999). Without a doubt, analysis of the factors provided by the snail cells that are essential for digenean parasites will revolutionize our views of digenean-snail interactions. For example, the recent demonstration of the production of cercariae in cultures initiated with primary sporocysts of *S. mansoni* indicate that subtle interactions with an intact molluscan brain and its secretions are not required to complete development.

The molecular study of digeneans, mostly of human-infecting schistosomes has yielded data concerning mRNA messages expressed by different stages of digenean life stages. These data are collected in both targeted fashions (Trottein *et al.*, 1992) and at large scale in random approaches such as EST projects (e.g., Adema *et al.*, 2000). The sequence information available from for instance SchistoDB (the *Schistosoma* Genome Database), the largest collection of digenetic trematode-derived sequences <www.nhm.ac.uk/ hosted_sites/ schisto/>, and other sources allows rapid identification and perhaps determination of functional context of newly obtained sequences. Molecular systematics help to clarify phylogenetic relationships between different digeneans (e.g., Morgan and Blair, 1995).

Similarly, molecular techniques have been applied to snail biology. A number of *B. glabrata* genes have been sequenced: actin (Lardans *et al.*, 1997), myoglobin (Dewilde *et al.*, 1998), 18S rDNA (Hanelt *et al.*, 1997), and heat shock protein 70 (Yoshino *et al.*, 1998), and of several factors that may function in defense contexts: signal transduction factor RACK (Lardans *et al.*, 1998), integrin that functions in the adherence of hemocyte (Davids and Yoshino, 1998, Davids *et al.*, 1999) and lectins such as selectin (Duclermortier *et al.*, 1999) and FREPs (Adema *et al.*, 1997b). More sequence information is available from an EST project (Knight *et al.*, 1998). Methods are being developed to produce transgenic snail cells or snails (Lardans *et al.*, 1996; Yoshino *et al.*, 1998). This would facilitate study of the role of putative snail defense factors towards resistance for digenean parasites.

7. Concluding Remarks

The immunobiology of echinostomes and snails provides a fascinating model to study parasite-host interactions. Solving the puzzle of how these organisms interact so intimately and continuously requires the application of new research techniques. This review has prompted us to reconsider the traditional description of the echinostome life cycle within the molluscan host, and from this one can question our present day understanding of several key points: the differing requirements and capabilities of different intramolluscan stages; the longevity and ultimate fate of individual larvae; the number of generations produced and how many are present at any one time; and the cues that trigger rediae to produce rediae versus cercariae. These issues should be revisited using modern approaches.

Molecular biology is already providing new information on the biology of echinostomes and snails. However, it is not sufficient to merely describe genes. Traditional approaches must be blended with new techniques to achieve a greater appreciation of digenean-snail interactions at a functional level. For example, the use of specific and characterized molecules derived from *in vitro* expression systems could be used in conjunction with traditional phagocytosis/encapsulation assays to evaluate the effect that these factors may have on the defense functions of hemocytes towards different non-self entities including digeneans.

In conclusion. the study of echinostomes will continue to contribute greatly to our understanding of parasite virulence and parasite-host specificity. Insights obtained from this model have the potential to illuminate several fields such as parasitology, cell biology, molecular biology of digeneans and snails, and comparative immunobiology.

8. Acknowledgments

Figure 1 is comprised of photographs taken by E.S. Loker with the help of the researchers and facilities of the Centre de Biologie et d'Ecologie Tropicale et Méditerranéenne, Université de Perpignan in France. We thank Dr Gennadi Ataev (Zoology, Faculty of Biology, Russian Pedagog University, St Petersburg, Russia) and Dr Christine Coustau (Centre de Biologie et d'Ecologie Tropicale et Méditerranéenne, Université de Perpignan, France) for providing Figure 2. This work received support from NIH grant AI24340 (ESL) and NSF grant MCB9974930 (CMA).

9. References

Adema, C. M., Arguello, D. F., Stricker S. A. and Loker, E. S. (1994b) A time-lapse study of interactions between *Echinostoma paraensei* intramolluscan larval stages and adherent hemocytes from *Biomphalaria glabrata* and *Helix aspersa*, *Journal of Parasitology* **80**, 719-727.

Adema, C. M., Hertel, L. A. and Loker, E. S. (1997a) Infection with *Echinostoma paraensei* (Digenea) induces parasite-reactive polypeptides in the hemolymph of the gastropod host *Biomphalaria glabrata*, in N. Beckage (ed.), *Parasite Effects on Host Physiology and Behavior*, Chapman Press, New York, pp. 77-99.

Adema, C. M., Hertel, L. A., Miller, R. D. and Loker, E. S. (1997b) A family of fibrinogen-related proteins that precipitates parasite-derived molecules is produced by an invertebrate after infection, *Proceedings of the National Academy of Sciences USA* **94**, 8691-8696.

Adema, C. M., Hertel, L. A. and Loker, E. S. (1999) Evidence from two planorbid snails of a complex and dedicated response to digenean (echinostome) infection, *Parasitology* **119**: In press.

Adema, C. M., Léonard, P. M., DeJong, R. J., Day, H., Edwards, D., Burgett, G., Hertel, L. A. and Loker, E. S. (2000) Analysis of messages expressed by *Echinostoma paraensei* miracidia and sporocysts, obtained by random EST sequencing, *Journal of Parasitology* **86**: In press.

Adema, C. M. and Loker, E. S. (1997) Specificity and immunobiology of larval digenean-snail associations, in B. Fried and T. K. Graczyk (eds.), *Advances in Trematode Biology*, CRC Press, Boca Raton Florida, pp. 230-263.

Adema, C. M., van Deutekom-Mulder, E. C., van der Knaap, W. P. W. and Sminia, T. (1994) Schistosomicidal activities of *Lymnaea stagnalis* hemocytes: the role of oxygen radicals, *Parasitology* **109**, 479-485.

Ataev, G. L. and Coustau, C. (1999) Cellular response to *Echinostoma caproni* infection in *Biomphalaria glabrata* strains selected for susceptibility/resistance, *Developmental and Comparative Immunology* **23**, 187-198.

Ataev, G. L., Dobrovolskij, A. A., Fournier, A. and Jourdane, J. (1997) Migration and development of mother sporocysts of *Echinostoma caproni* (Digenea, Echinostomatidae), *Journal of Parasitology* **83**, 444-453.

Ataev, G. L., Fournier, A. S., and Coustau, C. (1998) Comparison of *Echinostoma caproni* mother sporocyst development *in vivo* and *in vitro* using *Biomphalaria glabrata* snails and a *Biomphalaria glabrata* embryonic cell line, *Journal of Parasitology* **84**, 227-235.

Basch, P. F. and Diconza, J. J. (1975) Predation by echinostome rediae upon schistosome sporocysts *in vitro*, *Journal of Parasitlogy* **61**, 1044-1047.

Clapham, D. E. (1995) Calcium signaling, *Cell* **80**, 259-268.

Connors, V. A., Lodes, M. J. and Yoshino, T. P. (1991) Identification of a *Schistosoma mansoni* sporocyst excretory secretory antioxidant molecule and its effect on superoxide production by *Biomphalaria glabrata* hemocytes, *Journal of Invertebrate Pathology* **58**, 387-395.

Connors, V. A. and Yoshino, T. P. (1990) *In vitro* effect of larval *Schistosoma mansoni* excretory secretory products on phagocytosis-stimulated superoxide production in hemocytes from *Biomphalaria glabrata*, *Journal of Parasitology* **76**, 895-902.

Couch, L., Hertel, L. A. and Loker, E. S. (1990) Humoral response of the snail *Biomphalaria glabrata* to trematode infection: observations on a circulating hemagglutinin, *Journal of Experimental Zoology* **255**, 340-349.

Coustau, C., Ataev, G., Jourdane J. and Yoshino, T. P. (1997) *Schistosoma japonicum*: *in vitro* cultivation of miracidium to daughter sporocyst using a *Biomphalaria glabrata* embryonic-cell line, *Experimental Parasitology* **87**, 77-87.

Davids, B. J., Wu, X. J. and Yoshino, T. P. (1999) Cloning of a beta integrin subunit cDNA from an embryonic cell line derived from the freshwater mollusc, *Biomphalaria glabrata*, *Gene* **228**, 213-223.

Davids, B. J. and Yoshino, T. P. (1998) Integrin-like RGD-dependent binding mechanism involved in the spreading response of circulating molluscan phagocytes, *Developmental and Comparative Immunology* **22**, 39-53.

Degaffé, G. and Loker, E. S. (1998) Susceptibility of *Biomphalaria glabrata* to infection with *Echinostoma paraensei*: correlation with the effect of parasite secretory-excretory products on host hemocyte spreading, *Journal of Invertebrate Pathology* **71**, 64-72.

Dewilde, S., Winnepenninckx, B., Arndt, M. H. L., Nascimento, D. G., Santoro, M. M., Knight, M., Miller, A. N., Kerlavage, A. R., Geoghagen, N., Van Marck, E., Liu, L. X., Weber, R. E. and Moens, L. (1998) Characterization of the myoglobin and its coding gene of the mollusc *Biomphalaria glabrata*, *Journal of Biological Chemistry* **22**, 13583-13592.

Dikkeboom, R., van der Knaap, W. P. W., Meuleman, E. A. and Sminia, T. (1985) A comparative study on the internal defence system of juvenile and adult *Lymnaea stagnalis*, *Immunology* **55**, 547-553.

Dinnik, J. A. and Dinnik, N. N. (1956) Observations on the succession of redial generations of *Fasciola gigantica* Cobbold in a snail, *Zeitschrift zuer Tropenmedizin und Parasitenkunde* **4**, 397-418.

Dissous, C. and Capron, A. (1995) Convergent evolution of tropomyosin epitopes, *Parasitology Today* **11**, 45-46.

Doolittle, R. F. (1992) A detailed consideration of a principal domain of vertebrate fibrinogen and its relatives, *Protein Science* **1**, 1563-1577.

Duclermortier, P., Lardans, V., Serra, E., Trottein F. and Dissous, C. (1999) *Biomphalaria glabrata* embryonic cells express a protein with a domain homologous to the lectin domain of mammalian selectins, *Parasitology Research* **85**, 481-486.

DuPasquier, L. and Flajnik, M. (1999) Origin and evolution of the vertebrate immune system, in W. E. Paul (ed), *Fundamental Immunology, 4th edition*, Lippincott-Raven Publishers New York, pp. 605-650.

Fried, B. and Huffam, J. E. (1996) The biology of the intestinal trematode *Echinostoma caproni*, *Advances in Parasitology* **38**, 311-168.

Gibson, D. I. (1987) Questions in digenean systematics and evolution, *Parasitology* **95**, 429-446.

Gokudan, S., Muta, T., Tsuda, R., Koori, K., Kawahara, T., Seki, N., Mizunoe, Y., Wai, S. N., Iwanaga, S. and Kawabata, S. (1999) Horseshoe crab acetyl group-recognizing lectins involved in innate immunity are structurally related to fibrinogen, *Proceedings of the National Academy of Sciences USA* **96**, 10086-10091.

Hanelt, B., Adema, C. M., Mansour, M. H. and Loker, E. S. (1997) Detection of *Schistosoma mansoni* in *Biomphalaria* using nested PCR, *Journal of Parasitology* **83**, 387-394.

Hansen, E. L. (1976) A cell line from embryos of *Biomphalaria glabrata* (Pulmonata): establishments and characteristics. in K. Maramorosch (ed), *Invertebrate Tissue Culture: Research Applications*, Academic Press, New York pp. 75-99.

Hertel, L. A., Stricker, S. A. and Loker, E. S. (1999) Calcium dynamics of the gastropod *Biomphalaria glabrata*: effects of digenetic trematodes and selected bioactive compounds, *Invertebrate Biology*, In press.

Hertel, L. A., Stricker, S. A., Monroy, F. P., Wilson, W. D. and Loker, E. S. (1994) *Biomphalaria glabrata* hemolymph lectins: binding to bacteria, mammalian erythrocytes, and to sporocysts and rediae of *Echinostoma paraensei*, *Journal of Invertebrate Pathology* **64**, 52-61.

Horak, P. and van der Knaap, W. P. W. (1997) lectins in snail-trematode immune interactions : a review, *Folia Parasitologica* **44**, 161-172.

Hosoi, T., Imai, Y. and Irimura, T. (1998) Coordinated binding of sugar, calcium, and antibody to macrophage C-type lectin, *Glycobiology* **8**,791-798.

Ivanchenko, M. G., Lerner, J. P., McCormick, R. S., Toumadje, A., Allen, B., Fischer, K., Hedstrom, O., Helmrich, A., Barnes, D. W. and Bayne, C. J. (1999) Continuous *in vitro* propagation and differentiation of cultures of the intramolluscan stages of the human parasite *Schistosoma mansoni*, *Proceedings of the National Academy of Sciences U. S. A.* **96**, 4965-4970.

Janeway, C. A. Jr. (1989) Approaching the asymptote? Evolution and revolution in immunology, *Cold Spring Harbor Symposia on Quantitative Biology* **54**,1-13.

Jeong, K. H., Lie, K. J. and Heyneman, D. (1983) The ultrastructure of the amoebocyte-producing organ in *Biomphalaria glabrata*, *Developmental and Comparative Immunology* **7**, 217-228.

Johnson, P. T. J., Lunde, K. B., Ritchie, E. G. and Launer, A. E. (1999) The effect of trematode infection on amphibian limb development and survivorship, *Science* **284**, 802-804.

Kanev, I., Fried, B., Dimitrov, V. and Radev, V. (1995) Redescription of *Echinostoma trivolvis* (Cort, 1914) (Trematoda, Echinostomatidae) with a discussion on its identity, *Systematic Parasitology*, **302**, 61-70.

Knight, M., Miller, A. N., Geoghagen, N. S. M., Lewis, F. A. and Kerlavage, A. R. (1998) Expressed sequence tags (ests) of *Biomphalaria glabrata*, an intermediate snail host of *Schistosoma mansoni*: use in the identification of RFLP markers, *Malacologia* **39**, 175-182.

Kurachi, S., Song, Z. W., Takagaki, M., Yang, Q., Winter, H. C., Kurachi, K. and Goldstein, I. J. (1998) Sialic-acid-binding lectin from the slug *Limax flavus*: cloning, expression of the polypeptide, and tissue localization, *European Journal of Biochemistry* **254**, 217-222.

Kuris, A. M. and Warren, J. (1980) Echinostome cercarial penetration and metacercarial encystment as mortality factors for a 2nd intermediate host, *Biomphalaria-glabrata*, *Journal of Parasitology* **66**, 630-635.

Kurosawa, Y. and Hashimoto, K. (1996) The immunoglobulin superfamily: where do invertebrates fit in? in E. Cooper (ed), *Advances in Comparative and Environmental physiology*, Springer Berlin, **23** pp. 151-184.

Langand J., Jourdane, J., Coustau, C., Delay, B. and Morand, S. (1998) Cost of resistance, expressed as a delayed maturity, detected in the host-parasite system *Biomphalaria glabrata-Echinostoma caproni*, *Heredity* **80**, 320-325.

Langand J. and Morand, S. (1998) Heritable non-susceptibility in an allopatric host-parasite system: *Biomphalaria glabrata* (Mollusca) *Echinostoma caproni* (Platyhelminths, Digenea), *Journal of Parasitology* **84**, 739-742.

Lardans, V., Boulo, V., Duclermortier, P., Serra, E., Mialhe, E., Capron, A. and Dissous, C. (1996) DNA transfer in a *Biomphalaria glabrata* embryonic-cell line by DOTAP lipofection, *Parasitology Research* **82**, 574-576.

Lardans, V. and Dissous, C. (1998) Snail control strategies for reduction of schistosomiasis transmission, *Parasitology Today* **14**, 413-417.

Lardans, V., Ringaut, V., Duclermortier, P., Cadoret, J. P. and Dissous, C. (1997) Nucleotide and deduced amino-acid-sequences of *Biomphalaria glabrata* actin cDNA, *DNA Sequence* **7**, 353-356.

Lardans, V., Serra, E., Capron, A., Dissous, C. (1998) Characterization of an intracellular receptor for activated protein-kinase-c (RACK) from the mollusk Biomphalaria glabrata, the intermediate host for *Schistosoma mansoni*, *Experimental Parasitology* **88**, 194-199

Laursen, J. R. and Yoshino, T. P. (1999) *Biomphalaria glabrata* embryonic (Bge) cell line supports *in vitro* miracidial transformation and early larval development of the deer liver fluke, *Fascioloides magna*, *Parasitology* **118**, 187-194.

Lie, K. J. (1973) Larval trematode antagonism: principles and possible application as a control method, *Experimental Parasitology* **33**, 343-349.

Lie, K. J. (1982) Survival of *Schistosoma mansoni* and other trematode larvae in the snail B*iomphalaria glabrata* : a discussion of the interference theory, *Tropical and Geographical Medicine* **34**, 111-122.

Lie, K. J. and Basch, P. F. (1967) The life history of *Echinostoma paraensei* sp. n. (Trematoda: Echinostomatidae), *Journal of Parasitology* **53**, 1192-1199.

Lie, K. J, and Heyneman, D. (1976) Studies on resistance in snails. 3. Tissue reactions to *Echinostoma lindoense* sporocysts in sensitized and resensitized *Biomphalaria glabrata*, *Journal of Parasitology* **62**, 51-58.

Lie, K. J. and Heyneman, D. (1977a) *Schistosoma mansoni, Echinostoma lindoense,* and *Paryphostomum-segregatum* interference by trematode larvae with acquired-resistance in snails, *Experimental Parasitology* **42**, 343-347.

Lie, K. J. and Heyneman, D. (1977b) Capacity of irradiated echinostome sporocysts to protect *Schistosoma mansoni* in resistant B*iomphalaria glabrata*, *International Journal for Parasitology* **9**, 539-543

Lie, K. J., Heyneman, D. and Richards, C. S. (1977) Studies on resistance in snails : interference by nonirradiated echinostome larvae with natural resistance to *Schistosoma mansoni* in *Biomphalaria glabrata*, *Journal of Invertebrate Pathology* **29**, 118-125.

Lie, K. J., Heyneman, D. and Kostanian, N. (1975) Failure of *Echinostoma lindoense* to reinfect snails already harboring that species, *International Journal of Parasitology* **5**, 483-486

Lie, K. J., Jeong, K. H. and Heyneman D. (1981) Selective interference with granulocyte function induced by *Echinostoma paraensei* (Trematoda) larvae in *Biomphalaria glabrata*, *Journal of Parasitology* **67**, 790-796.

Lie, K. J., Jeong, K. H. and Heyneman D. (1983) Acquired-resistance in snails: induction of resistance to *Schistosoma mansoni* in *Biomphalaria glabrata*, *International Journal for Parasitology* **13**, 301-304.

Lim, H. K. and Heyneman, D. (1972) Intramolluscan inter-trematode anatagonism: a review of factors influencing the host-parasite system and its possible role in biological control, in B. Dawes (ed.), *Advances in Parasitology* **10**, Academic Press, London, pp. 191-268.

Lodes, M. J., Connors, V. A. and Yoshino, T. P. (1991) Isolation and functional-characterization of snail hemocyte-modulating polypeptide from primary sporocysts of *Schistosoma mansoni.*, *Molecular and Biochemical Parasitology* **49**, 1-10.

Lodes, M. J. and Yoshino, T. P. (1990) The effect of schistosome excretory secretory products on *Biomphalaria glabrata* hemocyte motility, *Journal of Invertebrate Pathology* **56**, 75-85.

Loker, E. S. (1994) On being a parasite in an invertebrate host : a short survival course, *Journal of Parasitology* **80**, 728-747.

Loker, E. S. and Adema, C. M. (1995) Schistosomes, echinostomes and snails: comparative immunobiology, *Parasitology Today* **11**, 120-124.

Loker, E. S., Bayne, C. J. and Yui, M. A. (1986) *Echinostoma paraensei* : hemocytes of *Biomphalaria glabrata* as targets of echinostome-mediated interference with host snail resistance to *Schistosoma mansoni*, *Experimental Parasitology* **62**,149-154.

Loker, E. S., Boston, M. E. and Bayne, C. J. (1989) Differential adherence of M-line *Biomphalaria glabrata* hemocytes to *Schistosoma mansoni* and *Echinostoma paraensei* larvae, and experimental manipulation of hemocyte binding, *Journal of Invertebrate Pathology* **54**, 260-268.

Loker, E. S., Cimino, D. F. and Hertel, L. A. (1992) Excretory-secretory products of *Echinostoma paraensei* sporocysts mediate interference with *Biomphalaria glabrata* hemocyte functions, *Journal of Parasitology* **78**, 104-115.

Loker, E. S., Cimino, D. F., Stryker, G. A. and Hertel, L. A. (1987) The effect of size of M-line *Biomphalaria glabrata* on the course of development of *Echinostoma paraensei*, *Journal of Parasitology* **73**, 1090-1098.

Loker, E. S., Couch, L. and Hertel, L. A. (1994) Elevated agglutination titers in plasma of *Biomphalaria glabrata* exposed to *Echinostoma paraensei*: characterization and functional relevance of a trematode-induced response, *Parasitology* **108**, 17-26.

Loker, E. S., Coustau, C., Ataev G. L. and Jourdane, J. (1999) *In vitro* culture of rediae of *Echinostoma caproni*, *Parasite - Journal de la Societé Française de Parasitologie* **6**, 169-174.

Loker, E. S. and Hertel, L. A. (1987) Alterations in *Biomphalaria glabrata* plasma induced by infection with the digenetic trematode *Echinostoma-paraensei*, *Journal of Parasitology* **73**, 503-513.

Matsushita, M., Endo, Y., Taira, S., Sato, Y., Fujita, T., Ichikawa, N., Nakata, M. and Mizuochi, T. (1996) novel human serum lectin with collagen- and fibrinogen-like domains that functions as an opsonin, *Journal of Biology and Chemistry* **271**, 2448-2454.

Millar, D. A. and Ratcliffe, N. A. (1994) Invertebrates, in R. J. Turner (ed), *Immunology: a comparative approach*, Wiley, New York, pp. 29-68.

Monroy, F. P., Hertel, L. A. and Loker, E. S. (1992) Carbohydrate-binding plasma-proteins from the gastropod *Biomphalaria glabrata*: strain specificity and the effects of trematode infection, *Developmental and Comparative Immunology* **16**, 355-366.

Monroy, F. P. and Loker, E. S. (1993) Production of heterogeneous carbohydrate-binding proteins by the host snail *Biomphalaria glabrata* following exposure to *Echinostoma paraensei* and *Schistosoma mansoni*, *Journal of Parasitology* **79**, 416-423.

Morgan, J. A. T. and Blair, D. (1995) Nuclear rDNA ITS sequence variation in the trematode genus *Echinostoma*: an aid to establishing relationships within the 37-collar-spine group, *Parasitology* **111**, 609-615.

Noda, S. (1992) Effects of excretory-secretory products of *Echinostoma paraensei* larvae on the hematopoietic organ of M-Line *Biomphalaria glabrata* snails, *Journal of Parasitology* **78**, 512-517.

Noda, S. (1993) The inhibitory effect of excretory-secretory products of *Echinostoma paraensei* primary sporocysts on M-line *Biomphalaria glabrata* hemocyte motility, *Japanese Journal of Parasitology* **42**, 479-483.

Noda, S. and Loker, E. S. (1989a) Effects of infection with *Echinostoma paraensei* on the circulating hemocyte population of the host snail *Biomphalaria glabrata*, *Parasitology* **98**, 35-41.

Noda, S. and Loker, E. S. (1989b) Phagocytic-activity of hemocytes of M-line *Biomphalaria glabrata* snails : effect of exposure to the trematode *Echinostoma paraensei*, *Journal of Parasitology* **75**, 261-269.

Paraense, W. L. and Correa, L. R. (1963) Variation in susceptibility of populations of *Australorbis glabratus* to a strain of *Schistosma mansoni*, *Revista do Instituto de Medecina Tropical de São* Paulo **5**, 15-22.

Rakondravao, A., Moukrin, P., Hourdin, P. and Rondelaud, D (1992) Redial generations of *Fasciola gigantica* in the pulmonate snail *Lymnaea truncatula*, *Journal of Helminthology* **66**, 159-166.

Richards, C. S. (1984) Influence of snail age on genetic variation in susceptibility of *Biomphalaria glabrata* for infection with *Schistosoma mansoni*, *Malacologia* **25**, 493-502.

Richards, C. S., Knight, M. and Lewis, F. A. (1992) Genetics of *Biomphalaria glabrata* and its effect on the outcome of *Schistosoma mansoni* infection, *Parasitology Today* **8**, 171-174.

Richards, E. and Renwrantz, L. R. (1991) Two lectins on the surface of *Helix pomatia* haemocytes: a Ca2+-dependent, GalNac-specific lectin and a Ca2+- independent, mannose 6-phosphate-specific lectin which recognizes activated homologous opsonins, *Journal of Comparative Physiology* **161**, 43-54.

Rondelaud, D, (1994) *Fasciola hepatica*: the infection rate and the development of redial generations in *Lymnaea truncatula* exposed to miracidia after experimental dissection and activation in water, *Journal of Helminthology* **68**, 63-66.

Rysavy, B., Ergens, R., Groschaft, J., Moravec, F., Yousif, F. and Elhassan, A. A. (1973) Preliminary report on the possibility of utilizing competition of larval schistosomes and other larval trematodes in the intermediate hosts for the biological control of schistosomiasis, *Folia Parasitologica* **20**, 293-296.

Schell S. C. (1970) *Handbook of trematodes of North America North of Mexico*, University Press of Idaho, Moscow.

Sapp, K. K. (1999) Mechanisms underlying digenean-snail specificity: a comparative approach, Ph.D. Dissertation Department of Biology, The University of New Mexico, Albuquerque NM.

Sapp, K. K., Meyer, K. A. and Loker, E. S. (1998) Intramolluscan development of the digenean *Echinostoma paraensei* : rapid production of a unique mother redia that adversely affects development of conspecific parasites, *Invertebrate Biology* **117**, 20-28.

Schluter, S. F., Bernstein, R. M. and Marchalonis, J. J. (1997) Molecular origins and evolution of immunoglobulin heavy-chain genes of jawed vertebrates, *Immunology Today* **18**, 543-549.

Sousa, W. P. (1992) Interspecific interactions among larval trematode parasites of freshwater and marine snails, *American Zoologist* **32**, 583-592.

Sorensen, R. E. and Minchella, D. J. (1998) Parasite influences on host life-history : *Echinostoma revolutum* parasitism of *Lymnaea elodes* snails, *Oecologia,* **115**, 188-195.

Strand, M. R. and Grbic, M. (1997) The life history and development of polyembryonic parasitoids, in N. Beckage (ed.), *Parasite Effects on Host Physiology and Behavior*, Chapman Press, New York, pp. 37-55.

Sullivan, J. T. (1988) Hematopoiesis in three species of gastropods following infection with *Echinostoma paraensei* (Trematoda,: Echinostomatidae), *Transactions of the American Microscopical Society* **107**, 355-361.

Tielens, A. G. M., Horemans, A. M. C., Dunnewijk, R., van der Meer, P. and van den Bergh, S. G. (1992) The facultative anaerobic energy-metabolism of *Schistosoma mansoni* sporocysts, *Molecular and Biochemical Parasitology* **56**, 49-58.

Trottein, F., Goding, G., Sellin, B., Gorillot, I., Samaio, M., Lecocq, J. P. and Capron, A. (1992) Inter-species variation of schistosome 28kDa glutathione-S transferases, *Molecular and Biochemical Parasitology* **54.**, 63-72.

Uchikawa, R. and Loker, E. S. (1991) Lectin-binding properties of the surfaces of *in vitro*-transformed *Schistosoma mansoni* and *Echinostoma paraensei* sporocysts, *Journal of Parasitology* **77**, 742-748.

Uchikawa, R, and Loker, E. S. (1992) *Echinostoma paraensei* and *Schistosoma mansoni*: adherence of unaltered or modified latex beads to hemocytes of the host snail *Biomphalaria glabrata*, *Experimental Parasitology* **75**, 223-232.

van der Knaap, W. P. W. and Loker, E. S. (1990) Immune mechanisms in trematode snail interactions, *Parasitology Today* **6**, 175-182.

Yamaguti, S. (1975) A synoptical review of life histories of digenetic trematodes of vertebrates, Academic Press of Japan, Kyoto.

Yoshino, T. P. and Boswell, C. A. (1986) Antigen sharing between larval trematodes and their snail hosts: how real a phenomenon is immune evasion? *Symposium of the Zoological Society of London* **56**, 221-238

Yoshino, T. P. and Laursen, J. R. (1995) Production of *Schistosoma mansoni* daughter sporocysts from mother sporocysts maintained in synxenic culture with *Biomphalaria glabrata* embryonic (Bge) cells, *Journal of Parasitology* **81**, 714-722.

Yoshino, T. P., Wu, X. J. and Liu, H. D. (1998) Transfection and heat-inducible expression of molluscan promoter-luciferase reporter gene constructs in the *Biomphalaria glabrata* embryonic snail cell-line, *American Journal of Tropical Medicine and Hygiene* **59**, 414-420.

Yoshino, T. P. and Vasta, G. R. (1996) Parasite-invertebrate host immune interactions, *Advances in Comparative and Environmental Physiology* **24**, 125-167.

Zelck, U. E., Becker, W. and Bayne C. J. (1995) The plasma proteins of *Biomphalaria glabrata* in the presence and absence of *Schistosoma mansoni*, *Developmental and Comparative Immunology* **19**, 181-194.

THE BEHAVIORAL BIOLOGY OF ECHINOSTOMES

WILFRIED HAAS

*Institute of Zoology I, University Erlangen-Nürnberg, D-91058 Erlangen,
Germany*

B. Fried and T.K. Graczyk (eds.),
Echinostomes as Experimental Models for Biological Research, 175–197.
© 2000 *Kluwer Academic Publishers. Printed in the Netherlands.*

1. Introduction

Worms that have adopted the parasitic way of life have often highly reduced sensory and behavioral capabilities and secondarily simplified or degenerated nervous systems. This is obvious when individual parasitic stages are compared with free-living organisms, and it may lead to the conclusion that behavior is unimportant for the fitness of parasitic worms. However, considering the life cycle of echinostomes as an example, we realize that nearly all developmental stages show behavioral patterns which play key roles for the survival of the species. Complex series of these patterns are responsible for the facts that: 1) miracidia hatch, find and invade their snail hosts, 2) sporocysts reach their habitats within the snail host, 3) rediae leave the sporocysts, move towards particular habitats, feed on the appropriate tissues and on the sporocysts and rediae of competing digeneans, 4) cercariae leave the rediae at a certain time, migrate through the snails, leave the snails, find their next intermediate hosts or habitats for encystation, enter the hosts and encyst in the appropriate sites within the hosts, 5) metacercariae leave their cysts at the appropriate site within the final hosts and select their particular habitats within the hosts, and 6) the adult worms select their sites within the hosts, feed properly, find their mate partners and successfully mate. This simplified review of digenean life cycle reveals that the behavior of the different stages are determining factors for the existence of these parasites. Behavior studies in helminths are hampered by many factors, including difficulties to maintain the parasitic stages *in vitro* and not easy access to the nervous system in the body cavity, filled with parenchyma, particularly when nerve cells need to be penetrated with micro-electrodes. Despite these difficulties modern technologies elaborated much information on the neuroanatomy and neurochemistry of digeneans and other flatworms, as well as on the ultrastructure of various putative sensory receptors, and these topics are presented in reviews (Halton *et al.*, 1990; 1992; 1994; 1997; Fairweather and Halton, 1991; Pax and Bennett, 1991; 1992, Gustafsson, 1992; Halton and Gustafsson, 1996; Pax *et al.*, 1996) and in chapter 11 of this book by Halton *et al*. However, presently it is not yet possible to integrate neurobiological results with studies on the behavior, as well as it is not possible to ascribe any sensory function to various receptor (except photoreceptor) structures.

Analyses of digenean behavior are particularly successful when it is possible to split the continuous flow of behavioral patterns into discrete units which may be determined by only few, or at the best, one neurophysiological mechanism. Then, the information may be obtained on the nature of the stimulating cues and the receptors involved, and hypotheses may be established on the neurophysiological background and on the adaptive function of the behavioral patterns. Although the mechanisms that control the behavior are not yet understood, it is possible to describe how internal control mechanisms and external stimuli may interact. For example, studies on the behavior of cercariae may allow to establish detailed flow diagrams describing the control of spontaneous swimming behavior and how it is modified by light, dark, and water turbulence stimuli (Haas, 1992).

Echinostomes are very suitable models for physiological behavioral studies and many data are already available. This chapter summarizes the findings and discusses the physiological background and the adaptive function of the various types of behavior. Each of the different stages in the echinostome life cycle has its particular function and they differ considerably in their behavior. Therefore, it is necessary to treat the behavior of each stage separately.

2. Behavior of Free-Living Stages: Miracidia and Cercariae

2.1. FUNCTIONS OF MIRACIDIAL AND CERCARIAL BEHAVIOR

Both miracidia and cercariae do not feed. They are provided with limited energy reserves, which allow them only for few hours of full activity. The only chance to transmit their genes is to invade appropriate hosts. It can be expected that they have evolved behavioral patterns to achieve this goal. Many studies have dealt with miracidial and cercarial behavior and, in fact, most of the described behavioral patterns can be interpreted as adaptations to successful invasion of the hosts (reviewed by Ulmer, 1971; Cable, 1972; Erasmus, 1972; MacInnis, 1976; Saladin, 1979; Christensen, 1980; Smyth and Halton, 1983; Sukhdeo and Mettrick, 1987; Haas, 1992; 1994; Combes et al., 1994; Haas et al., 1995a; Haberl, 1995; Haas and Haberl, 1997). The behavior of the actively host invading miracidia and cercariae may have functions in host-finding and host-recognition processes such as: 1) departure from the habitat of origin and dispersal, 2) localization of and remaining in the host's habitats, 3) host directed orientation, 4) attachment to the host, 5) enduring contact with the host, 6) creeping to entry sites, 7) host invasion, and 8) migration within the host. Other functions of behavioral patterns which may contribute to successful transmission have also been reviewed, for example, predator avoidance and longer survival (Haas 1994); however, this subject needs further investigation.

2.2. MIRACIDIAL BEHAVIOR

2.2.1. *Miracidial Hatching*
The hatching of digenean miracidia from the eggs is stimulated by different factors (reviewed by Smyth and Halton, 1983; Xu and Dresden, 1990). The major stimulus for *E. trivolvis*, *E. caproni*, and *E. paraensei* is light. *Echinostoma caproni* and *E. paraensei* miracidia hatch in a strict diurnal pattern between 11.00 and 16.00 hr. This pattern is of an unknown origin, and it is maintained even when the eggs are kept under constant illumination or in the dark (Behrens and Nollen, 1992; Markum and Nollen, 1996; Meece and Nollen, 1996). One selective advantage of this pattern might be that it enables the miracidia to use photo-orientation in their host-finding. However, *E. trivolvis* miracidia maintain their transmission without such a daily hatching pattern (Nollen, 1994). Certain chemical conditions may also stimulate the

hatching. For example, water conditioned with *B. glabrata* increased the number of hatching *E. caproni* miracidia and lipophilic components seem to contribute to this effect (Fried and Reddy, 1999).

2.2.2. *Miracidial Responses to Environmental Stimuli*
After hatching, the miracidia of at least some species seem to show a phase of dispersal from the site of origin, which may minimize overinfection of a small number of hosts and support distribution of miracidia. In the dispersal phase, the miracidia swim faster and more linear; they may show a reversed orientation to light and gravity, and a reduced response to host stimuli (reviewed by Saladin, 1979; Sukhdeo and Mettrick, 1987). The miracidia then enter a phase of microhabitat selection, in which they move to and remain in microhabitats frequented by their host snails. In this phase they respond to environmental stimuli such as light, gravity, temperature, magnetic fields, and physical boundaries (reviewed by MacInnis, 1976; Sukhdeo and Mettrick, 1987; Haberl, 1995). The miracidia of *E. trivolvis* and *E. caproni* (Table 1) show a negative geotaxis (they move to the top of containers also under dark conditions) which is dominated by a positive phototaxis (the miracidia move to the bottom when illuminated from below) (Behrens and Nollen, 1992; Nollen, 1994). This behavior may increase the chances for an encounter with the host snails, which tend to feed at the water surface. However, it is not understood why *E. paraensei* miracidia, which like *E. caproni* miracidia invade *B. glabrata*, show no distinct geo-orientation and only weak photo-orientation (Meece and Nollen, 1996).

2.2.3. *Miracidial Chemo-Orientation*
It is now generally accepted that miracidia approach their host snails in response to snail emitted excretory-secretory-products (ESP). Also the main mechanisms by which miracidia orientate towards their hosts are known (reviewed by Haas *et al.*, 1995a; Haas and Haberl, 1997). Most species studied (including *E. caproni*, Table 2) show an increased rate of change of direction (RCD) in increasing and a turnback response in decreasing concentration gradients of the ESP (Fig. 1). This chemokinesis (Fraenkel and Gunn, 1961) may be an adaptation to the particular swimming behavior of miracidia. The swimming mode seems to be optimized to achieve a wide dispersal by a high swimming speed (three times faster than the speed of the considerably bigger ciliate, *Paramecium caudatum*) and a rotation along their long axis supporting a relatively straight path of movement, even when cilia are damaged. During their chemokinetical orientation the miracidia need only information on an increase or decrease of the attractant concentration and this may be easily detected by the fast swimming and rotating organisms. On the other hand, *S. japonicum* miracidia swim directed along increasing concentration gradients and for this chemotaxis they obtain information on the direction of the concentration gradient in an unknown manner (Haas *et al.*, 1991).

Many studies have shown that miracidia respond to very different chemicals when offered as pure compounds (reviewed by MacInnis, 1976; Saladin, 1979; Haas *et al.*,

1995a; Haberl, 1995; Haas and Haberl, 1997), and this is also characteristic for echinostome miracidia (Table 2). However, pure chemicals, when presented to the miracidia in relatively high concentrations, may not be necessarily identical with the stimulating cues of the host ESP and tissues. This has been shown for host-finding responses of many species of arthropod parasites, several species of cercariae and is also true for echinostome miracidia. It is evident that host chemical signals, used by the miracidia for their host-finding, can only be identified when they stimulate the miracidia to respond at natural concentrations and mostly at natural pH-conditions. The miracidia of at least six digenean species (including *E. caproni*, Table 2) swim towards their snail hosts in response to snail-emitted macromolecular glycoproteins (reviewed by Haas *et al.*, 1995a; Haberl, 1995; Haas and Haberl, 1997). Only defined fractions of the snail ESP are recognized. The signaling fractions are extremely effective, e.g., the macromolecular host-signal of *Lymnaea stagnalis* is recognized by *Trichobilharzia ocellata* miracidia at concentrations of 1 mg in 10,000 liters of water (Kalbe, unpubl. results). The molecules signal species- and even strain-specificity of the snail hosts (Kalbe *et al.*, 1996; 1997). At first sight it seems paradoxical that aquatic snails signal their identity as the suitable hosts with such precision despite 400 million years of coevolution with their platyhelminth parasites. One hypothesis could be that the glycoprotein signals are pheromones which are needed by the snails for intraspecific communication and that the miracidia use this code for their host-identification. It is possible that host-finding via glycoprotein signals is a general principle in the miracidial host-finding process and that other echinostome species behave similarly to *E. caproni*.

2.2.4. *Miracidial Behavior after Contact with the Host*
The responses of miracidia after contact with the snail host have been described in detail for miracidia of *S. mansoni* (MacInnis, 1965). The behavioral patterns "repeated investigation" and "attachment" seem to be the typical responses upon contact with the suitable host snails. In schistosome miracidia these responses are mainly stimulated by macromolecular glycoconjugates of the snail's body surface mucus (Haas *et al.*, 1991; Haberl and Haas, 1992; Haberl *et al.*, 1995). No information is available for echinostome miracidia regarding the host cues which stimulate these responses.

2.3. CERCARIAL BEHAVIOR

2.3.1. *Cercarial Emergence*
Little is known about the behavior of cercariae when they leave the rediae, migrate through their snail host's tissues, and exit from the snail hosts. However, many studies investigate the daily cycles of emergence of cercariae from their molluscan hosts (reviewed by Jourdane and Théron, 1987; Shostak and Esch, 1990; Combes *et al.*, 1994). In most species, the photoperiod is the major synchronizer of the daily shedding pattern; however, other factors such as the thermoperiod may also play a

role.   The two species of echinostome cercariae studied, *E. caproni* (referred to as *Echinostoma liei*) and *Euparyphium albuferensis*, show a diurnal emergence rhythm (Christensen *et al.*, 1980; Toledo *et al.*, 1999).   In *E. albuferensis* the cercarial emergence rhythm is synchronized mainly by the light-dark alternation as shown in experiments with changed light-dark cycles (Toledo *et al.*, 1999).   In fact, in *E. trivolvis* cercariae the major stimulus for emergence was the exposure to light; however, temperature had also some effect (Schmidt and Fried, 1996).

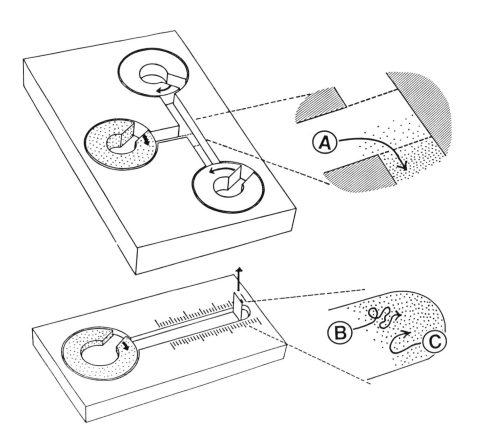

Figure 1. Mechanisms of chemo-orientation of echinostome miracidia and cercariae towards their snail hosts and the choice-chambers used to study their responses. The parasites are released from the chambers by opening the closure ring (spotted) and their swimming paths in increasing and decreasing concentration gradients of attractants recorded. A, chemotaxis, directed orientation along increasing gradients; B, increase of the rate of change of direction (RCD) in increasing gradients; C, turnback swimming in decreasing gradients. Note that miracidia use two orientation patterns (in 6 species studied) and cercariae one orientation pattern per species. (Modified after Haas *et al.*, 1995a, 1995b).

2.3.2. Mode of Swimming of Cercariae

The movement patterns of the cercariae, when they swim free in the water after having left the snail host, have been analyzed in several species (reviewed by Haas, 1992). The patterns of movement differ considerably among the species. However, they are all composed of movement cycles that are repeated in defined frequencies and some of the neurophysiological components that control the swimming movements are intrinsic to the tail. Although the mechanisms which control and modify cercarial swimming behavior are not yet understood in terms of neurophysiological events, behavioral analyses of cercarial swimming allowed researchers to describe how internal control mechanisms and external stimuli may interact (reviewed by Haas, 1992; 1994).

Echinostome cercariae show a continuous swimming mode, which contrasts to the intermittent swimming mode of many other cercariae (which show upward directed swimming bursts which alternate with phases of passive sinking). The swimming movements of *Echinoparyphium sp.* and *Himasthla secunda* were studied with cinematography (Graefe and Burkert, 1972; Chapman and Wilson, 1973). The base of the monocercous tail (i.e., a tail without furcae) of echinostome cercariae is directed upward and the complex movement patterns of the tail and the body allow locomotion in all directions. In the cercaria of *H. secunda,* the pattern of propulsion has been assigned to sequential contractions of muscle blocks in the tail and also cercarial turning has been described; the cercariae alter the symmetry and speed of their tail movements on 1 side of the body (Chapman and Wilson, 1973). However, it is not understood how the sensory input is processed in such modifications of the movement. The speed of swimming and the number of complete movement cycles depend mainly on the size of the cercariae and temperature. For example, at 25°C *Pseudechinoparyphium echinatum* swims 3.2 mm/sec (15 movement cycles/sec), *Echinoparyphium sp.* 2.5 mm/sec (10 cycles/sec) (Graefe and Burkert, 1972), *Echinostoma revolutum* 1.3 mm/sec (11 cycles/sec), and *Hypoderaeum conoideum* 1.6 mm/sec (16 cycles/sec) (Haas, 1994).

Cercarial swimming behavior is not just a sequence of automated self-regulated mechanisms. This behavior may be modified by environmental cues such as temperature, gravity, water currents, touch, chemicals, and direction and intensity of light radiation. Various receptors and the brain are taking part in the cercarial responses in an unknown manner. In contrast to many other cercarial species, the echinostome cercariae studied seem not to respond sensitively to water currents, touch, shadow, and light stimuli. Only gravity, light direction, light intensity, and chemical cues seem to modify their behavior in a manner which may serve to locate and recognize their hosts.

2.3.3. Cercarial Responses to Light and Gravity

Echinostome cercariae seem to disperse and to select particular habitats by responding mainly to the environmental cues of light intensity, light direction, and gravity. The available data indicate a very high diversity in the responses between different echinostome species to these environmental cues (Table 1). For example, photo-orientation in horizontal direction may be stimulated by the light intensity in addition to the direction of light radiation. *Pseudechinoparyphium echinatum* cercariae are poorly attracted by bright light, but strongly by low intensity of light (Fig. 2a) whereas freshly shed *E. revolutum*, *H. conoideum* and *Isthmiophora melis* cercariae are repelled by high and low intensity light. The photo-orientation in horizontal direction may also be modified with the age of the cercariae. For example, *H. conoideum* cercariae shift from a negative to a positive photo-orientation in the second half of their lifetime (Fig. 2a).

In particular, the vertical orientation of cercariae is dependent on interactions between the inducers gravity, direction, and intensity of light. For example, a negative geo-orientation may dominate the photo-orientation. This occurs in *I. melis* cercariae, which accumulate at the top of vertical cuvettes when illuminated from above (Fig. 2b), despite the fact that they show strong negative photo-orientation in horizontal direction. In other species, the geo-orientation may be induced mainly by the light intensity. For example, *E. revolutum* cercariae show increased negative geo-orientation with increased light intensity (Fig. 2b). As this occurs also when the light radiation comes from below, the light intensity must be the major inducer (Loy and Haas, unpublished results). Further species respond in their geo-orientation to both the direction and intensity of light. This is true for cercariae of *P. echinatum*. They are attracted to swim in upward direction by low light intensity and repelled by bright light coming from above (Fig. 2b). Because the reverse situation occurs due to illumination from below (low light intensity attracts the cercariae to swim downwards and high light intensity repels them upwards), the intensity and the direction of light are the inducers (Loy and Haas, unpublished results).

The mechanisms which control these types of orientation behavior are not understood in terms of neurophysiological events. The effect of light intensity on geo-orientation might be generated just by modification of the intensity of the upward directed tail-movements, and this orientation type would be then a kinesis (Fraenkel and Gunn, 1961). However, it is not known how the cercariae orientate along the direction of light radiation and we have no information on distinct photoreceptors in the species studied.

The function of the changing types of orientation may be to leave the habitats of the first intermediate host snails, to disperse and to select the habitats of the next hosts. For example, laboratory data suggest that *I. melis* searches for amphibian hosts in dark microhabitats near the water surface, and *P. echinatum* snail hosts in light microhabitats in deeper water levels. However, we do not have information whether the cercariae in fact distribute in the expected water depths and microhabitats in semi-field and field systems. Nor do we understand why various cercaria species select their habitats based on such different responses to environmental cues, even when they invade the same host genera.

(A) Horizontal photo-orientation

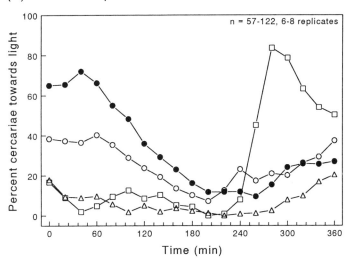

(B) Geo-orientation

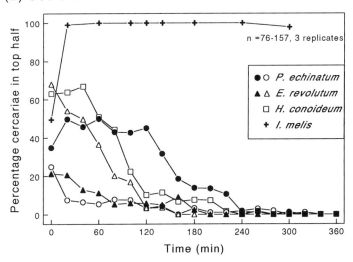

Figure 2. Effect of light and gravity on the orientation of echinostome cercariae at 25 °C. (A) Orientation towards light in horizontal direction. Percent cercariae in a 4.5 cm long section closest to the light source in a 18 cm long horizontal cuvette. (B) Effect of light intensity on geo-orientation, light radiation from above. Percent cercariae in the top half of a 18 cm high vertical cuvette. Light intensities: 500 W/m2 (blank symbols) and 5 W/m2 (filled symbols). (C. Loy, unpublished results).

2.3.4. *Chemo-Orientation of Swimming Cercariae*

A particular characteristic of swimming echinostome cercariae seems to be that they can orientate along chemical gradients of host emanations. Among the other species of cercariae studied, only cercariae of *Schistosoma mansoni* seem to show chemo-orientation; they tend to orientate towards still unknown compounds of human skin lipids (Stirewalt, 1971), and linolenic acid (Shiff *et al.*, 1993; Shiff and Graczyk, 1994). It was suggested that this chemo-orientation may enable the schistosomes to reach the skin surface when they are close to it, for example on hairs of furred mammals. It might be possible that the ability of echinostome cercariae for chemo-orientation is restricted to species which primarily invade snails, as stable concentration gradients of excretory-secretory products may establish only around slowly moving hosts. In fact, *I. melis* cercariae that invade fish and amphibia show much weaker chemo-orientation behavior than *P. echinatum* that invade snails (Matovu, 1971).

Chemo-orientation has been studied in five species of echinostome cercariae and some information is available on the attracting cues of the host, and on the cercarial orientation mechanisms (Fig. 1, Table 2). The mechanisms of snail host-finding of echinostome cercariae differ from those of miracidia. Whereas miracidia employ two orientation types, the cercariae show only one mechanism. Three cercaria species studied just turn back when the concentration gradients of chemical snail host ESP decrease (Fig. 1). This chemokinesis (Fraenkel and Gunn, 1961) seems to fit the swimming mode of the cercariae. They adjust their position in the water column by downstrokes of their upward directed tails and this movement type is probably adapted to minimize energetic costs. However, the tail movements create water currents which may disturb the detection of chemical gradients by the numerous sensory papillae on the cercarial body and tail. In fact, cercariae do not need information on the orientation of the stimulus gradient; a detection of the concentration of the stimulating chemicals is sufficient for their orientation. Nevertheless, cercariae of *H. conoideum* are able to swim directed chemotactically along increasing concentration gradients. It is not yet understood how they detect the direction of the gradient. As they swim more with their body in advance than other species, it was speculated (Haas *et al.*, 1995b) that they might detect the chemical gradient by comparison of simultaneous samples of stimulus by laterally positioned receptors (trophotaxis), or by sequential samples of stimulus during the oscillating movements of the body (klinotaxis). The selective benefits of the chemokinetic or the chemotactic orientation type can only be understood when much more information on function and energetics of cercarial swimming-movement is available.

Very different chemicals attract echinostome cercariae when offered as pure compounds (Table 2). However, like in miracidia, pure chemicals, when presented to the cercariae in relatively high concentrations may not be identical with the attractants of the hosts ESP. For example, cercariae of *H. conoideum* are highly attracted by the reduced form of glutathione (GSH) at concentrations of 10 μM. However, GSH does not contribute to the cercarial chemo-orientation towards snail-conditioned water

(SCW). Highly attractive SCW of *Lymnaea stagnalis* contains only 0.001 μM GSH, and SCW of *Biomphalaria glabrata* 0.01 μM and the cercariae are not attracted by GSH at these concentrations (Körner, 1998).

The chemical cues which attract cercariae of *E. revolutum* and *P. echinatum* towards their snail hosts have been analyzed in some detail (Haas *et al.*, 1995b; Körner, 1998; Körner and Haas, 1998a, 1998b). The attractive compounds of ESP of various aquatic animals are mainly the amino acids; urea and ammonia have also some effect. Water conditioned with diverse snail species and other aquatic animals showed different attractivity for the cercariae. As snails emit genotypic mixtures of amino acids when kept under constant conditions (Thomas and Eaton, 1993), it was suggested that the observed different species-specificities may be attributed to particular amino acid mixtures. However, when the samples of water conditioned with different species of snails, fish, tadpoles and leeches were diluted to the same total amino acid concentrations, they were equally attractive for the cercariae. Experiments with analogues and derivatives of amino acids showed that the cercariae could not distinctly differentiate between types of amino acids and that they responded only to certain common amino acid characteristics such as the primary α-amino group and the α-carboxyl group. Chemo-orientation to amino acids, ammonia and urea seems to reflect a strategy to locate a broad spectrum of aquatic hosts. A higher degree of host-specificity seems to occur in the chemo-orientation of *H. conoideum* cercariae, which are highly attracted by *Lymnaea stagnalis*, but weakly or not at all by other snails. It is not yet known whether the small peptides, which attract the cercariae towards *L. stagnalis*, signal particular host-specificities, as they disintegrate upon further analysis (Körner, 1998). Cercarial snail host-finding differs from the highly specific, glycoconjugate-mediated snail host-location of miracidia, demonstrating that snail host-finding by miracidia and cercariae have evolved independently.

Studies on cercarial chemo-orientation must take into account that the cercariae may show this behavior only in a certain phase of their lifetime. For example, *H. conoideum* cercariae start chemo-orientation towards their snail hosts not earlier than 1 hr after being shed (Haas *et al.*, 1995b). This may be interpreted as an adaptation to avoid metacercarial super-infection of the first intermediate hosts emitting the cercariae. Such a dispersal phase is also supported by the converted orientation of the cercariae towards light and gravity, and by a lower infectivity of cercariae in their immediate post-shedding period. For example, *E. recurvatum* cercariae show a submaximal infectivity for their second intermediate hosts within 2 hr after emission (Evans and Gordon, 1983).

2.3.5. Cercarial Attachment and Host Invasion

Most cercariae show particular behavioral patterns during their host-invasion phases of attachment to the host, enduring contact with the host, creeping to entry sites, penetration, gland secretion, and orientation within hosts tissues. Each of these invasion phases can be stimulated by separate physical and chemical host cues (reviewed by Haas, 1992; 1994; Haas and Haberl, 1997). In echinostome cercariae, studies on these behavioral patterns and their releasing cues were restricted to few species. *Isthmiophora melis* cercariae show higher attachment rates to fish and amphibian hosts than to other organisms; dissolved CO_2 and H_2CO_3/HCO_3^- participate in the stimulation of attachment (Motzel and Haas, 1985). The cercariae of *P. echinatum*, *E. revolutum* and *H. conoideum* respond intensively with attachment and enduring contact to various snail species and a frog, but with low intensity or not at all to a leech and fish. Both response types seem to be mainly stimulated by viscoelastic properties of the host's skin surface mucus in the three species of cercariae (Haas *et al.*, 1995b). No detailed information is available on echinostome cercarial behavior nor on the stimulating host cues when the cercariae penetrate the hosts and localize the sites within the hosts.

3. Behavior of Parasitic Stages: Sporocysts, Rediae, Metacercariae, and Adults

3.1. BEHAVIOR OF SPOROCYSTS AND REDIAE

Miracidia transform to sporocysts near their sites of entry and the sporocysts of many species move to particular locations within the snail hosts. For example, *E. caproni* sporocysts migrate from the various entry sites towards ventricle and aorta which are their final location. After moving through various tissues into the blood sinus system they finally reach the heart via the venous blood system (Ataev *et al.*, 1997). It is not known which behavior patterns and host cues are responsible for this site-finding, but it is suggested that ventricle and aorta must be recognized by the sporocysts. Sporocyst migration was also studied in *S. mansoni* daughter sporocysts which move from their mother sporocysts (located at various miracidial entry sites) to the digestive gland (Becker, 1970). They show nondirectional creeping movements until they enter a blood vessel. They then migrate fast within the vessels and are often passively transported to different highly vascular tissues. Finally, most of them are located in the digestive gland probably due to special conditions of the circulation of hemolymph and to their rapid growth in this organ. It is not known whether they in addition must recognize digestive gland tissue. They seem not to respond to the direction of blood flow; *in vitro* they show no rheotaxis (Becker, 1970).

Rediae are more active than sporocysts. They migrate within the snail's tissues and prefer particular habitats, mostly digestive gland and gonads of the snail. Rediae use their muscular pharynx to feed on host's tissues, but they may also attack and feed upon redia or sporocysts of other digeneans (reviewed by Lim and Heyneman, 1972). During direct antagonism, the rediae of dominant species search for rediae or sporocysts of subordinate species, kill, and consume them. The predaceous rediae accumulate around the victims; they must recognize them. Such rediae show, in fact, chemo-orientation towards each other, which has been demonstrated in *in vitro* experiments. *Echinostoma trivolvis* and *E. caproni* rediae moved towards individuals of the other species, but also towards conspecifics (Reddy and Fried, 1996). The attractants were contained in the fractions of free sterols and fatty acids of the redial ES-products (Reddy *et al.*, 1997b). This chemo-orientation can not yet be assigned to a particular type of redial behavior displayed within the snail host.

3.2. METACERCARIAL ACTIVATION AND EXCYSTATION

Echinostomes as gastrointestinal parasites enter their hosts passively by being ingested as encysted metacercariae. The metacercariae are passively carried with the ingesta to the intestine where the excystment processes are initiated. Much is known on the activation and excystment of metacercariae. The topic has been reviewed for various digeneans (Lackie, 1975; Smyth and Halton, 1983; Sommerville and Rogers, 1987; Sukhdeo and Mettrick, 1987; Fried, 1994; Irwin, 1997) and for echinostomes in particular by B. Fried in chapter 5 of this book and will not be treated here in detail. In several species metacercarial encystment seems to be passive, just by digestion of the cyst wall by host enzymes. However, in many species particular behavioral patterns of the metacercariae take part in the excystment. After activation the larvae may apply force to the cyst wall by coordinated body and sucker movements, they may secrete enzymes, e.g., with the contents of their intestinal caeca, and they may show complex movement patterns when they leave the cyst via the escape aperture. Very different host cues stimulating the metacercarial excystment behavior have been described. These include: mechanical effects, temperature, carbon dioxide, oxidation-reduction potentials, osmotic pressure, pH conditions, bile components, and various enzymes. Several observations indicate that the parasites even recognize the prospective escape aperture. It is very probable that the various host cues activate the coordinated emergence behavior via sensory receptors. For example, bile stimulates the emergence of *Fasciola hepatica* metacercariae only via glycocholic acid and various analogs of this molecule have no effect. The response to glycocholic acid is dose dependent with a sigmoidal log dose-effect curve, which is characteristic of the binding of a ligand to a receptor (Sukhdeo and Mettrick, 1986; 1987). The sequence of escape behavior patterns and their stimulating cues seem to assist the metacercariae of some species to select their particular habitats within the host.

3.3. HABITAT SELECTION WITHIN THE VERTEBRATE HOST

Most parasites select precise microhabitats in their hosts; however, little information is available on the mechanisms and stimulatory host cues which enable the parasites to find and to identify their microhabitats (reviewed by Ulmer, 1991; Crompton, 1973; Holmes, 1973; Sukhdeo and Mettrick, 1987; Sukhdeo, 1992; Sukhdeo and Sukhdeo, 1994; Sukhdeo and Bansemir, 1996). Different ultimate causes have been discussed for the restriction to specific habitats such as interspecific competition (Holmes, 1973), predation and hyperparasitism, facilitation of mating, reinforcement of reproductive barriers, and adaptations to the habitats complexity (Rohde, 1994). Also the various echinostome species select different intestine sections of their vertebrate hosts. Even within 1 species complex (*Echinoparyphium recurvatum*), two sympatric sibling species were totally separated in different intestine sections of their duck host (McCarthy, 1990). It is not known how echinostomes select their microhabitats in the intestine. When adult *E. trivolvis* (referred to as *E. revolutum*) were placed on the chorioallantoic membrane of chick embryos they aggregated in the area above the embryo which contained more blood vessels than the remaining chorioallantoic membrane (Fried and Diaz, 1987). However, it is not known what mechanisms and host cues govern this orientation. The microhabitat selection might also be influenced by the factors believed to attract adult echinostomes to each other.

3.4. MATE-FINDING AND MATING BEHAVIOR OF ADULT WORMS

The mate-finding behavior and the intra- and interspecific mating patterns were extensively studied in dioecious schistosomes (reviewed by Basch, 1991; Nollen, 1997) and in hermaphroditic digeneans. The act of copulation was rarely directly observed, but the mating behavior was indirectly determined by sperm radiolabeling, transplantation of labeled and unlabeled worms and subsequent autoradiography (reviewed by Nollen, 1983; 1997, and in chapter 7 of this book). The echinostome species studied (*E. caproni, E. trivolvis*, and *E. paraensei*) self-inseminated in single infections and cross- and self-inseminated in multiple infections. In concurrent infections the echinostomes showed interspecies mating, which did not occur when individuals of the same species were available. Thus, mating barriers are weak between the species, but the individuals can identify partners of their own species. Mate finding is facilitated by the fact that most of the species (also *E. paraensei* and *E. caproni*) select a small section of the relatively large intestinal tract. Chemical communication seems also to be important. Several studies showed that *in vitro* the worms find each other and that there must be chemical communication among the individuals (reviewed by Bone, 1982; Fried, 1986; Haseeb and Fried, 1988). In echinostomes creeping movements *in vitro* towards conspecifics were reported for *E. trivolvis* and *E. caproni* (Fried *et al.*, 1980; Fried, 1986; Fried and Haseeb, 1990). The worms of the two *Echinostoma*-species approached also the individuals of the other species, but the interspecific attraction seemed to be weaker than the intraspecific (Fried and Haseeb, 1990). The compounds which stimulate the worm-to-worm attraction in ESP of *E. trivolvis* were found to be lipophilic and the lipid fraction containing the free sterols. Cholesterol, however, which was the major free sterol in this fraction, had no effect as pure chemical (Fried *et al.*, 1980); the sterol fraction of *E. trivolvis* extracts contained nine additional sterols (Chitwood *et al.*, 1985). More information is needed on the nature of the attracting components and on the orientation mechanisms guiding the worms together. It was argued that helminths during their migration within hosts may not necessarily employ chemotactic orientation (Sukhdeo and Mettrick, 1987; Sukhdeo and Sukhdeo, 1994), and other non-directed activation-dependent or contact-dependent mechanisms may be more important. Another big gap of knowledge relates to the behavior of hermaphroditic worms during copulation. The very few direct observations of copulation (review: Nollen, 1997) suggest complex behavior sequences.

4. Concluding Remarks

This review shows that our knowledge on the behavioral patterns and their functions in the life-cycles of echinostomes and other digeneans is still very fragmentary. However, the available information indicates that the behaviors of the different stages in the life-cycles are determining factors for the fitness of the species and that they seem to be important determinants in the evolution of the life-cycles. When the different behavioral responses are considered, one may doubt that the parasitic digeneans should display a lower array of behavior responses than their free-living turbellarian relatives. Of course, it may be true that the behavior of the endoparasitic rediae, metacercariae, and adult worms may be simpler than those of their free-living turbellarian relatives, although they also must select their appropriate habitats, escape competing species, find their feeding sites and mating partners, and successfully mate. They may fulfill these tasks with simpler behavior responses to fewer cues, as they can make use of the extremely homeostatic and predictable conditions within the vertebrate and molluscan hosts. However, many digenean species produce, in addition to the endoparasitic stages, free-living miracidia and cercariae which are optimized to survive in the aquatic habitats and to find and invade two, often fully different, host types with complex behavior sequences. When all the behavioral patterns in a digenean life cycle are taken together, it seems possible that the digenean species base on similar numbers of genetically fixed behavioral responses to environmental and biotic cues that the free-living species.

It is self-evident that the behavior of parasitic worms merits attention for such topics as transmission dynamics, parasite-host relationships, and evolution of parasitism. Helminth behavior may even offer new approaches for transmission control, for example, by disguise of miracidia and cercariae (Haas et al., 1998).

Table 1. Orientation of echinostome miracidia and cercariae towards light and gravity during swimming (modified after Haas, 1994). In experiments with cercariae light sources were positioned lateral, above or below the organisms. In experiments with miracidia sections of chambers were darkened. Conditions of natural day-light-intensity and temperature were simulated. Abbreviations: pos, positive; neg, negative; -, no response; [], weak response or effect.

	Hosts	Photo-orientation		Geo-orientation		Movement in water column	Ref
		Horizontal	Vertical	Type	Inducer		
Miracidia							
Echinostoma trivolvis	Helisoma trivolvis	pos	pos	neg	not light	up	1
Echinostoma caproni	Biomphalaria sp.	pos	pos	neg	not light	up	2
Echinostoma paraensei	Biomphalaria glabrata	[pos]	[pos]	neutral	-	up, down	3
Cercariae							
Isthmiophora melis	Amphibians	neg	?	neg	not light	up	4
Echinostoma revolutum	Amphibians, freshwater snails	neg	[neg]	neg [+pos][a]	light	up [down]	5
Pseudechinoparyphium echinatum	Freshwater snails	neg, pos[a]	neg, pos[a]	pos [+neg][a]	light	down [up]	5
Hypoderaeum conoideum	Freshwater snails	neg, pos[a]	[neg]	neg [+pos][a]	light	up [down]	5

[a] Inserted periods of the reverse orientation type or response depends on light intensity. References: (1) Nollen, 1994; (2) Behrens and Nollen, 1992; (3) Meece and Nollen, 1996 (4) Motzel, unpublished results; (5) Loy and Haas, unpublished results.

Table 2. The nature of chemical cues, which are supposed to stimulate chemo-orientation towards the hosts by echinostome miracidia and cercariae. ESP, excretory-secretory products; RCD, rate of change of direction; SCW, snail conditioned water. Snail species include *Biomphalaria alexandrina, Biomphalaria glabrata, Helisoma trivolvis* CO strain (from Colorado) and PA strain (from Pennsylvania), *Lymnaea elodes, Lymnaea palustris, Lymnaea stagnalis, Planorbarius corneus.*

Echinostome species	Type of response	Nature of stimulating chemical cues	Ref
Miracidia *Echinostoma trivolvis*	Accumulation in phi-chamber	SCW of *H. trivolvis* and the not buffered pure chemicals glutamic acid, aspartic acid, H_2SO_4, HCl, acetic acid; not leucine, $MgSO_4$	1
Echinostoma caproni	Accumulation in phi-chamber. Chemo-orientation: increase of RCD in increasing and turnback swimming in decreasing concentration gradients	SCW of *B. glabrata* and the not buffered pure chemicals glutamic acid, aspartic acid, H_2SO_4; not acetic acid, leucine, HCl, $MgSO_4$. Solely stimulating components of SCW of *B. glabrata, B. alexandrina, L. stagnalis*: defined fraction of macromolecular glycoproteins	2 3
Cercariae *Echinostoma trivolvis*	Accumulation in petri dish assay.	Dialysate of *B. glabrata* ESP. ESP, dialysate and lipophilic extracts of ESP of *B. glabrata, H. trivolvis, L. elodes.* Pure chemicals, 5 mM in agar plugs: 14 amino acids, glutathione, fructose, maltose, valeric acid; not 10 carbohydrates and butyric acid	4 5 6

Echinostoma caproni	Accumulation in petri dish assay. Chemo-orientation: turnback swimming in decreasing concentration gradients	ESP and their lipophilic extracts of *B. glabrata*, *H. trivolvis*, *L. elodes*; dialysate of ESP of *B. glabrata*, *H. trivolvis* (CO strain), but **not** of *H. trivolvis* (PA strain), *L. elodes*. The stimulus in *B. glabrata* SCW is contained in the hydrophilic fraction < 3 kDa, mainly in the amino acids. SCW of *B. glabrata*, *L. stagnalis*, *P. corneus*, and ESP of a species of fish and frog are similarly attractive when adjusted to the same total content of amino acids	5 7
Echinostoma revolutum, *Pseudechinoparyphium echinatum*	Accumulation in choice-chambers and chemo-orientation: turnback swimming in decreasing concentration gradients	SCW of 6 species of freshwater snails, ESP of 2 species of fish, 1 species of leech and frog. The full stimulating activity of SCW is achieved by its content of amino acids, urea and ammonia. Quantitative differences in the attraction of ESP of different animals depend on different total content of amino acids therein. Mixtures of various amino acids and their analogues and derivatives show specific sensitivities of the cercariae to particular characteristics of amino acids	8 9 10

| Hypoderaeum conoideum | Accumulation in choice-chambers and chemo-orientation: chemotaxis in increasing concentration gradients | SCW of L. stagnalis, L. palustris, not of R. ovata, B. glabrata. The stimulating components of SCW from L. stagnalis are small molecular peptides | 7, 8 |

References: (1) Nollen, 1994; (2) Behrens and Nollen, 1992; (3) Haberl et al., unpublished results; (4) Fried and King, 1989; (5) Fried et al., 1997; (6) Reddy et al., 1997a; (7) Körner, 1998; (8) Haas et al., 1995b; (9, 10) Körner and Haas, 1998a, 1998b.

Table 3. Chemotaxis, increase of the rate of change direction, and turnback swimming reactions of echinostome miracidia and cercariae.

Stage	A Chemotaxis	B Increase of RCD[a]	C Turnback swimming	Ref
Miracidia				
Echinostoma caproni	no	yes	yes	1
Cercariae				
Echinostoma caproni	no	no	yes	2
Echinostoma revolutum	no	no	yes	3
Pseudechinoparyphium echinatum	no	no	yes	3
Hypoderaeum conoideum	yes	no	no	3

References: (1) Haberl et al., unpublished results; (2) Körner, 1998; (3) Haas et al., 1995b. [a] Rate of change direction.

5. References

Ataev, G.L., Dubrovolskij, A.A., Fournier, A. and Jourdane, J. (1997) Migration and development of mother sporocysts of *Echinostoma caproni* (Digenea: Echinostomatidae). *Journal of Parasitology* **83**, 444-453.

Basch, P.F. (1991) Schistosomes: *Development, Reproduction, and Host Relations*, Oxford University Press, Oxford.

Becker, W. (1970) Untersuchungen über die aus der Muttersporocyste auswandernden Tochtersporocysten von *Schistosoma mansoni*. II. Die Wanderung dieser Stadien zur Mitteldarmdrüse. *Zeitschrift für Parasitenkunde* **34**, 226-241.

Behrens, A.C. and Nollen, P.M. (1992) Responses of *Echinostoma caproni* miracidia to gravity, light, and chemicals. *International Journal for Parasitology* **22**, 673-675.

Bone, L.W. (1982) Reproductive chemical communication of helminths. I. Platyhelminthes. International *Journal of Invertebrate Reproduction* **5**, 261-268.

Cable, R.M. (1972) Behaviour of digenetic trematodes. Zoological Journal of the Linnean Society **51**, 1-18.

Chapman, H.D. and Wilson, R.A. (1973) The propulsion of the cercariae of *Himasthla secunda* (Nicoll) and *Cryptocotyle lingua*. *Parasitology* **67**, 1-15.

Chitwood, D.J., Lusby, W.R. and Fried, B. (1985) Sterols of *Echinostoma revolutum* (Trematoda) adults. *Journal of Parasitology* **71**, 846-847.

Christensen, N.O. (1980) A review of the influence of host- and parasite-related factors and environmental conditions on the host-finding capacity of the trematode miracidium. *Acta Tropica* **37**, 303-318.

Christensen, N.O., Frandsen, F. and Roushdy, M.Z. (1980) The influence of environmental conditions and parasite-intermediate host-related factors on the transmission of *Echinostoma liei*. *Zeitschrift für Parasitenkunde* **63**, 47-63.

Combes, C., Fournier, A., Moné, H. and Théron, A. (1994) Behaviours in trematode cercariae that enhance parasite transmission: patterns and processes. *Parasitology* **109**, S3-S13.

Crompton, D.W.T. (1973) The sites occupied by some parasitic helminths in the alimentary tract of vertebrates. *Biological Reviews* **48**, 27-83.

Erasmus, D.A. (1972) *The Biology of Trematodes*, Edward Arnold Limited, London.

Evans, N.A. and Gordon, D.M. (1983) Experimental studies on the transmission dynamics of the cercariae of *Echinoparyphium recurvatum* (Digenea: Echinostomatidae). *Parasitology* **87**, 167-174.

Fairweather, I. and Halton, D.W. (1991) Neuropeptides in platyhelminths. *Parasitology* **102**, S77-S92.

Fraenkel, G.S. and Gunn, D.L. (1961). *The orientation of Animals. Kineses, Taxes and Compass Reactions*, 2nd edn., Dover Publications, New York.

Fried, B. (1986) Chemical communication in hermaphroditic digenetic trematodes. *Journal of Chemical Ecology* **12**, 1659-1677.

Fried, B. (1994) Metacercarial excystment of trematodes. *Advances in Parasitology* **33**, 91-144.

Fried, B. and Haseeb, M.A. (1990) Intraspecific and interspecific chemoattraction in *Echinostoma caproni* and *Echinostoma trivolvis* adults *in vitro*. *Journal of the Helminthological Society of Washington* **57**, 72-73.

Fried, B. and Diaz, V. (1987) Site-finding and pairing of *Echinostoma revolutum* (Trematoda) on the chick chorioallantois. *Journal of Parasitology* **73**, 546-548.

Fried, B. and King, W. (1989) Attraction of *Echinostoma revolutum* cercariae to *Biomphalaria glabrata* dialysate. *Journal of Parasitology* **75**, 55-57.

Fried, B. and Reddy, A. (1999) Effects of snail-conditioned water from *Biomphalaria glabrata* on hatching of *Echinostoma caproni* miracidia. *Parasitology Research* **85**, 155-157.

Fried, B., Frazer, B.A. and Reddy, A. (1997) Chemoattraction and penetration of *Echinostoma trivolvis* and *E. caproni* cercariae in the presence of *Biomphalaria glabrata*, *Helisoma trivolvis*, and *Lymnaea elodes* dialysate. *Parasitology Research* **83**, 193-197.

Graefe, G.W. and Burkert, D.G. (1972) Zur Lokomotionsmechanik von Diplostomatiden- und Echinostomatiden-Cercarien (Trematoda). *Zoologischer Anzeiger, Leipzig* **188**, 366-369.

Gustafsson, M. K. S. (1992) The neuroanatomy of flatworms. *Advances in Neuroimmunology* **2**, 267-296.

Haas, W. (1992) Physiological analysis of cercarial behavior. *Journal of Parasitology* **78**, 243-255.

Haas, W. (1994) Physiological analyses of host-finding behaviour in trematode cercariae: adaptations for transmission success. *Parasitology* **109**, S15-S29.

Haas, W. and Haberl, B. (1997) Host recognition by trematode miracidia and cercariae, in B. Fried and T.K. Graczyk (eds.), *Advances in Trematode Biology*, CRC Press, Boca Raton, Florida, pp. 197-227.

Haas, W., Gui, M., Haberl, B. and Ströbel, M. (1991) Miracidia of *Schistosoma japonicum*: approach and attachment to the snail host. *Journal of Parasitology* **77**, 509-513.

Haas, W., Haberl, B., Kalbe, M. and Körner, M. (1995a) Snail-host finding by miracidia and cercariae: chemical host cues. *Parasitology Today* **11**, 468-472.

Haas, W., Körner, M., Hutterer, E., Wegner, M. and Haberl, B. (1995b) Finding and recognition of the snail intermediate hosts by 3 species of echinostome cercariae. *Parasitology* **110**, 133-142.

Haas, W., Haberl, B., Kalbe, M. and Stoll, K. (1998) Traps for schistosome miracidia/cercariae, in I. Tada, S. Kojima and M. Tsuji (eds.), *IX International Congress of Parasitology, ICOPA IX*. Monduzzi Editore, Bologna, pp. 359-363.

Haberl, B. (1995) Schistosome miracidial host-finding: synopsis and molecular approach. Doctoral Thesis, Naturwissenschaftliche Fakultät II, University Erlangen-Nürnberg.

Haberl, B. and Haas, W. (1992) Miracidium of *Schistosoma mansoni*: a macromolecular glycoconjugate as signal for the behaviour after contact with the snail host. *Comparative Biochemistry and Physiology* **101A**, 329-333.

Haberl, B., Kalbe, M., Fuchs, H., Ströbel, M., Schmalfuss, G. and Haas, W. (1995) *Schistosoma mansoni* and *S. haematobium*: miracidial host-finding behavior is stimulated by macromolecules. *International Journal for Parasitology* **25**, 551-560.

Halton, D.W., Fairweather, I., Shaw, C. and Johnston, C.F. (1990) Regulatory peptides in parasitic platyhelminths. *Parasitology Today* **6**, 284-290.

Halton D.W., Shaw, C., Maule, A.G., Johnston, C.F. and Fairweather, I. (1992) Peptidergic messengers: A new perspective of the nervous system of parasitic platyhelminths. *Journal of Parasitology* **78**, 179-193.

Halton D.W., Shaw, C., Maule, A.G. and Smart, D. (1994) Regulatory peptides in helminth parasites. *Advances in Parasitology* **34**, 163-227.

Halton, D.W., Maule, A.G. and Shaw, C. (1997) Trematode neurobiology, in B. Fried and T.K. Graczyk (eds.), *Advances in Trematode Biology*, CRC Press, Boca Raton, Florida, pp. 345-382.

Halton, D.W. and Gustafsson, M.K.S. (1996) Functional morphology of the platyhelminth nervous system. *Parasitology* **113**, S47-S72.

Haseeb, M.A. and Fried, B. (1988) Chemical communication in helminths. *Advances in Parasitology* **27**, 169-207.

Holmes, J.C. (1973) Site selection by parasitic helminths: interspecific interactions, site segregation, and their importance to the development of helminth communities. *Canadian Journal of Zoology* **51**, 333-347.

Irwin, S.W.B. (1997) Excystation and cultivation of trematodes, in B. Fried and T.K. Graczyk (eds.), *Advances in Trematode Biology*, CRC Press, Boca Raton, Florida, pp. 57-86.

Jourdane, J. and Théron, A. (1987) Larval development: eggs to cercariae, in D. Rollinson and Simpson, A.J.G (eds.), *The Biology of Schistosomes. From Genes to Latrines*, Academic Press, London, San Diego, pp. 83-113.

Kalbe, M., Haberl, B. and Haas, W. (1996) *Schistosoma mansoni* miracidial host-finding: species-specificity of an Egyptian strain. *Parasitology Research* **82**, 8-13.

Kalbe, M., Haberl, B. and Haas, W. (1997) Miracidial host-finding in *Fasciola hepatica* and *Trichobilharzia ocellata* is stimulated by species-specific glycoconjugates released from the host snails. *Parasitology Research* **83**, 806-812.

Körner, M. (1998) Chemoorientierung von Cercarien zum Schneckenwirt: Analyse der Wirtsignale. Doctoral Thesis, Naturwissenschaftliche Fakultät II, University Erlangen-Nürnberg.

Körner, M. and Haas, W. (1998a) Chemo-orientation of echinostome cercariae towards their snail-hosts: amino acids signal a low host-specificity. *International Journal for Parasitology* **28**, 511-516.

Körner, M. and Haas, W. (1998b) Chemo-orientation of echinostome cercariae towards their snail-hosts: the stimulating structure of amino acids and other attractants. *International Journal for Parasitology* **28**, 517-

525.

Lackie, A.M. (1975) The activation of infective stages of endoparasites of vertebrates. *Biological Reviews* **50**, 285-323.

Lim, H.K. and Heyneman, D. (1972) Intramolluscan inter-trematode antagonism: a review of factors influencing the host-parasite system and its possible role in biological control. *Advances in Parasitology* **10**, 191-268.

MacInnis, A.J. (1965) Responses of *Schistosoma mansoni* miracidia to chemical attractants. *Journal of Parasitology* **51**, 731-746.

MacInnis, A.J. (1976) How parasites find hosts: Some thoughts on the inception of host-parasite integration, in C.R. Kennedy (ed.), *Ecological Aspects of Parasitology*, North-Holland Publications, Amsterdam, pp. 3-20.

Markum, B.A. and Nollen, P.M. (1996) The effect of light intensity on hatching of *Echinostoma caproni* miracidia. *Journal of Parasitology* **82**, 662-663.

Matovu, D.B. (1971) Reizphysiologische Untersuchungen zur Chemotaxis bei Cercarien (Trematoda, Digenea). Doctoral Thesis, Naturwissenschftliche Fakultät, University Würzburg.

McCarthy, A.M. (1990) Speciation of echinostomes: evidence for the existence of two sympatric sibling species in the complex *Echinoparyphium recurvatum* (von Linstow 1873) (Digenea: Echinostomatidae). *Parasitology* **101**, 35-42.

Meece, J.K. and Nollen, P.M. (1996) A comparison of the adult and miracidial stages of *Echinostoma paraensei* and *E. caproni*. *International Journal for Parasitology* **26**, 37-43.

Motzel, W. and Haas, W. (1985) Studies on the attachment response of *Isthmiophora melis* cercariae (Trematoda: Echinostomatidae). *Zeitschrift für Parasitenkunde* **71**, 519-526.

Nollen, P.M. (1983) Patterns of sexual reproduction among parasitic platyhelminthes. *Parasitology* **86**, 99-120.

Nollen, P.M. (1994) The hatching behavior of *Echinostoma trivolvis* miracidia and their responses to gravity, light and chemicals. *International Journal for Parasitology* **24**, 637-642.

Nollen, P.M. (1997) Reproductive physiology and behavior of digenetic trematodes, in B. Fried and T.K. Graczyk (eds.), *Advances in Trematode Biology*, CRC Press, Boca Raton, Florida, pp. 117-147.

Pax, R.A. and Bennett, J.L. (1991) Neurobiology of parasitic platyhelminths: possible solutions to the problems of correlating structure with function. *Parasitology* **102**, S31-S39.

Pax, R.A. and Bennett, J.L. (1992) Neurobiology of parasitic flatworms: how much "neuro" in the biology? *Journal of Parasitology* **78**, 194-205.

Pax, R.A., Day, T.A., Miller, C.L. and Bennett, J.L. (1996) Neuromuscular physiology and pharmacology of parasitic flatworms. *Parasitology* **113**, S83-S96.

Reddy, A. and Fried, B. (1996) *In vitro* studies on intraspecific and interspecific chemical attraction in daughter rediae of *Echinostoma trivolvis* and *E. caproni*. *International Journal for Parasitology* **26**, 1981-1085.

Reddy, A., Frazer, B.A. and Fried, B. (1997a) Low molecular weight hydrophilic chemicals that attract *Echinostoma trivolvis* and *E. caproni* cercariae. *International Journal for Parasitology* **27**, 283-287.

Reddy, A., Frazer, B.A., Fried, B. and Sherma, J. (1997b) Chemoattraction of *Echinostoma trivolvis* (Trematoda) rediae to lipophilic excretory-secretory products and thin layer-chromatographic analysis of redial lipids. *Parasite* **4**, 37-40.

Rohde, K. (1994) Niche restriction in parasites: proximate and ultimate causes. *Parasitology* **109**, S69-S84.

Saladin, K.S. (1979). Behavioral parasitology and perspectives on miracidial host-finding. *Zeitschrift für Parasitenkunde* **60**, 197-210.

Schmidt, K.A. and Fried, B. (1996) Emergence of cercariae of *Echinostoma trivolvis* from *Helisoma trivolvis* under different conditions. *Journal of Parasitology* **82**, 674-676.

Shiff, C.J. and Graczyk, T.K. (1994): A chemokinetic response in *Schistosoma mansoni* cercariae. *Journal of Parasitology* **80**, 879-883.

Shiff, C.J., Chandiwana, S.K., Graczyk, T., Chibatamoto, P. and Bradley, M. (1993). A trap for the detection of schistosome cercariae. *Journal of Parasitology* **79**, 149-154.

Shostak A.W. and Esch, G.W. (1990) Photocycle-dependent emergence by cercariae of *Halipegus occidualis* from *Helisoma anceps*, with special reference to cercarial emergence patterns as adaptations for transmission. *Journal of Parasitology* **76**, 790-795.

Smyth, J.D. and Halton, D.W. (1983) *The Physiology of Trematodes*. 2nd edn. Cambridge: Cambridge University Press, Cambridge.

Sommerville, R.I. and Rogers, W.P. (1987) The nature and action of host signals. *Advances in Parasitology* **26**, 239-293.

Stirewalt, M.A. (1971) Penetration stimuli for schistosome cercariae, in T.C. Cheng, (ed.), *Aspects of the Biology of Symbiosis*, University Park Press, Baltimore, Maryland, pp. 1-23.

Sukhdeo, M.V.K. (1992) The behavior of parasitic flatworms *in vivo*: what is the role of the brain? *Journal of Parasitology* **78**, 231-242.

Sukhdeo, M.V.K. and Bansemir, A.D. (1996) Critical resources that influence habitat selection decisions by gastrointestinal helminth parasites. *International Journal for Parasitology* **26**, 483-498.

Sukhdeo, M.V.K. and Mettrick, D.F. (1987) Parasite behaviour: understanding platyhelminth responses. *Advances in Parasitology* **26**, 73-144.

Sukhdeo, M.V.K. and Sukhdeo, S.C. (1994) Optimal habitat selection by helminths within the host environment. *Parasitology* **109**, S41-S55.

Thomas, J.D. and Eaton, P. (1993) Amino acid medleys of snail origin as possible sources of information for conspecifics, schistosome miracidia and predators. *Comparative Biochemistry and Physiology* **106C**, 781-796.

Toledo, R., Munoz-Antoli, C. and Esteban J.G. (1999) Production and chronobiology of emergence of the cercariae of *Euparyphium albuferensis* (Trematoda: Echinostomatidae). *Journal of Parasitology* **85**, 263-267.

Ulmer, M.J. (1971) Site-finding behaviour in helminths in intermediate and definitive hosts, in A.M. Fallis, (ed.), *Ecology and Physiology of Parasites*, University of Toronto Press, Toronto, pp. 123-60.

Xu, Y.Z. and Dresden, M.H. (1990) The hatching of schistosome eggs. *Experimental Parasitology* **70**, 236-249.

PHYSIOLOGY AND BIOCHEMISTRY OF ECHINOSTOMES

JOHN BARRETT

Institute of Biological Sciences, University of Wales, Aberystwyth
Ceredigion, SY23 3DA, Wales, United Kingdom

B. Fried and T.K. Graczyk (eds.),
Echinostomes as Experimental Models for Biological Research, 199–212.
© 2000 *Kluwer Academic Publishers. Printed in the Netherlands.*

1. Introduction

In contrast to *Schistosoma* from the blood stream and *Fasciola* in the bile duct, little is known about the physiology and biochemistry of intestinal flukes, and even less about the metabolism of the intramolluscan stages. In looking at the biochemistry of helminths we are interested first, in the way that the metabolism of the parasite differs from that of its host and secondly, how the metabolism of the parasitic forms differs from that of their free-living relatives. Differences between host and parasite represent possible sites for chemotherapy, whilst the differences between parasitic and free-living forms may give some insight into the molecular basis of parasitism.

Echinostomes are an ideal model for studying the biochemistry and physiology of intestinal flukes: they are large worms (5-10 mg), the life cycle is relatively easy to maintain, they can be cultured *in vitro* and on egg chorioallantoic membranes as well as being transplanted between hosts (Jaw and Lo, 1974; Huffman and Fried, 1990; Loker *et al.*, 1999).

2. Synthetic Reactions

Compared to catabolic pathways synthetic reactions have been relatively little studied in parasitic helminths. The specific activities of synthetic enzymes are often very low and synthetic pathways are usually under tight metabolic control.

2.1. CARBOHYDRATES

The main storage carbohydrate in parastic helminths is glycogen, and values in digeneans range from 3% of the dry weight in *Schistosoma mansoni* females to 21% in *Fasciola hepatica*. In preovigerous *E. trivolvis* glycogen was 45% of the dry weight, the glycogen content of older, ovigerous worms ranged from 17 to 37% (Sleckman and Fried, 1984). In *E. trivolvis* glycogen occurs mainly in the musculature and the parenchyma. The main substrate for glyconeogenesis in parasitic helminths are dietary hexoses and although parasitic helminths have all the enzyme systems necessary to carry out glyconeogenesis from glycolytic or tricarboxylic acid cycle intermediates, there is little evidence that they do so. Glycogen resynthesis from glucose has been shown in *E. trivolvis* (Fried and Kramer, 1963). The synthesis of glycogen from glucose in helminths is similar to that in mammals, starting with glucose-6-phosphate and proceeding via UDP-glucose. The physical and chemical properties of helminth glycogen, such as the lengths of the external and internal branches are similar to those of mammalian glycogen.

2.2. LIPIDS

Compared with other helminths, digeneans contain relatively little lipid. Adult *E. trivolvis* contain approximately 15% of the dry weight as lipid (Fried and Boddorf,

1978). The main neutral lipids identified were free-sterols, with lesser amounts of triacylglycerols, sterol esters and free-fatty acids; carotene and lutein have also been identified. Phospholipids have been identified in *E. trivolvis* by TLC, the major fractions being phosphatidylcholine, phosphatidylethanolamine and phosphatidylserine (Barrett *et al.*, 1970; Fried and Shapiro, 1979; Fried *et al.*, 1980; Fried *et al.*, 1993a, 1993b; Lee *et al.*, 1998). A similar range of lipids was found in *E. malayanum* (Yusufi and Siddiqi, 1976). In *E. trivolvis* sterols comprise 15.8% of the total lipids, this is mostly cholesterol with a number of minor components (Chitwood *et al.*, 1985). The lipid composition of the rediae of *E. trivolvis* has also been analysed and is not appreciably different from the adult (Fried *et al.*, 1993a). There is evidence from *E. caproni* that host diet can influence lipid composition (Frazer *et al.*, 1997). Lipid droplets are found in the parenchyma and in and around the excretory canals (Fried and Morrone, 1970; Butler and Fried, 1977). A feature of lipid distribution in parasitic helminths, including digeneans is that much of the lipid in the egg occurs in the space between the embryo and the eggshell. When the eggs hatch, the lipid is discarded.

Parasitic helminths, in general, readily incorporate labelled precursors into all of their neutral and phospholipid classes, although different lipids become labelled to different extents, reflecting different intrinsic rates of turnover (Frayha and Smyth, 1983). In the few cases where the pathways of lipid synthesis have been studied in helminths, they appear to be similar to those found in mammals, for example the phosphatidic diacylglycerol pathway for triacylglycerol synthesis and the synthesis of phosphatidylcholine from GDP-choline and 1,2-diacylglycerol.

However, despite being able to synthesize complex lipids from simple precursors, parasitic helminths all appear unable to synthesize long chain fatty acids or the steroid nucleus *de novo*, nor can they desaturate preformed fatty acids. In mammals, fatty acid desaturation requires molecular oxygen and involves a microsomal oxidase.   In helminths fatty acid synthesis is limited to simple chain lengthening by the sequential addition of acetyl CoA. The mechanism in helminths is unknown, but may be similar to the mammalian mitochondrial fatty acid elongation system. Parasitic helminths, therefore, have a nutritional requirement for long chain, unsaturated fatty acids and sterols. Interestingly free-living platyhelminths appear to be unable to synthesise long chain, unsaturated fatty acids *de novo*, and one of the products which acoel turbellaria get from their algal symbionts are polyunsaturated fatty acids (Meyer *et al.*, 1970).

Steriod synthesis, like fatty acid desaturation, also has an absolute requirement for molecular oxygen in the cyclization of squalene to lanosterol, the reaction being catalysed by a microsomal oxidase system. Sterol synthesis has been shown to be absent in *E. revolutum* (Barrett *et al.*, 1970). Parasitic helminths seem to lack microsomal oxidase systems (Precious and Barrett, 1989). However, the block to sterol synthesis in helminths appears to be in squalene formation and not in its cyclization. Parasitic helminths, including schistosomes can metabolise dietary sterols to other derivatives and are also able to synthesize a range of polyisoprenoids. The inability to synthesize sterols, like the inability to synthesize long chain, unsaturated fatty acids may be a general feature of platyhelminths and not just a parasitic adaptation.

2.3. PROTEINS AND AMINO ACIDS

Helminths are capable of rapid protein synthesis and the incorporation of labelled amino acids into proteins has been demonstrated in a wide variety of species. Although not investigated in detail, protein synthesis in helminths seems similar to that in mammals and other organisms. Cell free systems capable of synthesizing proteins have been prepared from helminths and helminth *m*RNA has been translated in a number of heterologous systems. *In vivo,* proteins are constantly turning over, with half-lives ranging from minutes to several months. The protein content of *E. trivolvis* was found to be between 26-32% (Sleckman and Fried, 1984) and electrophoretic studies have been carried out on the water-soluble fractions of this and other echinostomes (Vassilev *et al.*, 1978, 1984). The protein patterns of the excretory/secretory products have also been analysed (Trouvé and Coustau, 1998). Chemical analyses are also available for *E. malayanum* (Haque and Siddiqi, 1982). Haemoglobin like proteins have been described from *E. trivolvis* and *E. caproni*, but they have not been characterized in any detail (Lee and Smith, 1965; Taft and Fried, 1968; Ross *et al.*, 1989). The distribution of acetylcholine esterase, which has been widely used as a marker for the nervous system has been extensively studied in adult and larval echinostomes (Nizami *et al.*, 1977; Krishna, 1981; Fried *et al.*, 1984; Zdarska and Nasincova, 1985; Krishna and Shimha, 1987).

Recently a wide range of neuropeptides have been demonstrated in the nervous systems of helminths. These peptides show cross reactivity with the vertebrate neuropeptide Y family, the gastrin/cholecystokinin family and the invertebrate FMRFamide family, with the latter predominating (Brownlee *et al.*, 1996). A range of such peptides has been found in echinostomes (see chapter 11).

Amino acids occur in tissues either as free-amino acids or as components of proteins and peptides. There have been numerous studies on the total amino acid composition of different helminths and in practically all cases the majority of the standard amino acids have been found. In general there are usually no qualitative differences between the total amino acid composition of different species and total amino acid composition has not proved particularly useful as an aid to taxonomy.

Compared with the protein amino acids, the free amino acid fraction in tissues is small, ranging from 0.1-2.6% of the fresh weight (Barrett, 1991). The free amino acid pool of *E. trivolvis* consists primarily of alanine, proline, serine and methionine (Bailey and Fried, 1977). Amino acid synthesis is a relatively neglected area of helminth biochemistry and the information available is often of uneven or uncertain quality. In addition to the ten essential amino acids for mammals (arginine, histidine, isoleucine, lysine, methionine, phenylalanine, threonine, tryptophan, valine) helminths probably also require a dietary source of tyrosine. Helminths seem to be able to synthesize the non-essential amino acids (alanine, aspartate, asparagine, cysteine, glutamate, glutamine, glycine, proline, hydroxyproline, serine) by pathways similar to those in mammals.

2.4. PURINES, PYRIMIDINES AND NUCLEIC ACIDS

Helminths, like other organisms, can synthesize DNA and RNA, often at high rates. However, there have been no detailed studies on nucleic acid synthesis in helminths. No novel bases have been described from parasitic helminths. The genomes of platyhelminths are AT rich, relatively small with up to 30% repeat sequences. Parasitic platyhelminths seem to have more complex genomes (in terms of their information content) than their free-living relatives (Searcy, 1970), this presumably reflects the complex nature of most parasite life cycles.

Purines and pyrimidines are precursors of DNA and RNA as well as a wide variety of other compounds including NAD^+, $NADP^+$ ATP, GTP, UTP, CTP and ITP. Adult *S. mansoni* are unable to synthesize purines *de novo* and rely entirely on exogenous purines (especially hypoxanthine) and salvage pathways (Coles, 1984). It was originally assumed that *de novo* pyrimidine synthesis was also lacking in parasitic helminths. However, *de novo* pyrimidine synthesis has been demonstrated in *S. mansoni,* where the ratio of *de novo* pyrimidine synthesis to salvage pathway is 1:2.

Like other synthetic pathways, purine and pyrimidine biosynthesis is under tight metabolic control and activity may be inhibited by exogenous nucleotides. Little is known about synthetic pathways in the free-living stages of digeneans. It is possible that in synthesis (as in catabolism) pathways that appear to be absent in the adult may function at other stages of the life cycle.

3. Catabolic Reactions

3.1. CARBOHYDRATE CATABOLISM

Adult parasitic helminths have an absolute dependency on carbohydrate as their energy source (either in the form of glucose or glycogen). In adult helminths there is no beta-oxidation of fatty acids and no significant catabolism of amino acids, nor is there any evidence for the co-fermentation of carbohydrate with amino acids or fatty acids. Characteristically, adult helminths break down carbohydrates to reduced organic acids or occasionally alcohols, which are then excreted (Barrett, 1984). The pathways of carbohydrate catabolism are essentially anaerobic, and on the basis of the end products produced parasites are divided into two broad groups (Barrett, 1981). First there are those that rely essentially on glycolysis and produce as end products of carbohydrate breakdown lactate or some other derivative of pyruvate (as in schistosomes). Secondly there are those that 'fix' carbon dioxide and whose primary end products of carbohydrate catabolism are succinate and pyruvate. These acids are usually further metabolised to short chain fatty acids such as acetate, propionate, 2-methylvalerate and 2-methylbutyrate. Helminths often show marked differences in the proportions of different end products produced under aerobic and anaerobic conditions.

Adult *E. caproni* catabolise glucose primarily to n-valerate, acetate, propionate, butyrate, n-hexanoate, lactate and succinate with traces of 2-methylbutyrate (Schaeffer *et al.*, 1977). Organic acid production is qualitatively similar under aerobic and anaerobic conditions, but quantitatively more organic acids are produced under anaerobic conditions suggesting a Pasteur effect. A number of glycolytic (phosphoglucomutase, phosphoglucose isomerase, lactate dehydrogenase), tricarboxylic acid cycle (isocitrate dehydrogenase, malate dehydrogenase, succinate dehydrogenase, fumarase) and associated enzymes (malic enzyme, mannose-phosphate isomerase, glucose-6-phophate dehydrogenase, 6-phosphogluconate dehydrogenase, alpha-glycerophosphate dehydrogenase, octopine dehydrogenase) have been demonstrated in echinostomes (Le Flore *et al.*, 1984; Gorchilova and Kanev, 1984; Kristensen and Fried, 1991; Ross *et al.*, 1989; Schaefer *et al.*, 1977; Sloss *et al.*, 1995; Voltz et al., 1985; Voltz *et al.*, 1988; Wright *et al.*, 1979). There have, however, been no isotopic studies on the pathways concerned.

The main energy store in digenean miracidia and cercariae is glycogen and they appear to be primarily aerobic (although they may be able survive anaerobically for varying periods of time). The intramolluscan stages of trematodes are facultative anaerobes, their metabolism seems to be different from that of cercariae and miracidia and may, to a certain extent, resemble that of the adult. However, unlike adult digeneans, sporocysts and rediae can catabolise both lipids and amino acids (Barrett, 1977).

3.2. CYTOCHROME CHAINS

Despite the essentially anaerobic nature of carbohydrate catabolism in parasitic helminths, they all utilize oxygen when it is available. That is, at least in air, they have a measurable oxygen consumption, and where investigated in detail they have all been shown to possess cytochromes and to be capable of oxidative phosphorylation. Oxygen consumption has been measured in adult *E. trivolvis* and *E. malayanum* and there is indirect evidence for the presence of cytochrome oxidase (Taft and Fried, 1968; Nizami and Siddiqi, 1982; Fujino *et al.*, 1995).

There are several reports of carbon monoxide reactive *b* type cytochromes (cytochrome *o*) and cyanide insensitive terminal oxidases occurring in helminth mitochondria.   However, the occurrence and physiological significance of these alternative oxidases, including cytochrome c peroxidase are still unclear at the molecular level.

3.3. LIPID CATABOLISM

Complex lipids are broken down by lipases to give fatty acids, glycerol and the various phospholipid polar groups.   In aerobic organisms the released fatty acids are then oxidized by the beta-oxidation sequence.   Non-specific esterases, capable of hydrolysing lipids have been demonstrated in a wide range of helminths, including echinostomes (Fried et al., 1984; Siddiqi et al., 1985; Sloss et al., 1995).   However, there is no good evidence that any adult parasitic helminth can catabolise its lipid stores, nor is there any evidence for a functional beta-oxidation sequence.

In contrast to the adult trematode, there is evidence that some cercariae and possibly miracidia can catabolise their endogenous lipid stores (Barrett, 1977).   However, there was no evidence for lipid utilization by the cercariae of E. trivolvis (Fried et al., 1998).

The absence of lipid catabolism in adult parasites poses the problem as to why they accumulate lipid if they are unable to catabolise it.   Some of the lipid in the adult may be destined for incorporation into the egg, since miracidia unlike the adults may be able to catabolize lipids.   Alternatively parasites may have to absorb large amounts of lipid to accumulate enough of a particular fatty acid or fat-soluble vitamin and the excess is stored rather than being excreted again.   Changes in lipid content in parasites is often correlated with the onset of the host's immune response and lipid accumulation may be related to the retention of potentially immunogenic compounds.

3.4. AMINO ACID CATABOLISM

Amino acids are not a significant energy source in parasitic helminths and only in digenean rediae and sporocysts is there evidence for appreciable amino acid catabolism (Barrett, 1991).   However, parasitic helminths all appear able to catabolise amino acids by pathways that appear to be identical with those found in mammals.   A number of enzymes (glutamate-oxaloacetic transaminase, glutamate-pyruvate transaminase, glutamate dehydrogenase) involved in amino acid metabolism have been demonstrated in echinostomes (Le Flore et al., 1984; Siddiqui and Siddiqi, 1990).   Despite the absence of microsomal oxidases parastic helminths have been shown to be able to carry out a number of oxidative reactions associated with amino acid metabolism including cysteine dioxygenase, proline hydroxylase, tyrosine hydroxylase and tryptophan hydroxylase.   Biogenic amines have been demonstrated in several species of echinostomes (Shishov and Kanev, 1986).

4. Transport Mechanisms

The lack of what appear to be a number of basic synthetic pathways in parasitic helminths has led to the development of extremely efficient uptake mechanisms. In digeneans there are two sites of uptake, the tegument and the gut. A survey of transport mechanisms in digeneans (principally *Fasciola* and *Schistosoma*) would suggest that glucose is taken up primarily through the tegument via an active mediated transport mechanism, while amino acids and short chain fatty acids are taken up by a mixture of mediated transport and diffusion. Purines and pyrimidines are taken up by diffusion alone. Sporocysts take up amino acids by passive diffusion, thymidine and adenine uptake has also been described in sporocysts but the mechanism is unknown (Barrett, 1981).

Many invertebrates, including the monogenean *Diclidophora merlangi*, appear to be able to absorb amino acids from sea water through their general body surface (Halton, 1978). The tegumental porter systems of parasitic platyhelminths may thus represent an elaboration of mechanisms present in free-living platyhelminths.

Several possible digestive enzymes have been demonstrated in echinostomes including proteases, aminopeptidases, esterases, lipases, nucleosidases and glucuronidases (Fried *et al.*, 1984; Siddiqui and Nizami, 1985; Siddiqi *et al.*, 1985; Mueller and Fried, 1999). A number of these enzymes may of course have an intracellular rather than digestive role. Acid and alkaline phosphatases occur widely in echinostomes and probably have a variety of functions including transport and digestion (Nizami *et al.*, 1975; Gorchilova and Kanev, 1984; 1994; Fried *et al.*, 1984; Siddiqui *et al.*, 1985).

5. Osmoregulation and Excretion

It is convenient to consider osmoregulation and excretion together as the same organs are often involved in both. Some 90% of excretory nitrogen comes from the alpha amino group of amino acids, the remainder from the breakdown of purines and pyrimidines. In digeneans, as in other parasitic helminths the major excretory product is ammonia, with small quantities of urea, uric acid and amino acids. There is no evidence for a functional urea cycle in parasitic helminths. The significance of amino acid excretion is uncertain. In many invertebrates and possibly helminths as well, amino acids are involved in regulating intracellular osmotic pressure. Much of the amino nitrogen excreted by digeneans probably comes from the partial products of digestion, but some may not be a normal excretory product, but the result of leakage from moribund worms. At least some of the amino acids excreted by digeneans are excreted via the protonephridial system and may represent true excretory products, and some digeneans in particular excrete large amounts of proline (Isseroff *et al.*, 1983).

It is possible that excreted amino acids may also have a signalling function.

Digeneans often excrete significant amounts of lipid, mostly via the protonephridial system. In *F. hepatica* lipid excretion can amount to as much as 2% of the wet weight per day. All classes of lipid appear to be excreted suggesting a general loss of lipid rather than a specialized secretion (Barrett, 1981). Adult *E. trivolvis* excrete mainly free sterols together with free-fatty acids (Fried and Appel, 1977; Fried et al., 1980; Gallo and Fried, 1984; Bennett and Fried, 1983; Rivas et al., 1998). The released lipids are also probably involved in chemical communications both in the adult and larval stages (Fried, 1986; Haseeb and Fried, 1988; Reddy et al., 1997). Echinostome ES products also affect the behavior of mollusc hemocytes (Noda, 1992; Loker et al., 1992; De Graffe and Loker, 1998; see also chapter 8).

Digeneans appear to be osmoconformers, swelling in dilute solutions and shrinking in concentrated ones, water and ions passing freely across the tegument. Adult digeneans show little if any volume regulation, however, *in vitro* they seem to be able to withstand hyper-osmotic conditions better than hypo-osmotic ones. Nothing is known about osmoregulation in larval digeneans, but during their life cycle they may have to cope with considerable variations in osmotic pressure, for example passing from the tissues of a mammal, to fresh water, to the tissues of a freshwater mollusc and back again.

6. Metabolic Control and Integration

In digeneans, as in other parasitic helminths, infection of the final host involves metabolic changes. During infection metabolic pathways are lost, new ones appear and enzymes are replaced by isoenzymes adapted to the new environmental conditions. New enzymes could be synthesized prior to infection, or they could be synthesized *de novo* at infection, $mRNA$ being transcribed from DNA; alternatively stable $mRNA$ could be involved. The infective stages of helminth parasites usually posses at least some of the enzyme systems characteristic of the next stage. So most post-infection changes will tend to involve the loss or reduction of enzyme systems, rather than the acquisition of new pathways. Vernberg (1961) has shown that larval digeneans are 'preadapted' to the body temperature of their next host and temperature change seems to be an important stimulus in initiating the metabolic switches associated with infection in parasites.

In addition to temperature, the pH, rH, pO_2, pCO_2 and osmotic pressure can all change during infection and invading stages are exposed to a range of host metabolites. Although many of these factors have been shown to be involved in initiating infection, it is not clear if the same stimuli are responsible for causing the switches in metabolic pathways. So far only temperature has been shown to have any direct effect on the metabolism of infective helminth stages (Barrett, 1968).

A final question is whether environmental stimuli, such as increase in temperature or change in pCO_2 affect infective stages via sense organs and the nervous and hormonal systems or whether they act directly on the target enzymes. An increase in pCO_2 for example might favour an increase in carboxylation reactions, whilst an increase in temperature could modify the behaviour of membrane bound enzymes or activate isoenzymes with different temperature characteristics.

7. Summary

In general, the biochemical information relating to echinostomes is very limited and they have not been extensively used in biochemical investigations, despite their obvious advantages. The development of new drugs requires model systems that are easy to manipulate. There have been a number of studies on anthelmintics using echinostomes (Krotov *et al.*, 1972; Leger and Notteghem, 1975; Leger, *et al.*, 1974; Notteghem *et al.*, 1979, 1980; Franco *et al.*, 1988; Maurer *et al.*, 1996; Bogh *et al.*, 1996; Schmidt, 1998), but echinostomes have not been widely adopted as a model for drug screening.

The potential of echinostomes for biochemical research has not yet been exploited. They could be used to study the control of synthetic pathways and transport mechanisms as well as the metabolism of intermediate stages. One of the fundamental questions in parasite biochemistry is how is flux partitioned between different metabolic pathways and how does this change during the life cycle? The relative importance of different metabolic pathways vary at different stages of the life cycle. Some stages are aerobic, using both glycolysis and the tricarboxylic acid cycle, in addition there may be an active beta-oxidation sequence. Other life cycle stages of the same parasite can be essentially anaerobic relying on glycolysis alone, or glycolysis plus part of the TCA cycle (Barrett, 1981). Similarly, synthetic pathways may be up-regulated in some stages and down-regulated in others. Echinostomes would be ideal models with which to study how these changes are controlled and integrated.

During their life cycle, echinostomes have at least four major environmental changes (entry and exit from the mollusc host and entry and exit from the vertebrate host). Changes in gene expression at different life cycle stages can be monitored with genomic micro-arrays, while post-translational modifications could be characterised by proteome analysis. Genetic mapping would reveal whether the genes, whose expression is restricted to particular life cycle stages are grouped together on the chromosome in functional 'cassettes'. Stage specific genes may have stage specific response elements associated with them. The identification of stage specific response elements and the corresponding stage specific transcriptional factors would allow all of the genes which are uniquely expressed at any one stage to be identified and tagged.

8. References

Barrett, J. (1968) The effect of physical factors on the rate of respiration of the infective larvae of *Strongyloides ratti* Sandground, 1925. *Parasitology* **58**, 641-651.

Barrett, J. (1977) Energy metabolism and infection in helminths. *Symposia of the British Society for Parasitology* **15**, 121-144.

Barrett, J. (1981) *Biochemistry of Parasitic Helminths,* Macmillan, London.

Barrett, J. (1984) The anaerobic end-products of helminths. *Parasitology* **88**, 179-198.

Barrett, J. (1991) Amino acid metabolism in helminths. *Advances in Parasitology* **30**, 39-105.

Barrett, J., Cain, G.D. and Fairbairn, D. (1970) Sterols in *Ascaris lumbricoides* (Nematoda), *Macracanthorhynchus hirudinaceus* and *Moniliformis dubius* (Acanthocephala), and *Echinostoma revolutum* (Trematoda). *Journal of Parasitology* **56**, 1004-1008.

Bailey, R.S. and Fried, B. (1977) Thin-layer chromatography analyses of amino acids in *Echinostoma revolutum* (Trematoda) adults. *International Journal for Parasitology* **7**, 497-499.

Bennett, S. and Fried, B. (1983) Densitometric thin-layer chromatographic analyses of free-sterols in *Echinostoma revolutum* (Trematoda) adults and their excretory-secretory products. *Journal of Parasitology* **69**, 789-790.

Bogh, H.O., Andreassen, J. and Lemmich, J. (1996) Anthelmintic usage of extracts of *Embelia schimperi* from Tanzania. *Journal of Ethnopharmacology* **50**, 35-42.

Brownlee, D.J.A., Fairweather, I., Holden-Dye, L. and Walker, R.J. (1996) Nematode neuropeptides: localization, isolation and functions. *Parasitology Today* **12**, 343-351.

Butler, M.S. and Fried, B. (1977) Histochemical and thin-layer chromatographic analysis of neutral lipids in *Echinostoma revolutum* metacercariae cultured *in vitro*. *Journal of Parasitology* **63**, 1041-1045.

Chitwood, D.J., Lusby, W.R. and Fried, B. (1985) Sterols of *Echinostoma revolutum* (Trematoda) adults. *Journal of Parasitology* **71**, 846-847.

Coles, G.C. (1984) Recent advances in schistosome biochemistry. *Parasitology* **89**, 603-637.

De Graffe, G. and Loker, E.S. (1998) Susceptibility of *Biomphalaria glabrata* to infection with *Echinostoma paraensei:* correlation with the effect of parasite secretory-excretory products on host hemocyte spreading. *Journal of Invertebrate Pathology* **71**, 64-72.

Franco, J., Huffman, J.E. and Fried, B. (1988) The effects of crowding on adults of *Echinostoma revolutum* (Digenea, Echinostomatidae) in experimentally infected golden hamsters, *Mesocricetus auratus*. *Journal of Parasitology* **74**, 240-243.

Frayha, G.J. and Smyth, J.D. (1983) Lipid metabolism in parasitic helminths. *Advances in Parasitology* **22**, 310-387.

Frazer, B.A., Reddy, A., Fried, B. and Sherma, J. (1997) Effects of diet on the lipid composition of *Echinostoma caproni* (Trematoda) in ICR mice. *Parasitology Research* **83**, 642-645.

Fried, B. (1986) Chemical communication in hermaphroditic digenetic trematodes. *Journal of Chemical Ecology* **12**, 1659-1677.

Fried, B. and Appell, A.J. (1977) Excretion of lipids by *Echinostoma revolutum* (Trematoda) adults. *Journal of Parasitology* **63**, 447.

Fried, B. and Boddorf, J.M. (1978) Neutral lipids in *Echinostoma revolutum* (Trematoda) adults. *Journal of Parasitology* **64**, 174-175.

Fried, B. and Kramer, M.D. (1963) Histochemical glycogen studies on *Echinostoma revolutum*. *Journal of Parasitology* **54**, 942-944.

Fried, B. and Morrone, L.J. (1970) Histochemical lipid studies on *Echinostoma revolutum*. *Proceedings of the Helminthological Society of Washington* **37**, 122-123.

Fried, B. and Shapiro, I. L. (1979) Thin-layer chromatographic analysis of phospholipids in *Echinostoma revolutum* (Trematoda) adults. *Journal of Parasitology* **65**, 243-245.

Fried, B., Tancer, R B. and Fleming, S.J. (1980) *In vitro* pairing of *Echinostoma revolutum* (Trematoda) metacercariae and adults, and characterization of worm products involved in chemoattraction. *Journal of Parasitology* **66**, 1014-1018.

Fried, B., Le Flore, W.B. and Bass, H.S. (1984) Histochemical localization of hydrolytic enzymes in the cercaria and excysted metacercaria of *Echinostoma revolutum* (Trematoda). *Proceedings of the Helminthological Society of Washington* **51**, 140-143.

Fried, B., Rao, K.S., Sherma, J. and Huffman, J.E. (1993a) Fatty acid composition of *Echinostoma trivolvis* (Trematoda) rediae and adults and of the digestive gland-gonad complex of *Helisoma trivolvis* (Gastropoda) infected with the intramolluscan stages of this Echinostome. *Parasitology Research* **79**, 471-474.

Fried, B., Beers, K. and Sherma, J. (1993b) Thin-layer chromatographic analysis of beta-carotene and lutein in *Echinostoma trivolvis* (Trematoda) rediae. *Journal of Parasitology* **79**, 113-114.

Fried, B., Eyster, L.S. and Pechenik, J.A. (1998) Histochemical glycogen and neutral lipid in *Echinostoma trivolvis* cercariae and effects of exogenous glucose on cercarial longevity. *Journal of Helminthology* **72**, 83-85.

Fujino, T., Fried, B. and Takamiya, S. (1995) Cytochemical localisation of cytochrome *c* oxidase activity in mitochondria in the tegument and tegumental and parenchymal cells of the trematodes *Echinostoma trivolvis, Zygocotyle lunata, Schistosoma mansoni, Fasciola gigantica* and *Paragonimus ohirai. Journal of Helminthology* **69**, 195-201.

Gallo, G. J. and Fried, B. (1984) Association of particular systems with the release of neutral lipids in *Echinostomum revolutum* (Trematoda) adults. *Journal of Chemical Ecology* **10**, 1065-1069.

Gorchilova, L. and Kanev, I. (1984) Enzymic characteristics of the tegument and intestinal wall of an *Echinostoma* marita with 37 collar spines. *Khelmintologiya* **18**, 31-36.

Gorchilova, L. and Kanev, I. (1994) *Echinostoma echinatum* (Zeder, 1803): ultrastructure and enzymocytochemical characteristic of the tegument and of the gut wall. *Helminthologia* **31**, 133-138.

Halton, D.W. (1978) Trans-tegumental absorption of L-alanine and L-leucine by a monogenean *Diclidophora merlangi. Parasitology* **76**, 29-37.

Haque, M. and Siddiqi, H.H. (1982) The biochemical composition of digenetic trematodes. *Indian Journal of Parasitology* **6**, 37-41.

Haseeb, M.A. and Fried, B. (1988) Chemical communication in helminths. *Advances in Parasitology* **27**, 169-207.

Huffman, J.E. and Fried, B. (1990) *Echinostoma* and echinostomiasis. *Advances in Parasitology* **29**, 215-269.

Isseroff, H., Bock, K., Owczarek, A. and Smith, K.R. (1983) Schistosomiasis: proline production and release by ova. *Journal of Parasitolgy* **69**, 285-289.

Jaw, C.Y. and Lo, C.T. (1974) *In vitro* cultivation of *Echinostoma malayanum* Leiper 1911. *Chinese Journal of Microbiology* **7**, 157-164.

Krishna, G.V.R. (1981) Esterase location and nervous system in *Echinostoma revolutum. Indian Journal of Parasitology* **5**, 191-193.

Krishna, G.V.R. and Simha, S.S. (1987) Esterase activity in the rediae and cercariae of *Echinostoma revolutum* (Trematoda-Digenea) with notes on the nervous system. *Rivista di Parassitologia* **1**, 139-143.

Kristensen, A.R. and Fried, B. (1991) A comparison of *Echinostoma trivolvis* (Trematoda: Echinostomatidae) adults using isoelectric focusing. *Journal of Parasitology* **77**, 496-498.

Krotov, A.I., Bekhli, A.F., Nutraeva, K.S., Khalilov, A.G., Gusel'nikova, L.M., Vorob'eva, Z.G. and Bayandina, D.G. (1972) Study of the anthelmintic and molluscicidal activity of piperazine salt of phenasal as compared with phenasal. *Meditsinskaya Parazitologiya i Parazitarnye Bolezni* **41**, 341-346 (In Russian).

Lee, D.L and Smith, M.H. (1965) Hemoglobins of parasitic animals. *Experimental Parasitology* **16**, 392-424.

Lee, M.S., Fried, B. and Sherma, J. (1998) HPLC determination of neutral lipids in *Echinostoma caproni, Echinostoma trivolvis* and *Zygocotyle lunata* (Platyhelminthes: Trematoda). *Journal of Planar*

Chromatography **11**, 105-107.

Le Flore, W.B., Fried, B. and Bass, H.S. (1984) Histochemical localization of dehydrogenases in the cercaria and excysted metacercaria of *Echinostoma revolutum* (Trematoda). *Comparative Biochemistry and Physiology* **77B**, 31-33.

Leger, N. and Notteghem, M.J. (1975) Study of the flukicide activity of a new compound, brotianide, on *Echinostoma caproni* Richard, 1964. *Annales Pharmaceutiques Francaises* **33**, 273-277 (In French).

Leger, N., Notteghem, M.J. and Cavier, R. (1974) Pharmacodynamic test for the trial of antifluke agents. Preliminary results. *Bulletin de la Societe de Pathologie Exotique* **66**, 732-736 (In French).

Loker, E.S., Cimino, D.F. and Hertel, L.A. (1992) Excretory-secretory products of *Echinostoma paraensei* mediate interference with *Biomphalaria glabrata* hemocyte functions. *Journal of Parasitology* **78**, 104-115.

Loker, E.S., Coustau, C., Ataev, G.L. and Jourdane, J. (1999) *In vitro* culture of rediae of *Echinostoma caproni*. *Parasitology* **6**, 169-174.

Maurer, K., Decere, M. and Fried, B. (1996) Effects of the anthelmintics clorsulon, rafoxanide, mebendazole and arprinocid on *Echinostoma caproni* in ICR mice. *Journal of Helminthology* **70**, 95-96.

Meyer, F., Meyer, H. and Beuding, E. (1970) Lipid metabolism in the parasitic and free-living flatworms, *Schistosoma mansoni* and *Dugesia dorotocephala*. *Biochimica et Biophysica Acta* **210**, 257-266.

Mueller, T.J. and Fried, B. (1999) Electrophoretic analysis of proteases in *Echinostoma caproni* and *Echinostoma trivolvis*. *Journal of Parasitology* **85**, 174-180.

Nizami, W.A. and Siddiqi, H.H. (1982) Studies on the oxygen consumption of digenetic trematodes. *Revista Iberica de Parasitologia* **42**, 63-72

Nizami, W.A., Siddiqi, A H. and Yusufi, N.K. (1975) Non-specific alkaline phosphomonoesterases of eight species of digenetic trematodes. *Journal of Helminthology* **49**, 281-287.

Nizami, W.A., Siddiqi, A.H. and Islam, M.W. (1977) Quantitative studies on acetylcholinesterase in seven species of digenetic trematodes. *Zeitschrift für Parasitenkunde* **52**, 275-280.

Noda, S. (1992) Effects of excretory-secretory products of *Echinostoma paraensei* larvae on the hematopoietic organ of M-line *Biomphalaria glabrata* snails. *Journal of Parasitology* **78**, 512-517.

Notteghem, M.J., Leger, N. and Cavier, R. (1979) A study of the anthelmintic activity of flubendazole on *Echinostoma caproni* Richard, 1964. *Annales Pharmaceutiques Francaises* **37**, 153-156 (In French).

Notteghem, M.J., Leger, N. and Forget, E. (1980) Comparison of the fluke-killing activity of some compounds derived from benzimidazole. *Annales Pharmaceutiques Francaises* **38**, 61-63 (In French).

Precious, W.Y and Barrett, J. (1989) The possible absence of cytochrome P-450 linked xenobiotic metabolism in helminths. *Biochimica et Biophysica Acta* **992**, 215-222.

Reddy, A., Frazer, B.A., Fried, B. and Sherma, J. (1997) Chemoattraction of *Echinostoma trivolvis* (Trematoda) rediae to lipophilic excretory-secretory products and thin layer chromatographic analysis of redial lipids. *Parasite* **4**, 37-40.

Rivas, F., Sudati, J., Fried, B. and Sherma, J. (1998) HPTLC analysis of neutral lipids in the intestinal mucose, serum, and liver of ICR mice infected with *Echinostoma caproni* (Trematoda) and of worm excretory-secretory products. *Journal of Planar Chromatography* **11**, 47-50.

Ross, G.C., Fried, B. and Southgate, V.R. (1989) *Echinostoma revolutum* and *Echinostoma liei:* observations on enzymes and pigments. *Journal of Natural History* **23**, 977-982.

Schaefer, F.W., Saz, H.J., Weinstein, P.P. and Dunbar, G.A. (1977) Aerobic and anaerobic fermentation of glucose by *Echinostoma liei*. *Journal of Parasitology* **63**, 687-689.

Schmidt, J. (1998) Effects of benzimidazole anthelmintics as microtubule-active drugs on the synthesis and transport of surface glycoconjugates in *Hymenolepis microstoma, Echinostoma caproni* and *Schistosoma mansoni*. *Parasitology Research* **84**, 362-368.

Searcy, D.G. (1970) Measurement by DNA hybridisation *in vitro* of the genetic basis of parasite reduction.

Evolution **24**, 207-219.

Shishov, B.A. and Kanev, I. (1986) Aminergic elements in the nervous system of echinostomatids and philophthalmids. *Parazitologiia* **20**, 46-52 (In Russian).

Siddiqui, A.A. and Nizami, W.A. (1985) 5'-Nucleosidase activity and substrate affinity in digenetic trematodes. *Journal of Helminthology* **59**, 263-266.

Siddiqui, J. and Siddiqi, A.H. (1990) Studies on glutamic-oxalacetic (GOT) and glutamic-pyruvic (GPT) transaminases of some digenetic trematodes. *Indian Journal of Parasitology* **14**, 227-230.

Siddiqui, A.A., Siddiqi, A.H. and Haque, M. (1985) Acid phosphatases of six species of digenetic trematodes. *Indian Journal of Parasitology* **9**, 49-53.

Sleckman, B.P. and Fried, B. (1984) Glycogen and protein content in adult *Echinostoma revolutum* (Trematoda). *Proceedings of the Helminthological Society of Washington* **51**, 353-356.

Sloss, B., Meece, J., Romano, M. and Nollen, P. (1995) The genetic relationships between *Echinostoma caproni, Echinostoma paraensei* and *Echinostoma trivolvis* as determined by electrophoresis. *Journal of Helminthology* **69**, 243-246.

Taft, J. and Fried, B. (1968) Oxygen consumption in adult *Echinostoma revolutum* (Trematoda). *Experimental Parasitology* **23**, 183-186

Trouvé, S. and Coustau, C. (1998) Differences in adult excretory-secretory products between geographical isolates of *Echinostoma caproni*. *Journal of Parasitology* **84**, 1062-1065.

Vasilev, I., Komandarev, S., Mikhov, L. and Kanev, I. (1978) Comparative electrophoretic studies of certain species of genus *Echinostoma* with 37-collar spines. *Khelmintologiya* **6**, 31-38 (In Bulgarian).

Vassilev, I., Michov, L., Kanev, I. and Fried, B. (1984) A comparative electrophoretic study of adult worms considered in Europe and the USA as identical with *Echinostoma revolutum* (Frohlich, 1802). *Khelmintologiya* **17**, 10-15 (In Bulgarian).

Vernberg, W.B. (1961) Studies on the oxygen consumption of digenetic trematodes. VI. The influence of temperature on larval trematodes. *Experimental Parasitology* **11**, 270-275.

Voltz, A., Richard, J., and Pesson, B. (1985) A genetic comparison between natural and laboratory strains of *Echinostoma* (Trematoda) by isoenzymatic analysis. *Parasitology* **95**, 471-478.

Voltz, A., Richard, J., Pesson, B. and Jourdane, J. (1988) Isoenzyme analysis of *Echinostoma liei:* comparison and hybridization with other African species. *Experimental Parasitology* **66**, 13-17.

Wright, C.A., Rollinson, D. and Goll, P.H. (1979) Parasites in *Bulinus senegalensis* (Mollusca: Planorbidae) and their detection. *Parasitology* **79**, 95-105.

Yusufi, A.N.K. and Siddiqi, A.H. (1976) Comparative studies on the lipid composition of some digenetic trematodes. *International Journal for Parasitology* **6**, 5-8.

Zdarska, Z. and Nasincova, V. (1985). Histological and histochemical studies on the cercaria and redia of *Echinostoma revolutum*. *Folia Parasitologica* **32**, 341-347.

NEUROMUSCULATURE – STRUCTURE AND FUNCTIONAL CORRELATES

J. E. HUMPHRIES, A. MOUSLEY, A.G. MAULE and D.W. HALTON
Parasitology Research Group, School of Biology & Biochemistry, Medical Biology Centre, The Queen's University of Belfast, Belfast BT9 7BL, UK

1. Introduction

The adult and developmental stages of most trematodes, including the echinostomes, are highly motile and often undergo quite elaborate behavioural patterns, mediated it is assumed by well-developed sensory modalities and neuromuscular control systems for host attachment, invasion and migration. The application of immunocytochemistry and the use of the phalloidin-fluorescence staining method in conjunction with confocal scanning laser microscopy have revolutionised the provision of detailed information on neuronal pathways and muscle organisation, respectively, in trematodes and other flatworms (Halton and Gustafsson, 1996; Mair *et al.* 1998a). Moreover, motility studies *in vitro* have provided some preliminary data on the somatic motor systems of these parasites, with respect to the actions of known neuroactive substances (Pax *et al.*, 1996; Halton *et al.*,1997). Research on the neuromusculature of trematodes holds the prospect of identifying pharmacologically important targets that may be exploited in the development of novel anthelmintics in the near future. Since echinostomes, such as *Echinostoma caproni*, are readily maintained in the laboratory (Fried and Huffman, 1996), they would seem to offer admirable model material for experimental study of the developing neuromuscular systems in trematodes. This fact has only just begun to be exploited, and the chapter herein presents what is largely unpublished and preliminary data on the structure and functional correlates of the nerve-muscle systems of adult and the developmental stages of *E. caproni*. Previous reports of neuroactive substances in echinostomes include those of Thorndyke and Whitfield (1987), Richard *et al.* (1989), and Riddel *et al.* (1991)

2. Muscle System

The gross anatomical features of the muscle system in *E. caproni* are beginning to emerge from the use of the phalloidin fluorescence technique. Fluoroscein isothiocyanate (FITC) labelled phalloidin serves as a site-specific probe for filamentous actin and, when applied to whole-mount preparations of the worm and then examined by confocal scanning laser microscopy (CSLM), can be used to reveal the three-dimensional organisation of the major muscle systems (Figs. 1-4). Thus, the body wall has been shown to comprise outer

B. Fried and T.K. Graczyk (eds.),
Echinostomes as Experimental Models for Biological Research, 213–227.
© 2000 *Kluwer Academic Publishers. Printed in the Netherlands.*

circular, intermediate longitudinal, and inner diagonal myofibres in a well-developed and highly organised lattice-like arrangement over the entire body. The three muscle types form distinct muscle layers. The circular fibres are mostly orientated at right angles to the longitudinal body axis, but are skewed in the mid-body to converge at the posterior tip. The intermediate longitudinal myofibres run parallel to the main body axis; they are particularly pronounced in the subtegumental region as a well-defined band of parallel fibres (Fig. 2), but less ordered in the deeper layers of the worm. Two sets of diagonal fibres run at angles of approximately 50° and 140°, respectively, to the longitudinal fibres (Figs. 3,4). A comparable arrangement of somatic muscle fibres has been described for the body wall of the digenean trematode, *Fasciola hepatica* by Mair *et al.* (1998b) and are believed to serve in motility (see 4 below) and in the maintenance of body shape. Somatic musculature provides the main means of body movements. Thus, contraction of the circular and longitudinal muscles, respectively, would extend and shorten the worm, while contraction of diagonal fibres would enable side-to-side and twisting movements. Dorso-ventral muscle fibres are also in evidence in the worm and are likely to be involved in maintaining structural integrity.

The musculature of the oral (Fig. 1) and ventral suckers derives in part from the body wall musculature, in that it incorporates the longitudinal muscle fibres as meridional fibres which serve to open the sucker concavity, and circular fibres that become the equatorially-arranged fibres of the sucker, whose contraction presumably creates suction to draw in a plug of host tissue. A third set of separate muscle fibres runs between the inner and outer faces of the sucker as radial fibres whose action closes the sucker into a cup-shape.

Comparable information is not yet available on the muscle fibre arrangement of the alimentary and reproductive tracts in an echinostome trematode. However, it is likely that, as in *F. hepatica*, the ducting of these systems will show a predominance of circular muscles, with only relatively few longitudinal fibres (Mair *et al.* 1998b). Peristaltic contractions of circular fibres presumably serve to move gut contents for alimentation purposes, and to help transport the components of eggs to the ootype and, when formed, to convey them through the uterus.

Plate 1. Muscle fibre arrangement in the body wall (subtegument) of adult *Echinostoma caproni*, as visualised by FITC-phalloidin staining of whole-mount preparations and examined by confocal scanning laser microscopy (CSLM). Fig. 1. Forebody showing musculature of the oral sucker (os), together with the characteristic collar of spines (arrow). Fig. 2. Lateral margin of the worm showing the well-ordered arrangement of evenly spaced longitudinal myofibres (between arrows) that form a pronounced intermediate muscle layer in the subtegument in this region. Fig. 3. Mid-region of the worm showing circular (cm), longitudinal (lm) and two distinct bands of diagonal muscle fibres (dm) in lattice-like arrangement. Fig. 4. Detail of the muscle fibres in Fig. 3, showing outer circular (cm), intermediate longitudinal (lm) and the two sets of inner diagonal (dm) muscle fibres.

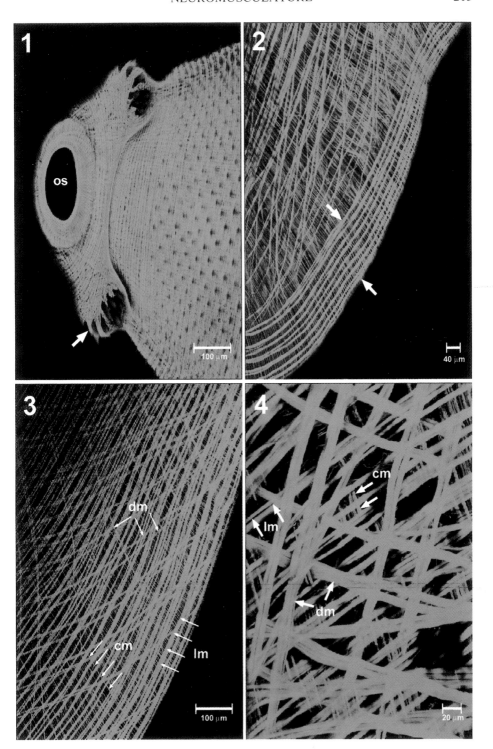

3. Nervous System

The basic organisation of the adult echinostome nervous system is typical of that of flatworm parasites (see review by Halton and Gustafsson, 1996). It is bilateral and differentiated into a central nervous system (CNS), consisting of a bilobed brain and fibrous commissure, from which emanate paired longitudinal nerve cords and associated cross connectives in a typical orthogonal arrangement; and a peripheral nervous system (PNS) that provides motor and sensory innervation to the musculature of the body wall, adhesive and feeding organs, and reproductive tract. The disposition of the longitudinal nerve cords is such that three pairs extend anteriorly from the brain to serve the oral sucker, pharynx and mouth region, whilst paired ventral, dorsal and lateral cords proceed posteriorly for the length of the body. Typically, the ventral pair are best developed as the main nerve cords and are fused posteriorly. All three nerve cords are cross-linked by transverse connectives and are in continuity with an extensive array of nerves that extend to the ventral sucker and somatic musculature. Here they divide and anastomose, giving rise to acetabular and subsurface plexuses, respectively. A similar arrangement of peripheral plexuses that are derived from the anterior nerve cords provide innervation to the muscle of the oral sucker and pharynx, and to the muscle associated with the copulatory apparatus and ducts of the reproductive system. The metacercarial and cercarial nervous systems are essentially similar to that of the adult but with the longitudinal nerve cords extending into the tail of the cercaria (Figs. 7,8). The nervous system of the redia is much less extensive than that of the cercaria and is concentrated largely in the anterior region, with paired cerebral ganglia, connecting commissure and nerve fibres extending down the body. The nervous system of the miracidium is also concentrated in the anterior portion of the body and is dominated by a single, large and compact ganglion from which nerve fibres extend the length of the body.

The application of immunocytochemical techniques has demonstrated substantial aminergic (serotonin, 5-hydroxytryptamine) and peptidergic immunoreactivities in both the central and peripheral nervous systems of *E. caproni* (Fig. 5). Controls employed usually consist of omission of primary antisera, use of nonimmune sera, and liquid-phase preincubation of the antisera with the appropriate antigen.

Plate 2. *Echinostoma caproni* immunostained and examined by CSLM to show the aminergic (5-HT) and peptidergic components of the nervous system. Fig. 5. Whole-mount of an adult worm dual labelled for 5HT (red, TRITC) and FaRPergic (green, FITC) immunoreactivities. Note the separate neuronal pathways and the relative sizes of aminergic (large arrow) and peptidergic neurons (small arrows). Fig. 6. Portion of the innervation of the egg chamber (ootype) immunostained to show it comprises aminergic neurons (green, arrow) and neurons immunoreactive for both 5-HT and FaRPs (yellow). Red structures are eggs. Fig. 7. Cercaria showing intense immunostaining for FaRPs (white) throughout the CNS (arrow). Note tail innervation (*). Fig. 4. Freshly excysted metacercaria showing little expression of FaRP-immunoreactivity (white) in the CNS (arrow).

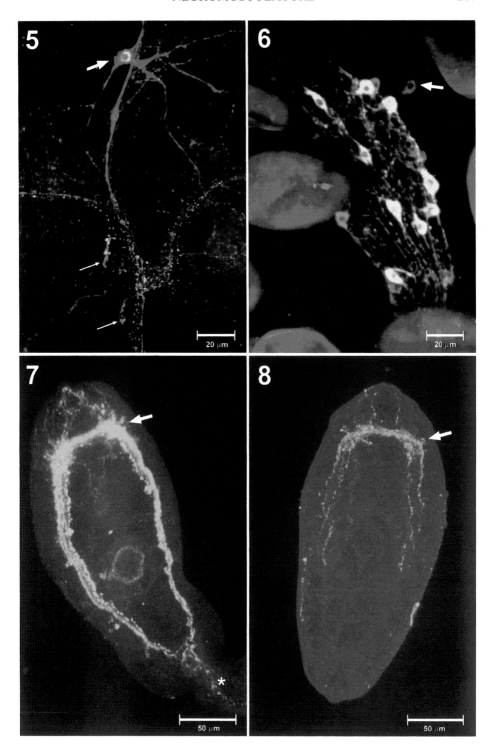

3.1. SEROTONINERGIC

Serotonin (5-HT)-immunoreactivity occurs in the nervous system of all of the stages of the life cycle studied.

3.1.1. *Adult*
In the adult worm, immunostaining for 5-HT has been shown to occur in the cerebral ganglia and throughout the length of the longitudinal nerve cords (LNCs), revealing each to be composed of at least four nerve fibres and associated cell bodies. At the level of the cirrus, a pair of neuronal cell bodies (c.19 x 12 μm) is positioned along the LNCs in close proximity to a larger pair of cell bodies (c.58 x 15μm) that project towards the cirrus sac. Fibres originating in the LNCs run behind the pharynx to form a post-pharyngeal loop containing cell bodies (c.21 x c.11μm), and which provides branches to innervate the pharyngeal muscle. No staining was apparent in the innervation of the oral sucker. In contrast, the acetabulum was immunoreactive. It is innervated by a pair of very broad (c.23 μm in width) parallel nerve fibres and by several finer nerve fibres (c. 1 μm in width), all of which are derived from the LNCs. Fine, immunoreactive nerve fibres occupy the wall of the uterus. The cirrus and cirrus sac are also innervated by fine nerve fibres, and a pair of large cell bodies (c.38 x 15μm) are positioned one either side of the cirrus sac. The ootype displays strong 5-HT-immunoreactivity in the form of a network of fine fibres and approximately 13 associated cell bodies (c.10 x 19μm) (see also 3.3 below).

3.1.2. *Miracidium*
Staining for 5-HT has revealed the brain comprises four, small immunoreactive neuronal cell bodies (c.3 x 3μm) aggregated into a single ganglionic mass from which paired LNCs of fine nerve fibres extend the length of the body.

3.1.3. *Redia*
In the anterior of the redia, the paired cerebral ganglia are seen as an accumulation of nerve fibres, with three or four large neuronal cell bodies (c.6 x 6μm x c.6 x 12μm) in each ganglion. Nerve fibres emanate from the ganglia some of which extend anteriorly to innervate the pharynx while others run posteriorly as the LNCs, being joined at irregular intervals by transverse connectives. Three pairs of immunoreactive cell bodies (c.6 x c.12μm) occur in parallel along the paired LNCs. The PNS consists essentially of a subtegumental plexus of immunoreactive nerve fibres.

3.1.4. *Cercaria*
Immunostaining for 5-HT highlights the paired cerebral ganglia and shows them to consist of accumulations of fibres and two or three neuronal cell bodies (c.6 x 6μm) per ganglion, connected by a dorsal commissure which is positioned just posterior to the oral sucker. Fibres extend anteriorly from the ganglia and provide strong immunoreactive innervation to the oral sucker. Each of the LNCs is a composite of three or more nerve fibres that extend the length of the body, proceeding into the tail where they appear as four clearly defined caudal nerves. The LNCs are joined by transverse connectives at

approximately four points. At least six pairs of cell bodies (c.6 x 6μm) are arranged in parallel symmetry along the LNCs in the main body, and stain more intensely than those in the cerebral ganglia. Two pairs are present in the tail, one at the tail junction and the other in the anterior third of the tail. Fibres branch from the LNCs to provide innervation to the acetabulum which is in the form of a rich plexus of fibres and associated cell bodies. Lateral nerve cords run parallel either side of the main LNCs, and are linked to the LNCs at intervals via cross-connective fibres. Two nerve fibres originate either side of the oral sucker and extend posteriorly between the LNCs, terminating in what appear to be duct-like structures anterior to the acetabulum.

3.1.5. *Metacercaria*
Excysted metacercariae display only weak immunoreactivity for 5-HT. The staining occurs in the paired cerebral ganglia, commissure and in the LNCs that run to the posterior end of the body and fuse at a point marked by a single neuronal cell body. Six pairs of cell bodies (c.6 x 6μm) are positioned symmetrically at intervals along the nerve fibres that run between the oral sucker and acetabulum. As in the cercaria, a pair of nerve fibres originating at the level of the oral sucker runs posteriorly to innervate duct-like structures just in front of the acetabulum.

3.2. PEPTIDERGIC

Immunoreactivity for neuropeptides, as demonstrated using antisera to the flatworm neuropeptides, neuropeptide F (NPF), GNFFRFamide and GYIRFamide, has been found to be similar in all of the life cycle stages of *E. caproni* that have been examined, although variations in the immunostaining were observed in the staining patterns for NPF and the two FaRPs.

3.2.1. *Adult*
Strong peptide-immunoreactivity delineates the paired cerebral ganglia, revealing it to consist of a dense accumulation of varicose nerve fibres and approximately 3-4 neuronal cell bodies (c.8-12μm x 12-19μm), cross-linked by a dorsal commissure. From the ganglia, fibres run anteriorly to innervate the muscle of the oral sucker where they give rise to a ring of fine nerves around the opening of the sucker. Nerve fibres and three associated cell bodies (c.12 x 15μm) extend from the cerebral ganglia and LNCs to innervate the pharyngeal muscle, and also dorsally to form a post-pharyngeal loop. The acetabulum is innervated by an extensive assemblage of anastomosing fibres derived from the LNCs, alongside nerves that have been shown to be reactive for 5-HT. Lateral nerve cords proceed either side of the main LNCs and at intervals are interconnected by transverse connectives. The LNCs extend the length of the worm and become progressively less well defined, displaying decreasing immunoreactivity towards the posterior region. The paired LNCs merge at a point behind the testes and then branch to innervate the sub-tegumental region. At least two pairs of cell bodies (c. 12-19 x 19-46μm) are located in parallel along the LNCs between the pharynx and the acetabulum. The muscle associated with the cirrus sac and genital pore is innervated by FaRPergic nerve fibres originating from the LNCs. The ootype displays GYIRFamide-

immunoreactivity in a complex network of fine nerve fibres and approximately eight associated cell bodies (c. 8 x 14μm) (see also 3.3 below). NPF-immunoreactivity was evident in the PNS in the form of an extensive subtegumental plexus of nerve fibres; immunostaining was most intense in the anterior region, in the innervation of the oral sucker, and in that of the cirrus and acetabulum.

3.2.2. *Miracidium*

As with 5-HT, neuropeptide-immunoreactivity is localised most intensely in the anterior neural mass or cerebral ganglion, showing it to consist of a dense accumulation of nerve fibres and several associated cell bodies. At least two pairs of LNCs originate from the ganglion and run the length of the miracidium to the posterior end; they are joined along their length by delicate transverse connectives. NPF-immunoreactive cell bodies (c.1.5 x 1.5μm) have been identified at the posterior tip of the larva.

3.2.3. *Redia*

Intense peptide-immunoreactivity marks the paired cerebral ganglia as a compact accumulation of nerve fibres and cell bodies (c.4 x 4 μm), and staining is also evident in the fibrous commissure. Nerve fibres extend from the ganglia, decreasing posteriorly both in number and in intensity of staining, and anteriorly fibres innervate the pharynx. NPF-immunoreactive cell bodies (c.6 x 6 μm) occupy a region dorsal to the ganglia. A GYIRFamide-immunoreactive cell body (c.6 x 9μm) projects from each ganglion; and GYIRFamide-immunoreactive nerve fibres with two associated cell bodies innervate the excretory pore in the posterior of the larva.

3.2.4. *Cercaria*

The paired cerebral ganglia of nerve fibres and neuronal cell bodies (c.3 x 6μm) and associated fibrous commissure all display intense neuropeptide-immunoreactivity (Fig. 7). Neurofibres emanate from the ganglia to innervate the oral sucker. Anteriorly, the LNCs consist of a broad tract of several immunoreactive nerve fibres but these become progressively narrower towards the posterior end of the main body and in the tail. A pair of cell bodies (c.6 x 9μm) are positioned symmetrically along the LNCs between the pharynx and the acetabulum. A longitudinal nerve fibre innervates the sub-tegumental region of the main body. Innervation of the acetabulum was via nerve fibres derived from the LNCs but these have only been observed using antisera to NPF. A pair of NPF-immunoreactive cell bodies (c.3 x 3μm) is located at the tail junction. Immunostaining for NPF was also observed in the PNS and consisted of plexuses of fine fibres innervating the oral sucker and acetabulum.

3.2.5. *Metacercaria*

Immunostaining for neuropeptides revealed the cerebral ganglia as an accumulation of nerve fibres and cell bodies (c.4 x c.4μm), with the intervening commissure showing only weak staining. The LNCs display a symmetrical arrangement of neuronal cell bodies and their fibres extend the length of the metacercaria. Immunoreactivity was at best moderate in the LNCs, and decreased towards the posterior end where staining was negligible or absent (Fig. 8). GYIRFamide-immunoreactive nerve fibres extend from the LNCs to innervate the acetabulum. Peptide-immunoreactivity in the PNS was largely confined to the innervation of the oral sucker and acetabulum via a sub-tegumental plexus of NPF- and GYIRFamide-immunoreactive nerve fibres.

3.3. DUAL-LABELLING

Using dual detectors in CSLM and the two fluorophores, TRITC (tetramethylrhodamine isothiocyanate) and FITC (fluorescein isothiocyanate), it has been possible to tag, respectively, 5-HT-immunoreactivity as red fluorescence and neuropeptide GYIRFamide-immunoreactivity as green fluorescence, and where colocalised, immunoreactivity as yellow fluorescence. With this approach, it has been found that in the CNS at least, the serotoninergic system appears to be quite distinct from the peptidergic system, with separate neuronal pathways for what are larger and fewer aminergic cell bodies and their smooth-sided fibres compared to the smaller and more numerous peptidergic cell bodies and their varicose or beaded fibres (Fig. 5). However, in parts of the PNS some elements of the serotoninergic and peptidergic systems of the adult worm show evidence of overlap, indicating co-localisation. One notable example is in the innervation of the ootype or egg chamber. Here immunoreactivities for both 5-HT and GYIRFamide neuropeptide have been recorded in a plexus of cell bodies and associated fibres that innervate the ootype (Fig. 6). Here a majority of the cell bodies appeared yellow, marking the presence of both 5-HT- and GYIRFamide-immunoreactivities. However, a subpopulation of 6-8 cell bodies out of a total of 20 neurons appeared green, indicating the presence of GYIRFamide peptide without 5-HT; of all of these cells none was seen to be immunopositive solely for 5-HT (see also 5 below).

4. Physiology and Pharmacology of the Neuromuscular System

Knowledge and understanding of the physiology of the neuromuscular systems of trematodes is meagre. This is not surprising in view of their usual small size, generally poor viability *in vitro*, and tegument that is often impermeable externally applied test compounds. On the other hand, echinostomes would seem better than average as subjects for motility studies *in vitro*. Thus, when recovered from laboratory mice, the vast majority (95%) of specimens of *E. caproni* exhibited strong spontaneous contractility and retained this activity in warm Hanks' balanced salt solution under experimental conditions for up to 8 h. Adult *E. caproni* are relatively large, measuring up to 12 mm in length, and quite robust, enabling them to be easily handled for recording purposes, such that they can be secured by suction pipettes for attachment under tension to a force transducer in an optoisolator system.

4.1. ACTIONS OF CLASSICAL TRANSMITTERS

Acetylcholine and the muscarinic agonists, arecoline and carbachol, each completely inhibited the spontaneous motility of *E. caproni*, causing a rapid flaccid paralysis involving complete abolition of frequency and amplitude (Fig. 9). Arecoline (at 1 μM) had more potent effects on worm motility than carbachol (30 μM), and also decreased baseline tension. These results confirm the presence of an inhibitory cholinergic system in trematodes, as previously demonstrated for *F. hepatica* (Chance and Mansour, 1953; Holmes and Fairweather, 1984), *Schistosoma mansoni* (Pax *et al.*, 1984) and *Haplometra cyclindracea* (McKay *et al.*, 1989).

In common with other trematode species examined thus far (see above), the addition of 5-hydroxytryptamine (5-HT, 1-10 μM) induced an increase in muscle activity by stimulating both frequency and amplitude of contraction; concentrations of 10 μM or greater also increased baseline tension (Fig. 10). As in the other trematodes (cited above) where intact preparations have been used, it is not known if the 5-HT acts presynaptically on the nervous system or postsynaptically on the musculature. What is interesting is that although there is strong evidence for an adenylate cyclase-activating 5-HT receptor in trematodes (see Day *et al.*, 1994), 5-HT does not induce contraction independently on isolated muscle fibres. With respect to *E. caproni*, and in contrast to all other adult trematodes examined immunocytochemically, little or no 5-HT immunostaining was found associated with the oral sucker.

4.2. ACTIONS OF NEUROPEPTIDES

As yet, no authentic endogenous trematode neuropeptide is known since none has been successfully isolated and structurally characterised. Therefore, where physiological studies have been conducted on trematodes (*Schistosoma*, *Fasciola*) they have been done largely using neuropeptides identified from free-living flatworms (turbellarians) and the cyclophyllidean cestode, *Moniezia expansa* (Day *et al.*, 1994; Marks *et al.*, 1996, 1997; Graham *et al.*, 1997). The total inventory of flatworm neuropeptides to date is 6. These are neuropeptide F (NPF) from *M. expansa* and the land turbellarian, *Artioposthia triangulata* (Maule *et al.*, 1991; Curry *et al.*, 1992), and the FMRFamide related peptides (FaRPs), GNFFRFamide from *M. expansa* (Maule *et al.*, 1993), RYIRFamide from *A. triangulata* (Maule *et al.*, 1994), GYIRFamide from the marine and freshwater turbellarians, *Dugesia tigrina* and *Bdelloura candida* (Johnston *et al.*, 1995, 1996), and YIRFamide from *B. candida* (Johnston *et al.*, 1996). Preliminary tests in our laboratory showed that the exogenous addition of any one of these peptides to intact specimens of *E. caproni* was without effect, necessitating the worms being cut just posterior to the acetabulum in order to facilitate peptide penetration. Using these trimmed preparations of *E. caproni*, all of the FaRPs tested brought about a significant decrease in the base-line tension (Figs. 11,12). The order of potency was: GYIRFamide > RYIRFamide = YIRFamide > GNFFRFamide. The results are in accord with those reported for other flatworms, including *F. hepatica*, where the same peptides as were used with *E. caproni* were tested on muscle-strip preparations of the worm (Marks *et al.*, 1996).

Figs. 9-12. Effects of carbachol, 5-HT and flatworm FaRPs on the motility of *E. caproni*. Arrows show period of test compound addition, and tension (mg) and time (min) are given at right of each figure.

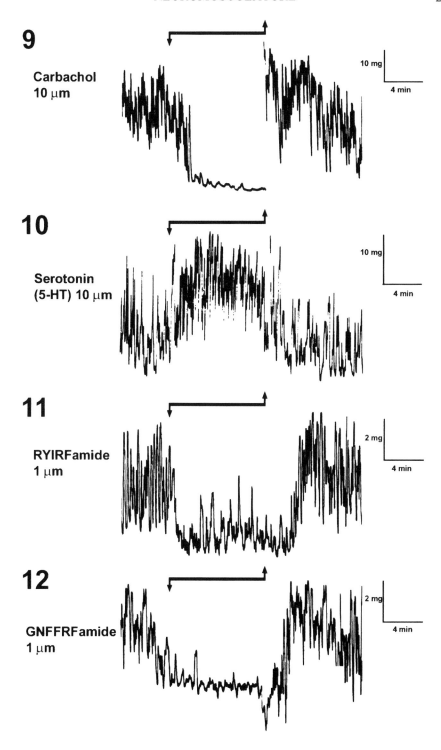

9

Carbachol
10 μm

10 mg

4 min

10

Serotonin
(5-HT) 10 μm

10 mg

4 min

11

RYIRFamide
1 μm

2 mg

4 min

12

GNFFRFamide
1 μm

2 mg

4 min

5. Conclusions and Future Developments

The immunolocalisation and distribution patterns of neuroactive substances can provide clues as to the putative roles of specific neuromediators. Thus, serotoninergic and peptidergic innervations of muscle implicate 5-HT and FaRPs in muscle control, a fact borne out in motility experiments *in vitro* (see 4 above). The CNS in *E. caproni* appears to be largely peptidergic in nature, as evidenced by the more extensive immunoreactivity for NPF and FaRPs than 5-HT, whereas serotoninergic elements predominate over peptidic substances in both male and female reproductive systems. In the ootype, there is evidence of 5-HT and peptide co-localisation in a subpopulation of neurons which may be involved in controlling sphincter muscles that regulate the entry or exit of oocytes or eggs, respectively. Armstrong *et al.* (1997) showed in the monogenean trematode, *Polystoma nearcticum*, where egg production is discontinuous, that expression of FaRPs in the ootype innervation only occurs when eggs are being produced in the ootype.

Both 5-HT- and neuropeptide-immunoreactivities have been shown to be weaker in the metacercaria, compared to all other developmental stages of the worm. This perhaps reflects the fact that the metacercaria is a quiescent resting stage and the least motile of all the forms. Thus, while the metacercarial nervous system is as well developed in structure as that of the cercaria or indeed the adult, it is not as functionally active and shows relatively little expression of 5-HT or neuropeptide. Metacercariae develop into juveniles then adults, and it is interesting to note that serotonin expression is evident in the innervation of the oral sucker of the cercaria but is absent from the metacercarial and adult stages. In contrast, peptidic innervation is present in all three stages. This could mean an active role for 5-HT in the host penetration process, its production being switched off following encystment.

Although a preliminary study, this is the first immunocytochemical insight into the neurochemistry of the developing nervous system of a trematode parasite. There are a few other immunocytochemical descriptions of the nervous system in larval trematodes, for example, for schistosomes (Skuce *et al.*, 1990; Solis-Soto and De Jong Brink, 1994) and for three marine trematodes (Pan *et al.*, 1994). However, the use of an echinostome trematode has proved potentially rewarding in that all successive stages in development can be readily maintained in the laboratory, and immunostained preparations examined by confocal microscopy and compared. A more detailed immunocytochemical study and analysis, using double or even triple labelling, should enable neuronal activity to be better monitored and correlated with the complex morphological and behavioural changes that accompany trematode development, following exposure to different environmental triggers and hosts.

Future developments in trematode neuromuscular biology are likely to centre around (a) the use of isolated muscle fibre assays to help elucidate the postsynaptic effects of putative transmitters and modulators, including neuropeptides and their analogues (see Day *et al.*, 1997), and (b) the application of molecular biology techniques as a means of isolating and characterising helminth neuromuscular receptors and ion channels. Available information suggests that helminth receptors and channels have a unique pharmacology, differing quite considerably from those of vertebrates. Their identification and characterisation using expression cloning procedures should enable *in vitro* expression and physiological analysis to be performed, such as has been achieved in identifying a *Shaker*-related voltage-gated potassium channel from *S. mansoni* (Kim *et al.*, 1995).

Immunoprophylaxis of helminth infections remains elusive in the face of effective immune-evasion strategies by parasites and the general absence of vaccines for human and animal parasite disease. Unfortunately, chemotherapy often involves toxicity problems and there is also growing resistance to many current anthelmintics, necessitating the need for rational development of novel antiparasitic drugs. The neuromuscular system of helminths presents an attractive target for pharmacological intervention in that it is central to locomotory movement and attachment and serves an essential role in alimentation and reproduction. In this respect, echinostomes offer considerable potential as model parasites for experimental study of trematode neuromuscular systems in future drug development programmes.

6. Acknowledgement

The authors thank Dr Bernard Fried, Lafayette College, Easton, PA 18042, USA for kindly providing cysts of *E. caproni* to our laboratory in Belfast, and for his help and advice in maintaining experimental infections of the parasite.

7. References

Armstrong, E.P., Halton, D.W., Tinsley, R.C., Cable, J., Johnston, R.N., Johnston, C.F. and Shaw, C. (1997) Immunocytochemical evidence for the involvement of a FMRFamide -related peptide (FaRP) in egg production in the flatworm parasite, *Polystoma nearcticum, Journal of Comparative Neurology* 377, 41-48.

Chance, M.R.A. and Mansour, T.E. (1953) A contribution to the pharmacology of movement in the liver fluke, *British Journal of Pharmacology* 8, 134-138.

Curry, W.J., Shaw, C., Johnston, C.F., Thim, L. and Buchanan, K.D. (1992) Neuropeptide F: primary structure from the turbellarian, *Artioposthia triangulata, Comparative Biochemistry and Physiology* 101C, 269-274.

Day, T.A., Maule, A.G., Shaw, C. and Pax, R.A. (1997) Structure-activity relationships of FMRFamide-related peptides contracting *Schistosoma mansoni* muscle, *Peptides* 18, 917-921.

Day, T.A., Maule, A.G., Shaw, C., Halton, D.W., Moore, S., Bennett, J.L. and Pax, R.A. (1994) Platyhelminth FMRFamide-related peptides (FaRPs) contract *Schistosoma mansoni* (Trematoda: Digenea) muscle fibres *in vitro, Parasitology* 109, 445-459.

Day, T.A., Chen, G.-Z., Miller, C., Tian, M., Bennett, J.L. and Pax, R.A. (1996) Cholinergic inhibition of muscle fibres isolated from *Schistosoma mansoni* (Trematoda: Digenea), *Parasitology* 113, 55-61.

Fried, B. and Huffman, J.E. (1996) The biology of the intestinal trematode *Echinostoma caproni, Advances in Parasitology* 38, 311-368.

Graham, M.K., Fairweather, I. and McGeown, J.G. (1997) The effects of FaRPs on the motility of isolated muscle strips from the liver fluke, *Fasciola hepatica*. *Parasitology* **114**, 455-465.

Halton, D.W. and Gustafsson, M.K.S. (1996) Functional morphology of the platyhelminth nervous system, *Parasitology* **113**, S47-S72.

Halton, D.W., Maule, A.G. and Shaw, C. (1997) Trematode neurobiology, in B. Fried and T.K Graczyk (eds.), *Advances in Trematode Biology*, CRC Press, Boco Raton, pp. 345-382.

Holmes, S.D. and Fairweather, I. (1984) *Fasciola hepatica*: the effects of neuropharmacological agents upon in *vitro* motility, *Experimental Parasitology* **58**, 194-208.

Johnston, R., Shaw, C., Halton, D.W., Verhaert, P., Baguña, J. (1995) GYIRFamide: a novel FMRFamide-related peptide (FaRP) from the triclad turbellarian, *Dugesia tigrina*, *Biochemical and Biophysical Research Communications* **209**, 689-697.

Johnston, R., Shaw, C., Halton, D.W., Verhaert, P., Blair, K.L., Brennan, G.P., Price, D.A. and Anderson, P.A.V. (1996) Isolation, localisation and bioactivity of the FMRFamide-related neuropeptides GYIRFamide and YIRFamide from the marine turbellarian, *Bdelloura candida*, *Journal of Neurochemistry* **67**, 814-821.

Kim, E., Day, T.E., Bennett, J.L. and Pax, R.A. (1995) Cloning and functional expression of a *Shaker*-related voltage-gated potassium channel gene from *Schistosoma mansoni* (Trematoda: Digenea), *Parasitology* **110**, 171-180.

Mair, G.R., Halton, D.W., Maule, A.G. and Shaw, C. (1998a) Muscling in on parasitic flatworms, *Parasitology Today* 14, 73-76.

Mair, G.R., Maule, A.G., Shaw, C., Johnston, C.F. and Halton, D.W. (1998b) Gross anatomy of the muscle systems of *Fasciola hepatica* as visualized by phalloidin-fluorescence and confocal microscopy, *Parasitology* **117**, 75-82.

Marks, N.J., Johnson, S., Maule, A.G., Halton, D.W., Shaw, C. Geary, T.G., Moore, S. and Thompson, D.P. (1996) Physiological effects of platyhelminth RFamides on muscle strip preparations of *Fasciola hepatica* (Trematoda: Digenea), *Parasitology* **113**, 393-401.

Marks, N.J., Maule, A.G., Halton, D.W., Geary, T.G., Shaw, C. and Thompson, D.P. (1997) Pharmacological effects of nematode FMRFamide-related peptides (FaRPs) on muscle contractility of the trematode, *Fasciola hepatica*, *Parasitology* **114**, 531-539.

Maule, A.G., Shaw, C., Halton, D.W., Thim, L., Johnston, C.F., Fairweather, I. and Buchanan, K.D. (1991) Neuropeptide F: a novel parasitic flatworm regulatory peptide from *Moniezia expansa* (Cestoda: Cyclophyllidea), *Parasitology* **102**, 309-316.

Maule, A.G., Shaw, C., Halton, D.W. and Thim, L. (1993) GNFFRFamide: a novel FMRFamide-immunoreactive peptide isolated from the sheep tapeworm, *Moniezia expansa*, *Biochemical and Biophysical Research Communications* **193**, 1054-1060.

Maule, A.G., Shaw, C., Halton, D.W., Curry, W.J. and Thim, L. (1994) RYIRFamide: a turbellarian FMRFamide-related peptide (FaRP), *Regulatory peptides* **50**, 37-43.

McKay, D.M., Halton, D.W., Allen, J.M. and Fairweather, I. (1989) The effects of cholinergic and serotoninergic drugs on motility in vitro of *Haplometra cylindracea* (Trematoda: Digenea), *Parasitology* **99**, 241-252.

Pan, J.-Z., Halton, D.W., Shaw, C., Maule, A.G. and Johnston, C.F. (1994) Serotonin and neuropeptide immunoreactivities in the intramolluscan stages of three marine trematode parasites, *Parasitology Research* **80**, 388-395.

Pax, R.A., Siefker, C. and Bennett, J.L. (1984) *Schistosoma mansoni*: differences in acetylcholine, dopamine, and serotonin control of circular and longitudinal parasite muscles, *Experimental Parasitology* **58**, 314-324.

Pax, R.A., Day, T.A., Miller, C.L. and Bennett, J.L. (1996) Neuromuscular physiology and pharmacology of parasitic flatworms, *Parasitology* **113**, S83-S96.

Richard, J., Klein, M.J. and Stoeckel, M.E. (1989) Neural and glandular localisation of substance P in *Echinostoma caproni* (Trematoda-Digenea), *Parasitology Research* **75**, 641-648.

Riddell, J.H., Whitfield, P.J., Thorndyke, M.C. and Balogun, M.A. (1991) FMRFamide-like peptides in the nervous and endocrine systems on the digenean helminth *Echinostoma liei*, *Acta Zoologica* **72**, 1-5.

Skuce, P.J., Johnston, C.F., Fairweather, I., Halton, D.W. and Shaw, C. (1990) A confocal scanning laser microscope study of the peptidergic and serotoninergic components of the nervous system in larval *Schistosoma mansoni*, *Parasitology* **101**, 227-234.

Solis-Soto, J.M. and De Jong Brink, M. (1994) Immunocytochemical study on biologically active neurosubstances in daughter sporocysts and cercariae of *Trichobilharzia ocellata* and *Schistosoma mansoni*, *Parasitology* **108**, 301-311.

Thorndyke, M.C. and Whitfield, P.J. (1987) VIP-like-immunoreactive tegumental cells in the digenean helminth *Echinostoma liei*: possible role in host-parasite interactions, *General and Comparative Endocrinology* **68**, 202-207.

IMMUNOBIOLOGY AND IMMUNODIAGNOSIS OF ECHINOSTOMIASIS

THADDEUS K. GRACZYK
*Department of Molecular Microbiology and Immunology, and
Department of Environmental Health Sciences, School of Hygiene and
Public Health, Johns Hopkins University, Baltimore, Maryland 21205,
USA*

B. Fried and T.K. Graczyk (eds.),
Echinostomes as Experimental Models for Biological Research, 229–244.
© 2000 *Kluwer Academic Publishers. Printed in the Netherlands.*

1. Introduction

Intestinal trematodes, such as echinostomes, elicit detectable immune responses in vertebrate hosts (Simonsen and Anderson, 1986; Andersen *et al.*, 1989; Simonsen *et al.*, 1990; Agger *et al.*, 1993; Graczyk and Fried, 1994; 1995). Much of the information on echinostome immunobiology is derived from research which utilizes rodent models, i.e., mouse, hamster, or rat (Huffman and Fried, 1990; Fried and Huffman, 1996). Various strains of mice, such as ICR, C3H, Swiss Webster, CBA, NMRI, SCID, RAG, C3H/HeN, and athymic have been used as experimental hosts of echinostomes (Bindseil and Christensen, 1984; Hosier and Fried, 1986; Odaibo *et al.*, 1988; Fujino and Fried, 1993a; 1993b; Baek *et al.*, 1996; Frazer *et al.*, 1999).

The humoral (based on antibodies [Ab]) and cellular (restricted to cell surfaces) immune responses of vertebrate hosts are stimulated by specific antigens (Ag) expressed by juvenile and adult echinostomes (Andersen *et al.*, 1989; Simonsen and Anderson, 1986; Simonsen *et al.*, 1990; Agger *et al.*, 1993; Graczyk and Fried, 1994; 1995). Antigenic stimulation can also be provided by excretory-secretory (ES) products derived from adult echinostomes and echinostome metacercarial cysts after excystation (Graczyk and Fried, 1994; 1995; Trouve and Coustau, 1998). The rodent model for echinostomes (Fried *et al.*, 1988; Fried and Huffman, 1996) provides an extensive set of details on their immunobiology in the vertebrate host (Huffman and Fried, 1990)

The purpose of this chapter is to present a broad spectrum of the immune interactions developed between echinostomes and their definitive hosts, i.e., homothermous vertebrates, and the means by which these interactions can be immunologically evaluated and measured.

2. Pathophysiology of Echinostome Infections

2.1. HUMAN ECHINOSTOMIASIS

2.1.1. *Clinical Disease, Diagnosis, and Treatment*

Human disease occurs focally, is linked to fresh- or brackish water habitats, and is associated with common socio-cultural practices of eating raw or insufficiently cooked mollusks, fish, crustaceans, and amphibians or using "night soil" (human excrement collected from latrines) for fertilization of fish ponds (Graczyk and Fried, 1998). Morbidity and mortality due to echinostomiasis are difficult to assess in endemic areas because of a prolonged latent phase, a short acute phase, and asymptomatic presentations; also because of a similarity of clinical symptoms to other intestinal helminthiasis (Shekhar, 1991). Clinical symptoms are related to parasite load (Miyamoto *et al.*, 1984; Huffman and Fried, 1990). In light to moderate infections anaemia, headache, dizziness, stomachache, gastric pain, and loose stools have most often been reported (Ujiie, 1936; Chai *et al.*, 1994). Heavy infections are associated with eosinophilia, abdominal pain, profuse watery diarrhoea, anemia, edema, and anorexia (Ujiie, 1936; Huffman and Fried, 1990; Chai *et al.*, 1994; Graczyk and Fried, 1998). Pathologically, echinostomes damage the intestinal mucosa and cause extensive intestinal and duodenal erosions, and catarrhal inflammation (Chai *et al.*, 1994).

Diagnosis of human and animal echinostomiasis is routinely done by finding characteristic operculate, unembryonated, ellipsoidal, yellow to yellow-brown eggs in fecal specimens (Graczyk and Fried, 1988). As egg size varies among species of echinostomes (Huffman and Fried, 1990), species identification can be done based on morphology of adult worms following anthelmintic treatment. Infections may be readily terminated by mebendazole (Cross *et al.*, 1986), albendazole (Punkpark *et al.*, 1984), praziquantel (Seo *et al.*, 1985) or niclosamide (Yomesan) (Rim, 1982). Diagnosis and treatment programs focused on human echinostomiasis should also include domestic animal reservoirs (Graczyk and Fried, 1998).

2.2. ANIMAL ECHINOSTOMIASIS

Pathophysiological responses to echinostome infections in rodents are modulated by the intensity of infection (Huffman and Fried, 1990; Fried and Huffman, 1996). Echinostomiasis in laboratory rodents is associated with clinical signs such as progressive unthriftiness, weight loss, lack of weight gain, poor body condition, inappetence, watery diarrhoea, and blood lost into the intestinal tract (Hufmann *et al.*, 1986; Huffman and Fried, 1990; Fried and Huffman, 1996). Serum chemistry analysis usually shows hypoalbuminemia and decreased total protein concentration in the serum (Hufmann *et al.*, 1986). Pathological changes in the intestines of rats infected with *E. malayanum* included edema of the lamina propria and epithelial hyperplasia (Mohandas and Nadakal, 1978). Histopathological response of hamsters to *E. trivolvis* include erosion of intestinal villi with lymphocytic infiltration and focal hepatic necrosis in livers (Hufmann *et al.*, 1986). In birds, echinostome infections cause hemorrhagic inflammation of the intestinal mucosa and severe enteritis (Huffman and Fried, 1990). Heavy infections with *Echinoparyphium echinatum* caused death of ducklings and goslings (Huffman and Fried, 1990).

3. Basic Principles of Vertebrate Immune Responses to Echinostomes

3.1. ANTIBODY RESPONSES

Immune responses to intestinal trematodes, such as echinostomes, are initiated by two categories of lymphocytes: B-cells which stimulate cells to produce Ab, and T-cells that bind to the Ag. Serum-circulating Ag is captured by Ag-presenting cells which release interleukins (IL) and present the Ag epitope to a subset of T-cells, i.e., T-helper (T_H) cells, which stimulate (together with IL) production of B-cells (see review in Graczyk, 1997). B-cells stimulate production of large numbers of plasma cells which secrete various amounts of systemic Ab that bind specifically to the presented Ag epitope. The amount of the circulating Ab is measured as Ab titer. Antibodies are glycoproteins constituting plasma protein immunoglobulins (Ig) which have specific binding activity to Ag epitopes. There are 5 Ig isotypes: IgA, IgD, IgE, IgG, and IgM. After presentation of the same Ag (re-infection, challenge, or vaccine) secondary (anamnestic) responses occur rapidly because B-cells have already produced at primary challenge a long lived memory cell (see review in Graczyk, 1997). The Ab that binds to the echinostome surface Ag (Simonsen et al., 1990) induce Ab-dependent cell mediated cytotoxicity (ADCC) with eosinophils, i.e., effector cells, which may damage the trematode tegument (see review in Graczyk, 1997).

3.2. CELL-MEDIATED RESPONSES

In cell-mediated responses, T-cells respond to Ag epitopes presented by macrophages, and those with Ag-specific receptors proliferate and release soluble lymphokines (i.e. interferon [IFN], or IL) which activate neutrophils and natural killer cells (see review in Graczyk, 1997). Cell-mediated responses are based on macrophages, eosinophils, neutrophils and mast cells. Cell-mediated immunity may show secondary response due to the memory T-cells. Delayed type hypersensitivity reaction is a cell-mediated process (immediate hypersensitivity is based on Ab) that occurs more than 24 hr after Ag presentation (see review in Graczyk, 1997).

3.3. EOSINOPHILS AND MAST CELLS

Eosinophils are minor components of white blood cells; however, their numbers increase after echinostome infection (Huffman and Fried, 1990; Fried and Huffman, 1996). Blood eosinophilia and eosinophil infiltration of the inflamed intestinal lamina propria are the most characteristic features of echinostome infection (Fujino et al., 1998a; 1998b). The increase of eosinophils is caused by an increase of eosinophil production in hemopoietic tissues (bone marrow, spleen, and liver), migration of eosinophils to the inflammation site, and mobilization of eosinophils from the production site to the circulation (Byram et al., 1978; Borojevic et al., 1981; Nawa et al., 1994). Eosinophils attach to the surface of the trematodes which are coated with Ab, complement, and Ab-Ag complexes, and release neurotoxins or proteolytic enzymes which affect the tegument of the trematode (Butterworth et al., 1979; Incani and McLaren, 1983; Graczyk, 1997).

Mast cells (and mastocytosis) are divided into mucosal type (MMC) and connective tissue type (CTMC) (Nawa et al., 1994). Mucosal type mast cells are present during echinostome infection (intestinal mastocytosis) not only in the lamina propria, but also in the intestinal epithelial cells (Fujino et al., 1998b). However, a significant reduction of mastocytosis was observed in C3H/HeN mice infected with E. trivolvis and treated intramuscularly with immunosuppressive agent FK506 (Fujino *et al.*, 1998b). Intestinal mastocytosis is one of the factors responsible for expulsion of *Echinostoma caproni* from the intestine of ICR mice (Huffman and Fried, 1990). The growth and differentiation of MMC is T-cell-dependent (Nawa *et al.*, 1994), whereas CTMC showed differences in T-cell dependency (Nawa *et al.*, 1994).

3.4. ANTIBODY-DEPENDENT (ADCC) AND COMPLEMENT-DEPENDENT CELL-MEDIATED CYTOTOXICITY (CDCC)

Complement is a set of plasma proteins which forms a cascade of triggering hydrolytic enzymes reacting with the tegument of intestinal helminths, e.g., echinostomes (Nawa *et al.*, 1994). The complement system can be activated by the Ag-Ab complex, or by the echinostome-derived surface Ag (see review in Graczyk, 1997). The biological role of complement during echinostome infection includes opsonization (coating of Ag-complex) (Simonsen *et al.*, 1990) and chemoattraction of neutrophils and eosinophils (Nawa *et al.*, 1994). The surface of the echinostome tegument is covered by Ab, complement, and Ab-Ag complexes which facilitate attachment of the inflammatory cells (Simonsen *et al.*, 1990; Nawa *et al.*, 1994). After attachment, the inflammatory cells release a variety of substances such as neurotoxins or proteolytic enzymes, which are directly toxic to the trematode or indirectly react with the trematode tegument (Nawa *et al.*, 1994).

4. Immunobiology of Echinostome Infections

4.1. HOMOLOGOUS AND HETEROLOGOUS RESISTANCE

Primary 20-or 28-day-old *E. caproni* infection in mice was sufficient to generate complete resistance to homologous challenge (Sirag *et al.*, 1990). However, the resistance to *E. caproni* challenge infection on day 21 following a primary mouse infection with 25 *E. caproni* metacercariae was affected by zinc deficiency to which the mice were subjected (Baek *et al.*, 1996). Cross-protection of mice against *E. caproni* was also induced by patent *Schistosoma mansoni* or *E. trivolvis* infections, and the resistance persisted after anthelmintic removal of the worms (Sirag *et al.*, 1990). However, resistance was not observed after single-sex *S. mansoni* infection indicating that egg Ag is essential in the generation of resistance (Christensen *et al.*, 1981). The ICR mice developed resistance against an homologous infection with either *E. trivolvis* or *E. caproni* (Hosier *et al.*, 1988). Christensen *et al.* (1986) demonstrated equal level of resistance following oral infection and intraduodenal implantation of newly excysted worms of *E. caproni*. Simonsen and Andersen (1986) reported that circulating Ab to the surface of juvenile *E. caproni* appeared in the serum of infected mice at the same time as resistance to reinfection is developed.

Hooded lister rats were resistant to secondary infection with *E. caproni* after anthelmintic treatment of a 7-day-old primary infection (Huffman and Fried, 1990). Golden hamsters however, were unable to acquire a resistance to homologous *E. trivolvis* infection (Mabus *et al.*, 1988).

4.2. INTESTINAL EXPULSION

Echinostoma caproni is retained in the intestines of ICR mice for months (Hosier and Fried, 1986), whereas the allopatric species, *E. trivolvis*, is expelled within 3 wk post-infection (PI) (Weinstein and Fried, 1991). The expulsion of *E. trivolvis* is related to an increase in the number of goblet cells in the mouse small intestine (Weinstein and Fried, 1991), loss of acetabular attachment by the worms (Gavet and Fried, 1994), retraction of worm collar spines, and mouse sera toxicity factors (Fujino and Fried, 1993a). Expulsion of *E. trivolvis* from the intestine of SCID mice was induced by the goblet cells that markedly increase in numbers, and suppression of the goblet cell hyperplasia delayed rejection of the worms (Fujino *et al.*, 1998a). The lack of suppression of goblet cell hyperplasia was directly related to expulsion of *E. trivolvis* from the intestine of C3H/HeN mice (Fujino *et al.*, 1998b). In murine hosts, increased mucus production associated with goblet cell hyperplasia is primarily involved in the expulsion of *E. trivolvis* (Fujino *et al.*, 1998b). Increased mucin secretion from hyperplastic goblet cells following *E. trivolvis* infections is characteristic for mice, whereas it does not occur in *E. trivolvis*-infected golden hamsters (Fujino and Fried, 1996). Activated glucose metabolism in the mouse (C3H strain) intestines infected with *E. trivolvis* results in an increased rate of mucin synthesis which mediates expulsion of *E. trivolvis* (Fujino and Fried, 1993a). Ultrastructural and immunocytochemical observations of *E. trivolvis* collar spines recovered from ICR mice or golden hamsters indicate that factors present in mouse serum are involved in spinal retraction enhancing worm expulsion (Fujino *et al.*, 1994). In contrast, most of the collar spines of *E. trivolvis* maintained in a defined medium supplemented with hamster serum were extended (Fujino *et al.*, 1994). Baek *et al.*, (1996) demonstrated that zinc deficiency delayed worm expulsion from the intestine of CBA mice. Rapid expulsion of *E. trivolvis* and *E. caproni* from C3H/HeN mice was observed after infection with *Nippostrongylus brasiliensis* (Fujino *et al.*, 1996). Dexamethasone treatment of C3H/HeN mice infected with *E. trivolvis* suppressed the increase in goblet cell numbers and caused an inhibition of worm expulsion (Fujino *et al.*, 1997). Spontaneous expulsion of *E. hortense* occurred in experimentally infected mice (Tani and Yoshimira, 1988). The worm expulsion began 3 wk PI and was associated with an increase in intestinal goblet cells, but not with the kinetics of the mast cells (Tani and Yoshimira, 1988).

4.3. INTESTINAL SITE SEGREGATION AND WORM CROWDING

Interactive site segregation was described based on the results of concurrent infection of golden hamsters with *E. caproni* and *Echinoparyphium recurvatum* (Barus *et al.*, 1974). Gravid *E. recurvatum* were found in the anterior half of the small intestine whereas *E. caproni* were found in the posterior part (Barus *et al.*, 1974). Extraintestinal infections of *E. trivolvis* (bile ducts, liver, gall-bladder, and pancreas) in golden hamsters resulted from worm crowding in heavy infection (Huffman *et al.*, 1986). Under crowded conditions, *E. trivolvis* produced approximately one-half of the number of eggs when compared with the worms from uncrowded sites (Fried and Freeborne, 1984). In general, crowding conditions enhance intraspecific competition and encrease the spatial distribution of the echinostomes in the intestine (Franco *et al.*, 1988).

4.4. HOST SPECIFICITY OF ADULT ECHINOSTOMES

Host specificity is an adaptation of echinostomes to physico-chemical conditions offered by the vertebrate host which allows the parasite to infect the host, and to establish and extend the infection (Huffman and Fried, 1990; Fried and Huffman, 1996; Haseeb and Fried, 1997). Echinostome specificity resulting from phylogenetic, physiologic, and ecologic accommodations between the fluke and its host an is an evolutionary dynamic process in which evasion of host immune responses serve an important function (Parkhouse, 1984). For adult echinostomes, the main factor influencing phylogenetic host specificity is host behavior, particularly the feeding habits of vertebrate hosts (Huffman and Fried, 1990; Fried and Huffman, 1996; Graczyk and Fried, 1998). Phylogenetic host-specificity is affected by a wide spectrum of ecological and physiological factors (Chowdhury *et al.*, 1994). The host digestive system which directly affects excystment of echinostome metacercarial cysts, constitutes an example of a physiological factor (Fried, 1994).

4.5. EVASION OF HOST HUMORAL RESPONSES

Evasion of the host immune response is an accommodation of the echinostomes to the host immune recognition system which allows the reproduction of the parasite and propagation to a new host (Parkhouse, 1984; Graczyk, 1997). Echinostomes modify host immune responsiveness by Ab cleavage, inactivation of complement, formation of Ag-Ab immunocomplexes, and shedding and renewal of surface Ag (Parkhouse, 1984; Andersen *et al.*, 1989; Simonsen *et al.*, 1990; Graczyk, 1997). The surface of mature *E. caproni* which were freshly recovered from mouse intestines was covered with Ab-Ag immunocomplexes (Simonsen *et al.*, 1990). These complexes were lost after 24 hr of culture in vitro; however, in vitro culture in a medium with mouse serum resulted in reformation of the Ab-Ag immunocomplexes (Simonsen *et al.*, 1990). Juvenile (newly excysted metacercariae) and young *E. caproni* which had never been exposed to mice Ab, obtained a layer of Ab-Ag immunocomplexes on their surface after incubation in vitro with immune mouse sera (Simonsen *et al.*, 1990). Immunologic studies utilizing SDS-PAGE, ELISA, and Western blot analysis demonstrated that juvenile, i.e., newly excysted metacercariae, and adult *E. caproni* shed solubilized surface Ag and that the Ag turn-over rate was very high, i.e., half-life of 8-15 min (Andersen *et al.*, 1989).

5. Immunodiagnosis of Echinostomiasis

5.1. MURINE HOSTS

Diagnosis of echinostome infection can be accomplished by recovering eggs from the stools; however, the eggs may not be present if worms are preovigerous (Huffman and Fried, 1990; Fried, 1997). Immunodiagnostic tests based on an indirect ELISA were developed for detection of anti-*E. caproni*, and anti-*E. trivolvis* serum IgG (Graczyk and Fried, 1994; 1995). Extraction of the outer membrane of the glycocalyx tegumental surface of adult *E. caproni* and *E. trivolvis* proved to be an effective method to obtain highly-reactive Ag that can be used for serodiagnosis of echinostomiasis in mice (Graczyk and Fried, 1994; 1995). With the glycocalyx membrane Ag from adult echinostomes used in the Ab ELISA, it was possible to detect humoral responses (IgG class) to *E. caproni* and *E. trivolvis* in ICR mice on day 8 PI (Graczyk and Fried, 1994), and on day 10 PI (Graczyk and Fried, 1994), respectively. A concentration of 10.0 μg/ml of Ag was used, and it was possible to detect anti-*E. caproni* and *E. trivolvis* and IgG at a dilution of 1/3,200 (Graczyk and Fried, 1994; 1995). The IgG level varied over a 40-day period showing a time-dependent pattern with a peak on day 16 PI (Graczyk and Fried, 1994). The results on reciprocal cross-reactivity indicate that adult *E. trivolvis* and *E. caproni* share at least some of the surface Ag, and express species-specific antigenic determinants in ICR mice at similar, but not significantly different levels (Graczyk and Fried, 1995). In another study, humoral responses to *E. caproni* in NMRI mice were detected with ELISA utilizing crude *E. caproni* Ag on day 14 PI (Agger *et al.*, 1993). The anti-*E. trivolvis* IgG titer in ICR mice (Graczyk *et al.*, 1995) was considerably lower than anti-*E. caproni* IgG titer in the same mouse strain (Graczyk and Fried, 1994), and anti-*E. caproni* Ab titer in NMRI mice (Agger *et al.*, 1993). Zinc deficiency induced in CBA mice resulted in a prolonged IgM response, a delayed IgG response, and an increased IgA response (Baek *et al.*, 1996). Considering the low IgG response of ICR mice against *E. trivolvis*, it is unlikely that the expulsion of the worms is a result of humoral-mediated immunity. A similar conclusion was reached by Fujino *et al.*, (1998) after expulsion of *E. trivolvis* from the intestine of SCID mice. The differences between *E. caproni* and *E. trivolvis* Ag can be a potential source of differences in the IgG levels. However, expulsion of *E.*

trivolvis prevented the use of 40-day-old worms for the Ag preparation (Graczyk and Fried, 1995). Heterogeneity of the Ag intestinal flukes over time was demonstrated by Rivera-Marrero *et al.* (1988), and change in surface Ag as the parasites grow was shown for *S. mansoni* (Ramalho-Pinto *et al.*, 1978). ELISA-detected heterogeneity of surface Ag related to the age of trematodes has been reported in human blood flukes (Ramalho-Pinto *et al.*, 1978), and intestinal trematodes (Hockley and McLaren, 1973; Rivera-Marrero *et al.*, 1988). However, the preparation of the Ag from various-age worms in the studies of Graczyk and Fried (1994; 1995) reduced the potential age-related heterogeneity of the Ag.

Golden hamsters mounted considerably weaker systemic Ab responses to *E. caproni* than mice (Simonsen and Andersen, 1986). The lack of humoral response in golden hamsters (*Mesocricetus auratus*) to *E. trivolvis* (referred to as *E. revolutum* in that study) was demonstrated by Mabus *et al.* (1988). The serum Ab of golden hamsters infected with *E. caproni* was examined by ELISA, SDS-PAGE, Western blot, and IFAT (Simonsen *et al.*, 1991). All hamsters developed Ab responses by week 11-13 PI; the responses were considerably weaker when compared to mouse Ab responses to *E. caproni* (Simonsen *et al.*, 1991).

5.2. THE HUMAN HOST

Detection of circulating Ab specific to echinostomes does not provide direct evidence of active infection but suggests exposure of the host to the fluke (Graczyk and Fried, 1994; 1995). For this reason, serum Ab tests may represent low values in epidemiological surveillance of human echinostomiasis in endemic areas (Graczyk and Fried, 1998). Also, although immunoassays possess a superior sensitivity for diagnosis of echinostome infections compared to conventional microscopical techniques, they may display low specificity due to cross-reactivity of the immunoreagents (Ahuja and Tsuji, 1984). ELISA can be applied to diagnostic surveys of exposure to echinostomes. The ELISA-wells that gave absorbance of 0.12 or higher (the cutoff level was 0.14) at 405 nm wavelength can be clearly distinguished visually (particularly on a white background) from the negative wells (Graczyk and Fried, 1994; 1995). Thus, the need for an automated ELISA-reader is eliminated, making this method suitable for field surveys.

5.3. SEVERE COMBINED IMMUNODEFICIENT (SCID) MICE

Severe combined immunodeficient mice (SCID) lacking functional T- and B-lymphocytes have been used to study infections with echinostomes. SCID mice have been experimentally infected with *E. trivolvis* and given daily intramuscular injections with dexamethasone (Fujino *et al.*, 1998a). In the dexamethasone-treated mice, goblet cell hyperplasia was significantly suppressed and the worms were retained in the intestines until day 15 PI (Fujino *et al.*, 1998a).   In control mice, goblet cell hyperplasia peaked at day 12 PI, and by day 15 PI, all worms were expelled (Fujino *et al.*, 1998a).   ELISA showed no marked rise in titers of circulating IgM, IgA, and IgG in both experimental and control mice (Fujino *et al.*, 1998a).

A strain of mice that fails to express the recombinase activating gene (RAG-2) protein (Shinkai *et al.*, 1992) has been used in experimental infection of *E. caproni* (Frazer *et al.*, 1999).   These mice exhibit a complete lack of mature B- and T-cells (Shinkai *et al.*, 1992).   The intestinal circumference of infected RAG mice was significantly greater that than of uninfected mice (Frazer *et al.*, 1999).   A significant goblet cell hyperplasia occurred at 2 wk PI, but the response was not effective in eliminating *E. caproni* from RAG mice (Frazer *et al.*, 1999).

6. Selected Protocols

6.1. ANTIGEN PREPARATION

6.1.2. *Surface Glycocalyx Membrane Crude Antigen of Adult Echinostomes*
The Ag is prepared from the pool of various age, e.g., 14-day-old and 40-day-old, adult echinostomes in order to avoid any potential age-related Ag heterogeneity (Hockley and McLaren, 1973; Rivera Marrero *et al.*, 1988).   Following mouse necropsy, the echinostomes are collected and stored frozen at -70° C.   The flukes are thawed, pooled (approximately 25 worms per pool) and each pool is suspended in 1.0 ml of phosphate-buffered saline (PBS) (pH 7.4) to which 1.0 ml of 85% phenol is added.   The incubation, centrifugation and dialyzing procedures follow the protocols of Caufield *et al.* (1987), and Nanduri *et al.* (1991).   The concentration of the glycocalyx-containing fraction should be determined by a spectrophotometer at ꞓ 280 nm wavelength or by bovine serum albumin (BSA) assay (Graczyk *et al.*, 1995; 1996).   The final preparation should be diluted 1/10 with PBS, stored at -70° C in 1.5-ml vials, and then thawed and aliquoted prior to the ELISA trials (Graczyk and Fried, 1994; 1995; Graczyk *et al.*, 1995; 1996).   The glycoprotein-enriched fractions of *E. caproni* and *E. trivovis* for use in the indirect ELISA can be obtained by concanavalin agglutinin (Con A) - Sepharose 4B affinity column chromatography (Johnson et al., 1989; De Repentigny *et al.*, 1991; Graczyk *et al.*, 1996).   To determine the purity of Ag preparation, the components of the Con A-bound fractions can be separated by SDS-PAGE with a 15% gel run at 100 V constant voltage for 7 hr at 21° C in electrode buffer (De Repentigny *et al.*, 1991; Graczyk *et al.*, 1996) and stained with Coomassie blue (Laemmli, 1970).

6.2. BLOOD SAMPLE COLLECTION

Mouse blood (500 μl) samples can be collected (starting at day 5 PI) by calibrated heparinized capillary tubes from the retro-orbital sinus (Graczyk and Fried, 1994; 1995), and stored on filter paper (Graczyk and Fried, 1994; 1995). Filter paper storage does not diminish IgG binding capacity (Graczyk et al., 1993; 1994). Prior to the ELISA trials, the blood is eluted as hemolysate from the filters (Graczyk et al. 1993; 1994).

6.3. INDIRECT ENZYME-LINKED IMMUNOSORBENT ASSAY (ELISA)

ELISA plate is coated with 100 μl of Ag (10.0 μg/ml) and incubated overnight at room temperature (RT). Eight wells on the plate are not coated to determine nonspecific background values (NBV) of absorbance. A low NBV of absorbance (less than 0.074) should be observed (Graczyk and Fried, 1994; 1995). Following coating, the plate is post-coated by incubating each well for 3 hr at RT with 200 μl of casein blocking buffer (CBB)/0.05% Tween (TW)-20 (Graczyk et al., 1993; 1994). Following post-coating the wells are emptied and filled in triplicate with 100 μl of test blood eluants diluted (1/50) with PBS (pH 7.4). Negative control (NC) is a pool of the blood from uninfected mice and their absorbance vary from 0.11 to 0.12; mean = 0.11 $\pm$ 0.01), with the cutoff level of 0.14. The positive control (PC) is prepared by dilution (1/50 with PBS, pH 7.4) of the blood of mice from which the trematodes are recovered. Negative and PC are prepared at the same time in several 1.0-ml vials, frozen at -70° C, and a single sample is thawed for the ELISA, then discarded after use. This procedure prevents changes in the background caused by repeated freeze-thaw cycles. Eight wells are filled with NC, and 8 wells are filled with PC. After 3 hr incubation at 37° C, the plates are washed 6 times with PBS/0.05% TW-20, 100 μl of anti-mouse IgG coupled to alkaline phosphatase diluted in PBS (pH 7.4) (1/500) is added to each well, and the plate is processed as described above. The absorbance is read (405 nm) by an ELISA microplate reader; most of such commercially available devices are controlled by computer software which subtract the NBV from all values on the plate, giving the corrected absorbance values used for analysis. The positive cutoff level is established as units of Ab unit greater than the mean of absorbance of NC $\pm$ 3 SD (conservative estimate), or NC $\pm$ 2 SD (liberal estimate).

6.3.1. Comparison Plate-to-Plate Results
The adjustment method of Schwartz et al. (1991) is used to compare absorbance values from different ELISA plates. The absorbance values can be comparable from plate to plate because the same NC and PC is used in each plate. The mean adjustment from plate to plate should not exceed 10.0% of the mean absorbance values of PC (Schwartz et al. 1991; Graczyk et al., 1993; 1994).

6.3.2. Testing of ELISA Reproducibility and Reliability
The reproducibility test for the indirect ELISA can be performed 2 wk later with the

same blood samples, and following the same protocol. The absorbance values of the ELISA reproducibility trial should be dispersed within 2 SD of the mean values obtained in the original ELISA trial.

To determine ELISA reliability, the coded trial with multiple, i.e., more than 30 blood samples, should be carried out. The pool of tested samples should contain at least 10 NC samples. The samples should be coded by a person not involved in the ELISA trial. ELISA should be carried out without knowledge of the sample identity. All NC samples should be identified as negative after revealing their status.

6.3.3. *Determination of ELISA Sensitivity and Specificity*

To determine the lower detection limits of the indirect ELISA, 8 wells are filled with 100 μl of following PBS dilutions of PC: 1/50, 1/100, 1/200, 1/400, 1/800, 1/1,600, 1/3,200, 1/6,400, and 12,800 (Graczyk and Fried, 1994; 1995; Graczyk et al., 1996). The test is run according the indirect ELISA protocol. Dilution of the PC that gives an absorbance value located within the mean of absorbance of NC $\pm$ 3 SD, determines the ELISA sensitivity. Usually it is possible to detect anti-*E. trivolvis* and *E. caproni* IgG at a dilution of 1/3,200 (Graczyk and Fried, 1994; 1995).

To determine the specificity of the indirect ELISA, the plate is coated with 10.0 μg/ml of antigen from other trematodes, e.g., *S. mansoni*, *S. japonicum*, *S. haematobium*, *Fasciola hepatica*, *H. dorsopora*, *L. learedi*, and *C. hawaiiensis* (Graczyk et al., 1995) obtained in the same manner as from *E. caproni* and *E. trivolvis*. The use of mouse NC and PC follow the protocol for the indirect ELISA described above and should be followed.

7. References

Agger, M.K., Simonsen, P.E. and Vennervald, B.J. (1993) The antibody response in serum, intestinal wall and intestinal lumen of NMRI mice infected with *Echinostoma caproni*. *Journal of Helminthology* **67**, 169-178.

Ahuja, S.P. and Tsuji, M. (1984) Immunodiagnosis of helminthic infection, in N. Chowdhury and I. Tada (eds.), *Helminthology,* Springer-Verlag, Narosa Publishing House, New York, pp. 283-298.

Andersen, K., Simonsen, P.E., Andersen, B.J. and Birch-Andersen, A. (1989) *Echinostoma caproni* in mice: shedding of antigens from the surface of an intestinal trematode. *International Journal for Parasitology* **19**, 111-118.

Baek, J., Simonsen P.E., Friis, H. and Christensen, N.Ø. (1996) Zinc deficiency and host responses to helminth infection: *Echinostoma caproni* infection in CBA mice. *Journal of Helminthology* **70**, 7-12.

Barus, V., Moravec, F. and Rysavy, B. 1974 Antagonistic interaction between *Echinostoma revolutum* and *Echinoparyphium recurvatum* (Trematoda) in the definitive host. *Folia Parasitologica* **21**, 155-161.

Bindseil, E. and Christensen, N.Ø. (1984) Thymus-independent crypt hyperplasia and villous atrophy in the small intestine of mice infected with the trematode *Echinostoma revolutum*. *Parasitology* **88**, 431-438.

Borojevic, R., Srocker, S. and Grimaud, J.A. (1981) Hepatic eosinophil granulocytopoiesis in murine experimental schistosomiasis mansoni. *British Journal of Experimental Pathology* **62**, 480-486.

Butterworth, A.E., Wassom, D.L., Gleich, G.J., Loegering, D.A. and David, J.R. (1979) Damage to schistosomula of *Schistosoma mansoni* induced directly by eosinophil major basic protein. *Journal of Immunology* **122**, 221-229.

Byram, J.E., Imohiosen, E.A.E. and Von Lichtenberg, F. (1978) Tissue eosinophil proliferation and maturation in schistosome-infected mice and hamsters. *American Journal of Tropical Medicine and Hygiene* **27**, 267-272.

Caufield, J.P., Cianti, C.M.L., McDiarmid, S.S., Suyemitsu, T. and Schmid, K. (1987) Ultrastructure, carbohydrate, and amino acid analysis of two preparations of the cercarial glycocalyx of *Schistosoma mansoni*. *Journal of Parasitology* **73**, 514-522.

Chai, J.Y., Hong, S.T., Lee, S.H., Lee, G.C. and Min, Y.I. (1994) A case of echinostomiasis with ulcerative lesions in the duodenum. *Korean Journal of Parasitology* **32**, 201-204.

Chowdhury, N., Sood, M.L. and O'Grady, R.T. (1994) Evolution, parasitism and host specificity in helminths, in N. Chowdhury and I. Tada (eds.), *Helminthology*, Springer-Verlag, Narosa Publishing House, New York, pp. 1-19.

Christensen, N.Ø., Nydal, R., Frandsen, F., Sirag, S. and Nansen P. (1981) Further studies on resistance to *Fasciola hepatica* and *Echinostoma revolutum* in mice infected with *Schistosoma* sp. *Zeitschrift für Parasitenkunde* **65**, 293-298.

Christensen, N.Ø., Knudsen, J. and Andreassen, J. (1986) *Echinostoma revolutum*: resistance to secondary and superimposed infections in mice. *Experimental Parasitology* **61**, 311-318.

Cross, J.H. and Basaca-Sevilla, V. (1986) Studies on *Echinostoma ilocanum* in the Philippines. *Southeast Asian Journal of Tropical Medicine and Public Health* **17**, 23-27.

DeRepentigny L., Kilanowski, E., Pendenault, L. and Boushira, M. (1991) Immunoblot analyses of the serologic response to *Aspergillus fumigatus* antigens in experimental invasive aspergillosis. *The Journal of Infectious Diseases* **163**, 1305-1311.

Franco, J., Huffman, J.E. and Fried, B. (1988) The effects of crowding on adult of *Echinostoma revolutum* (Digenea: Echinostomatidae) in experimentally infected golden hamsters, *Mesocriceus auratus*. *Journal of Parasitology* **74**, 240-243.

Frazer, B.A., Fried, B., Fujino, T. and Sleckman, B.P. (1999) Host-parasite relationships between *Echinostoma caproni* and RAG-2-deficient mice. *Parasitology Research* **85**, 337-342.

Fried, B. (1994) Metacercarial excystment of trematodes. *Advances in Parasitology* **33**, 91-144.

Fried, B. (1997) An overview of the biology of trematodes, in B. Fried and T.K. Graczyk, T. K. (eds.), *Advances in Trematode Biology*, ICR Press, Inc., Boca Raton, Florida, pp. 1-23.

Fried, B. and Freeborne, N.C. (1984) Effects of *Echinostoma revolutum* (Trematoda) adults on various dimensions of the chicken intestine, and observations on worm crowding. *Proceedings of the Helminthological Society of Washington* **51**, 297-300.

Fried, B. and Huffman, J.E. (1996) The biology of intestinal trematode, *Echinostoma caproni*. *Advances in Parasitology* **38**, 311-368.

Fried, B., Huffman, J.E. and Franco, J. (1988) Single-and five-worm infections of *Echinostoma revolutum* (Trematoda) in the golden hamster. *International Journal for Parasitology* **18**, 413-441.

Fujino, T. and Fried, B. (1993a) Expulsion of *Echinostoma trivolvis* (Cort, 1914) Kanev, 1985 and retention of *E. caproni* Richard, 1964 (Trematoda: Echinostomatidae in C3H mice: pathological, ultrastructural, and cytochemical effects on the host intestine. *Parasitology Research* **33**, 286-292.

Fujino, T. and Fried, B. (1993b) *Echinostoma caproni* and *E. trivolvis* alter the binding of glycoconjugates in the intestinal mucosa of C3H mice as determined by lectin histochemistry. *Journal of Helminthology* **67**, 179-188.

Fujino, T. and Fried, B. (1996) The expulsion of *Echinostoma trivolvis* from C3H mice: difference in glycoconjugates in mouse versus hamster small intestinal mucosa during infection. *Journal of Helminthology* **70**, 115-121.

Fujino, T., Fried, B. and Hosier, D.W. (1994) The expulsion of *Echinostoma trivolvis* (Trematoda) from ICR mice: extension/retraction mechanisms and ultrastructure of the collar spines. *Parasitology Research* **80**, 581-587.

Fujino, T., Ichikawa, H. and Fried, B. (1998b) The immunosuppressive compound FK506 does not affect expulsion of *Echinostoma trivolvis* in C3H mice. *Parasitology Research* **84**, 519-521.

Fujino, T., Ichikawa, H., Fried, B. and Fukuda, K. (1997) The expulsion of *Echinostoma trivolvis*: worm kinetics and intestinal reactions in C3H/HeN mice treated with dexamethasone. *Journal of Helminthology* **71**, 257-259.

Fujino, T. Ichikawa, H., Fukuda, K. and Fried, B. (1998a) The expulsion of *Echinostoma trivolvis* caused by goblet cell hyperplasia in severe combined immunodeficient (SCID) mice. *Parasite* **5**, 219-222.

Fujino, T., Yamada, M., Ichikawa, H., Fried, B., Arizono, N. and Tada, I. (1996) Rapid expulsion of the intestinal trematodes *Echinostoma trivolvis* and *E. caproni* from C3H/HeN mice after infection with *Nippostrongylus brasiliensis*. *Parasitology Research* **82**, 577-579.

Gavet, M.F. and Fried, B. (1994) Infectivity, growth, distribution and acetabular attachment of a one-hundred metacercarial cysts inoculum of *Echinostoma trivolvis* in ICR mice. *Journal of Helminthology* **68**, 131-134.

Graczyk, T.K. (1997) Trematode immunobiology in the vertebrate host, in Graczyk, T. K. and Fried, B. (eds.) *Advances in Trematode Biology*, ICR Press, Boca Raton, Florida, pp. 383-404.

Graczyk, T.K. and Fried, B. (1994) ELISA method for detecting anti-*Echinostoma caproni* (Trematoda: Echinostomatidae) antibodies in experimentally infected ICR mice. *Journal of Parasitology* **80**, 544-549.

Graczyk, T.K. and Fried, B. (1995) An ELISA for detecting anti-*Echinostoma trivolvis* (Trematoda) immunoglobulins in experimentally infected ICR mice: cross-reactivity with *Echinostoma caproni*. *Parasitology Research* **81**, 710-712.

Graczyk, T.K. and Fried, B. (1998) Echinostomiasis: a common but forgotten food-borne disease. *The American Journal of Tropical Medicine and Hygiene* **58**, 501-504.

Graczyk, T.K., Cranfield, M.R. and Shiff, C.J. (1993) ELISA method for detecting anti-*Plasmodium relictum* and anti-*Plasmodium elongatum* antibody in duckling sera using *Plasmodium falciparum* antigens. *Journal of Parasitology* **79**, 879-885.

Graczyk, T.K., Alonso, A.A. and Balazs, G.H. (1995) Detection by ELISA of circulating anti-blood fluke (*Carettacola, Hapalotrema,* and *Learedius*) immunoglobulins in Hawaiian green turtles (*Chelonia mydas*). *Journal of Parasitology* **81**, 416-421.

Graczyk, T.K., Cranfield, M.R., Skjoldager, M.L. and Shaw M.L. (1994) An ELISA for detecting anti-*Plasmodium* spp. antibodies in African black-footed penguins (*Spheniscus demersus*). *Journal of Parasitology* **80**, 60-64.

Graczyk, T.K., Brossy, J.J., Sanders, M.L., Dubey, J.P., Plos, A. and Stoskopf, M.K. (1996) Immunological survey of Babesiosis (*Babesia peircei*) and toxoplasmosis in Jackas penguins in South Africa. *Parasite* **4**, 313-319.

Haseeb, M.A. and Fried, B. (1997) Modes of transmission of trematode infections and their control, in B. Fried and T.K. Graczyk (eds.), *Advances in Trematode Biology*, ICR Press, Boca Raton, Florida, pp. 31-56.

Hockley, D.J. and McLaren, D.J. (1973) *Schistosoma mansoni*: changes in the outer membrane of the tegument during development from cercaria to adult worm. *International Journal for Parasitology* **3**, 13-25.

Hosier, D.W. and Fried, B. (1986) Infectivity, growth, and distribution of *Echinostoma revolutum* in Swiss Webster and ICR mice. *Proceedings of the Helmintological Society of Washington* **53**, 173-176.

Hosier, D.W. and Fried, B. (1991) Infectivity, growth, and distribution of *Echinostoma caproni* (Trematoda) in the ICR mouce. *Journal of Parasitology* **77**, 640-642.

Hosier, D.F., Fried, B. and Szewczak, J.P. (1988) Homologous and heterologous resistance of *Echinostoma revolutum* and *E. liei* in ICR mice. *Journal of Parasitology* **74**, 89-92.

Huffman, J.E. and Fried, B. (1990) *Echinostoma* and echinostomiasis. *Advances in Parasitology* **29**, 215-269.

Huffman, J.E., Iglesias, D. and Fried, B. (1988) *Echinostoma revolutum*: pathology of extraintestinal infection in the golden hamster. *International Journal for Parasitology* **18**, 873-874.

Hufmann, J.E., Michos, C. and Fried, B. (1986) Clinical and pathological effects of *Echinostoma revolutum* (Digenea: Echinostomatidae) in the golden hamster, *Mesocricetus auratus*. *Parasitology* **93**, 505-515.

Incani, R.N. and McLaren, D.J. (1983) Ultrastructural observation on the in vitro interaction of rat neutrophils with schistosomula of *Schistosoma mansoni* in the presence of antibody and/or complement. *Parasitology* **86**, 345-349.

Johnson, T.M., Kurup, V.P., Resnik, A., Ash, R.C. and Fink, J.N. (1989) Detection of circulating *Aspergillus fumigatus* antigen in bone marrow transplant patients. *Journal of Laboratory and Clinical Medicine* **114**, 700-707.

Laemmli, U.K. (1970) Cleavage of the structural proteins during the assembly of the head of bacteriophage. *Nature* **227**, 680-685.

Mabus, J., Huffman, J.E. and Fried B. (1988) Humoral and cellular response to infection with *Echinostoma revolutum* in the golden hamster, *Mesocricetus auratus*. *Journal of Helminthology* **62**, 127-132.

Miyamoto K., Nakao, M., Ohnishi, K. and Inoaka, T. (1984) Studies on zoonoses in Hokkaido, Japan. 6. Experimental human echinostomiasis. *Hokkaido Journal of Medical Sciences* **59**, 696-700.

Mohandas, A. and Nadakal, A.M. (1978) In vivo development of *Echinostoma malayanum* Leiper, 1911, with notes on effect on population density, chemical composition and pathogenicity and in vitro excystment of the metacercaria (Trematoda: Echinostomatidae). *Zeitschrift für Parasitenkunde* **55**, 139-152.

Nanduri, J., Dennis, J.E., Rosenberf, T.L., Mahmoud, A.A.F. and Tarlakoff, A.M. (1991) Glycocalyx of bodies versus tails of *Schistosoma mansoni*. *The Journal of Biological Chemistry* **266**, 1311-1347.

Nawa, Y., Abe, T. and Owhashi, M. (1994) Host responses to helminths with emphasis on eosinophils and mast cells, in N. Chowdhury and I. Tada (eds.), *Helminthology,* Springer-Verlag, Narosa Publishing House, New York, pp. 243-269.

Odaibo, A.B., Christensen, N.O. and Ukoli, F.M.A. (1988) Establishment, survival, and fecundity in *Echinostoma caproni* infections in NMRI mice. *Proceedings of the Helminthological Society of Washington* **55**, 192-198.

Parkhouse, R.M.E. (1984) Parasite evasion of the immune response. *Parasitology* **88**, 517-122.

Punkpark, S., Bunnag, D. and Harinasuta, T. (1984) Albendazole in the treatment of opisthorchiasis and concomitant intestinal helminthic infection. *Southeast Asian Journal of Tropical Medicine and Public Health* **15**, 44-50.

Ramalho-Pinto, F.J., McLaren, D.J., and Smithers, S.R. (1978) Complement-mediated killing of schistosomula of *Schistosoma mansoni* by rat eosinopile in vitro. *Journal of Experimental Medicine* **147**, 147-156.

Rim, H.J. (1982) CRC handbook series in zoonoses, in G.V. Hillyer and C. Hopla, (eds.), CRC Press, Inc., Boca Raton, Florida, pp. 53-69.

Rivera-Marrero, C.A., Santiago, N. and Hillyer, G.V. (1988) Evaluation of immunodiagnostic antigens in the excretory-secretory products of *Fasciola hepatica*. *Journal of Parasitology* **74**, 646-652.

Schwartz, B.S., Ford, D.P., Childs, J.E., Rothman, N. and Thomas, R.J. (1991) Anti-tick saliva antibody: A biologic marker of tick exposure that is a risk factor for Lyme disease seropositivity. *American Journal of Epidemiology* **134**, 86-95.

Seo, B.S., Hong, S.T., Chai, J.Y., Hong, S.J. and Lee, S.H. (1985) Studies on intestinal trematodes in Korea. XVII. Development and egg laying capacity of *Echinostoma hortense* in albino rats and human experimental infection. *Korean Journal of Parasitology* **23**, 24-32.

Shekhar, K.C. (1991) Epidemiological assessment of parasitic zoonoses in Malaysia. *Southeast Asian Journal of Tropical Medicine and Public Health* **22**, 337-339.

Shinkai, Y., Rathbun, G., Lam, K., Oltz, E., Stewart, V., Mendelhson, M., Charron J., Datta, M., Young, F., Stahl, A. and Alt, F. (1992) RAG-2-deficient mice lack mature lymphocytes owing the inability to initiate V(D)J rearrangement. *Cell* **68**, 855-867.

Simonsen, P.E. and Andersen, B.J. (1986) *Echinostoma revolutum* in mice; dynamics of the antibody attack to the surface of an intestinal trematode. *International Journal for Parasitology* **16**, 475-482.

Simonsen, P.E., Vennervald, B.J. and Birch-Andersen, A. (1990) *Echinostoma caproni* in mice: ultrastructural studies on the formation of immune complexes on the surface of n intestinal trematode. *International Journal for Parasitology* **20**, 935-941.

Simonsen, P.E., Estambale, B.B. and Agger, M. (1991) Antibodies in the serum of golden hamsters experimentally infected with the intestinal trematode *Echinostoma caproni*. *Journal of Helminthology* **65**, 239-247.

Sirag, S.B., Christensen, N.Ø., Frandsen, F., Monrad, J. and Nansen, P. (1990) Homologous and heterologous resistance in *Echinostoma revolutum* infections in mice. *Parasitology* **80**, 479-486.

Tani, S. and Yoshimira, K. (1988) Spontaneous expulsion of *Echinostoma hortense* Asada, 1926 (Trematoda: Echinostomatidae) in mice. *Parasitology Research* **74**, 495-497.

Trouve, S. and Coustau, C. (1998) Differences in adult excretory-secretory products between geographical isolates of *Echinostoma caproni*. *Journal of Parasitology* **84**, 1062-1065.

Ujiie, N. (1936) On the structure and development of *Echinochasmus japonicus* and its parasitism in man. *Journal of Medical Association Formosa* **35**, 525-546.

Weinstein, M.S. and Fried, B. (1991) The expulsion of *Echinostoma trivolvis* and the retention of *Echinostoma caproni* in the ICR mouse: pathological effect. *International Journal for Parasitology* **21**, 255-257.

MOLECULAR BIOLOGY OF ECHINOSTOMES

JESSICA A. T. MORGAN[a], AND DAVID BLAIR[b]
[a]Department of Microbiology and Parasitology, University of Queensland, Brisbane Queensland, 4072 Australia, and [b]Department of Zoology and Tropical Ecology, James Cook University of North Queensland, Townsville, Queensland, 4811 Australia

B. Fried and T.K. Graczyk (eds.),
Echinostomes as Experimental Models for Biological Research, 245–266.
© 2000 *Kluwer Academic Publishers. Printed in the Netherlands.*

1. Introduction to Molecular Methods and their History in Echinostome Research

1.1. EVOLUTION OF MODERN GENETICS AND THE APPLICATION OF
   MOLECULAR TECHNIQUES

Although most molecular studies on echinostomes have focused on species identification and systematics, molecular techniques are not limited to these fields. Over the past decade molecular technology has advanced at a rapid rate; techniques are now available to investigate gene expression (Schechtman *et al.*, 1995 working on an mRNA protein in *Schistosoma*), to resolve protein structure and function (Davis, 1997 studying spliced leader RNAs in flatworms), to determine genome organisation (Boore, 1999 reviewed the mitochondrial genome) and for species diagnostics and vaccine development (Bergquist, 1995 looking at possible vaccines for schistosomiasis).

Early molecular studies involved counting chromosomes and comparing isoenzyme patterns (Churchill, 1950, Voltz *et al.*, 1987). These techniques were limited by low resolution due to a lack of variation (Mutafova, 1994). Modern molecular tools, such as DNA sequencing, the characterisation of microsatellites and the expression of genes, are able to target regions of high or low variation.

Although publications on echinostome biology appeared well over a century ago (Frölich, 1802), the majority of molecular studies have been published in the past decade. Some of the earliest studies involved karyotyping or chromosome counting and comparisons. The karyotype of *Echinostoma trivolvis* (referred to as *E. revolutum*) was described in 1950 by Churchill (Churchill, 1950). Other early studies used methods such as isoelectric focusing and enzyme electrophoresis to compare adults (Vasilev *et al.*, 1978; Voltz, Richard and Pesson, 1987; Ross *et al.*, 1989). With the advent of PCR technology in the late 1980s questions requiring higher resolution could be answered. In 1995 DNA sequencing and RAPD analysis (Randomly Amplified Polymorphic DNA) were used to compare a number of echinostome species (Morgan and Blair, 1995; Fujino *et al.*, 1995). DNA sequences are currently being used to confirm the identity of echinostome species and strains, and to match adult and larval stages.

Using molecular techniques it is possible to compare animals across a scale that ranges from the level of kingdom through to individual differences (Pace *et al.*, 1986; Hillis and Dixon, 1991). The remainder of this section will discuss the various molecular methods that are currently being used to study echinostomes. They can be broken into two categories according to whether or not they use PCR technology.

1.2. KARYOTYPING AND ENZYME ELECTROPHORESIS

Molecular techniques that do not require PCR technology include karyotyping and enzyme electrophoresis. Compared to DNA sequencing both of these methods are cheap and large numbers of samples can be processed. However, they provide correspondingly low levels of information.

A karyotype is defined as the chromosome complement and of a cell and its structural characteristics (Lincoln *et al.*, 1982). Cells are microscopically examined while undergoing mitotic division and the chromosomes are stained and photographed. The karyotype is determined by arranging pairs of mitotic chromosomes in order of size. Karyotypes are compared using morphometric characters such as chromosome number, arm length and symmetry, centromere location and heterochromatin banding patterns. Within the genus *Echinostoma* most karyotype differences are limited to C-banding patterns (stained regions of constitutive heterochromatin which are dense at centromeric regions (Lawrence, 1989; Mutafova, 1994).

Genetic differences within and among species can be quantified using enzyme electrophoresis. Soluble enzymes which catalyse steps in major metabolic pathways are used in enzyme electrophoresis, as they are soluble in the cytoplasm and can be released into solution by rupturing the cell membrane. Allozymes (different allelic forms of the same enzyme) are separated in an electric field through a matrix of starch or polyacrylamide. The molecules migrate through the medium according to their charge, shape and size then remain in position when the current is removed. In gels with a single pH the proteins move constantly through the gel. In gels with a pH gradient the proteins move until they reach their isoelectric point where they stop (Hillis *et al.*, 1996). This second method is referred to as isoelectric focusing. The gel is then stained to detect differences in banding patterns for specific proteins. The method assumes that if the banding pattern of a homologous protein differs between two conspecific individuals then the alleles encoding this protein must also differ (Hillis *et al.*, 1996).

Problems can arise when interpreting electrophoretic patterns. Gene duplication, or organisms with polyploid origins, have multiple loci and complex patterns for which it is difficult to assign a genetic basis. Artefact bands also complicate interpretations as they represent enzyme activity that does not fit simple Mendelian and biochemical models. Further weaknesses of electrophoretic methodology are that it cannot detect differences in non-coding regions, silent mutations or amino acid substitutions that have no effect on the proteins charge (Hoelzel, 1992). To detect the true level of genetic variation among organisms requires comparisons at the nucleotide level.

1.3. DEVELOPMENT OF PCR BASED METHODS

One advantage of working with DNA rather than proteins is that DNA can be stored indefinitely at -20°C while proteins do not store well as they tend to degrade over time. Protein studies also require large amounts of tissue compared to DNA studies. This requirement limits research if parasite samples are small or rare. Prior to the development of the polymerase chain reaction (PCR), attempts to study nucleotide differences were hindered by the difficulties associated with obtaining sufficient quantities of pure DNA. PCR technology enabled small quantities of specific regions of DNA to be amplified (Hoelzel, 1992).

PCR is a powerful technique used to amplify double stranded target regions of DNA between known oligonucleotide primers. A primer anneals upstream of the target region on each strand of the denatured DNA template. In the presence of available, unbound nucleotides and a DNA polymerase the primers are extended by the addition of complementary bases. The original DNA template can be copied at an exponential rate by repeatedly cycling through the three temperature steps required for denaturation, primer annealing and primer extension. This process is now fully automated (Hillis *et al.*, 1996). The amplified template is then ready for DNA sequencing which will be discussed in the next section.

RAPD (random amplification of polymorphic DNA) analysis takes the PCR reaction a step further. This molecular tool requires no prior knowledge of sequence information (normal PCR requires prior sequence knowledge for primer design). The technique employs a single short primer of random sequence, usually 10-12 bases long, to amplify segments of DNA. During PCR the primer anneals in both forward and reverse directions amplifying fragments up to 4 kb long. A 10 base primer usually amplifies 5-10 fragments which can be viewed on an agarose gel. Different banding patterns are obtained if the intervening DNA varies in length between organisms. The weakness of RAPD analysis is that results are difficult to replicate due to the amplification step being sensitive to thermocycling conditions (Hillis *et al.*, 1996).

A risk associated with techniques based on PCR amplification is that low levels of contaminants may be amplified exponentially during the reaction producing spurious results (Hoelzel, 1992).

1.4. DNA SEQUENCING

DNA sequences display heritable variation at the highest resolution. Sequencing involves determining the order of nucleotide bases in a specific region of DNA. Although a number of sequencing methods have been used in the past, the majority of molecular laboratories now use fluorescence-based cycle sequencing technology (Hillis *et al.*, 1996). Reactions can be carried out on single or double stranded DNA templates, PCR fragments or clones. The sequencing reaction amplifies the template in a similar fashion to PCR. However the reaction mix also contains terminator nucleotides labelled with fluorescent dyes. Thermocycling the mixture produces fragments of varying length and known termination site. The fragments are separated using gel electrophoresis. As the fragments migrate through the gel a tuneable laser records them, sending the results to a computer where they are interpreted into a DNA sequence (Hillis *et al.*, 1996).

Different regions of the genome have evolved at different rates. Thus it is important to select a region that displays the appropriate level of variation for the questions being asked. At least two genomes, mitochondrial and nuclear, exist within metazoan cells. DNA within these genomes displays diverse characteristics making it appropriate for molecular studies at different levels. Mitochondrial DNA (mtDNA) is small (around 16kb), present in multiple copies within a cell and compared to nuclear ribosomal coding regions, displays a high level of variability (Hillis *et al.*, 1996; Boore, 1999). For this reason mtDNA sequences are well suited to questions investigating intraspecific variability or very closely related species (Moritz *et al.*, 1987).

In contrast the nuclear genome is large; it ranges from one million nucleotide pairs in bacteria to close to 100 billion nucleotide pairs in some amphibians and plants (Alberts *et al.*, 1989). In trematodes the nuclear genome has been estimated at 270 million base pairs for *Schistosoma mansoni* (Simpson *et al.*, 1982). Nuclear DNA is passed on from both parents and is subject to recombination. A tiny fraction of the nuclear genome has been used in systematic studies (Hillis *et al.*, 1996). Most investigations on trematodes have concentrated on ribosomal genes and their associated spacers (Blair *et al.*, 1996). Ribosomal DNA (rDNA) typically consists of several hundred tandemly repeated units. Each unit is subdivided into three coding and four non-coding regions or spacers. Functional constraints within the ribosomal unit result in varying degrees of sequence conservation with the spacers being the most variable regions (Hillis and Dixon, 1991).

Despite each rDNA unit being present in multiple copies within a cell, they seldom vary from each other in primary sequence. The units also tend to homogenise among individuals within a species. This effect is a consequence of concerted evolution combined with molecular drive. Concerted evolution homogenises sequences within an individual through gene conversion and unequal crossing over (Arnheim, 1983). Molecular drive fixes a variant throughout a species due to random genetic drift and natural selection (Dover, 1982). This property of rDNA sequences is particularly useful for comparative studies at a species level, as sample sizes can be small, because a single individual should be representative of its species (Blair *et al.*, 1996).

The most conserved of the ribosomal genes is the small subunit (18S) gene. It has been sequenced extensively to study relationships among Platyhelminthes, reviewed in Blair *et al.* (1996). Comparisons among familial groupings have concentrated on the variable V4 domain within this gene. The D1 domain of the 28S coding gene has also been sequenced for a number of digeneans. This domain was useful for investigating relationships among species, genera and closely related families (Blair *et al.*, 1996).

Sequences from the rDNA internal transcribed spacers (ITS1 and 2) are commonly used to investigate species level relationships. Digenean internal transcribed spacers were first sequenced by Luton *et al.* (1992). They found considerable interspecific difference occurred between two congeneric species of *Dolichosaccus*. Since then ITS2 sequences have been shown to be sensitive enough indicators to discriminate among species of *Schistosoma* (Després *et al.*, 1992, Bowles *et al.*, 1995b), species of *Fasciola* (Adlard *et al.*, 1993), species of *Paragonimus* (Blair *et al.*, 1997a) and species of *Echinostoma* (see below).

DNA sequencing has proved a highly sensitive tool with a range of applications in biology. However, sequencing also has its limitations, for example, it is expensive and time consuming, it cannot be applied to fossil specimens or formalin preserved tissue (some exceptions exist) and we still have little understanding of genetic interpretation.

2. Case Studies Involving Echinostomes

2.1. APPLICATION TO SOLVING SPECIES IDENTITY PROBLEMS IN ECHINOSTOMES

The family Echinostomatidae consists of 61 to 141 species (Skryabin, 1979; Kanev, 1985; see Chapter 2 for an update on species in the family Echinostomatidae). Most research has focused on the genus *Echinostoma* although karyotyping has been completed for a range of species in the family including *Echinostoma trivolvis*, *E. caproni*, *E. echinatum*, *E. barbosai*, *E. miyagawai*, *E. cinetorchis*, *Echinoparyphium recurvatum*, *Echinoparyphium aconiatum*, *Hypoderaeum conoideum* and *Neoacanthoparyphium echinatoides* (Churchill, 1950; Terasaki *et al.*, 1982; Richard and Voltz, 1987; Baršiene and Kiseliene, 1990, 1991; Mutafova, 1994). With the exception of *Echinostoma hortense*, species in the genus *Echinostoma* appear to have a diploid chromosome number of 22. Species in the remaining genera posess 2n = 20 chromosomes. Terasaki *et al.* (1982) reported 2n = 20 chromosomes in *Echinostoma hortense*.

Within the genus *Echinostoma*, species identification problems have fallen into two categories: A. Misidentification, especially among 37 collar-spine species *E. revolutum*, *E. caproni* and *E. trivolvis*; and B. Descriptions of isolates as new species, which subsequently turned out to be junior synonyms of known species.

Identifying members within the 37 collar-spine group is difficult because only one consistently different morphological character exists and it is only visible in a larval stage of the life cycle (first reported by Kanev *et al.*, 1993). Morphological identification should not be based on adults alone.

A comparison of chromosome structural and C-banding patterns could not distinguish among *Echinostoma* species in the 37 collar-spine group (Mutafova, 1994). Enzyme electrophoresis proved more successful with fixed differences detected among *E. revolutum*, *E. paraensei*, *E. caproni* and *E. trivolvis* (Ross *et al.*, 1989; Kristensen and Fried, 1991; Sloss *et al.*, 1995).

RAPD analysis has also been used to investigate the taxonomic status of 37 collar-spine species. *Echinostoma trivolvis* and *E. caproni* were shown to be genetically distinct by Fujino *et al.*, (1995). A later study by Petrie *et al.* (1996) distinguished *E. caproni* from *E. paraensei*. Unfortunately the banding patterns obtained from RAPD analysis cannot be used to study systematic relationships because co-migrating bands do not necessarily represent homologous regions of DNA (Petrie *et al.*, 1996).

DNA sequencing has proved a useful tool for distinguishing among 37 collar-spine species. Morgan and Blair (1995) sequenced 1,000 nucleotide bases, of non-coding ribosomal DNA (ITS), from each of six echinostome species. The five 37 collar-spine species studied, *E. paraensei*, *E. caproni*, *E. trivolvis*, *E. revolutum* and an African isolate *E. sp.I.* differed from each other by 2.2% on average. The same species displayed greater sequence divergence across their mitochondrial DNA genes (Morgan and Blair, 1998a). Among the 37 collar-spine species the CO1 gene averaged 8% divergence (257 bases sequenced) and ND1 averaged 14% divergence (530 bases sequenced). In both studies Morgan and Blair found intraspecific variation to be much lower than interspecific. Sorensen *et al.* (1998) detected higher levels of intraspecific variation among ITS sequences from isolates of *E. revolutum* (0.9%) and *E. trivolvis* (0.6%). Their results highlight a potential problem with using sequence data alone for taxonomic diagnosis.

2.2. POPULATION DIFFERENCES AND IDENTIFYING SIBLING SPECIES

Echinostome species appear to have broad distributions and low host specificity. This has led to geographically distinct isolates being described as new species. A classic example concerns the *E. caproni* complex. Several synonyms exist for *E. caproni* (Madagascar) including *E. liei* (Egypt), *E. togoensis* (Africa) and *E.* sp. (an African isolate from Cameroon currently maintained by J. Jourdane, France). The identity of each of these isolates as *E. caproni* strains was confirmed using isoenzyme analyses (Voltz *et al.*, 1987, 1988) and DNA sequencing (Morgan and Blair, 1995, 1998a). Cross breeding experiments also support this conclusion (Voltz *et al.*, 1988).

Morgan and Blair (1998a, 1998b) found mtDNA ND1 sequences more useful than nuclear rDNA spacers for studying Echinostome populations. Mitochondrial DNA sequences from the *E. caproni* isolate originating in Madagascar differed from the Cameroon sample. However, the level of divergence was 5-fold lower than that detected between recognised species (less than 2.5% between populations compared to 12.5% between species). The same individuals were indistinguishable through the ribosomal ITS. Similar results were obtained for *E. revolutum* populations originating from Germany and Australia and *E. paraensei* isolates from Brazil and Australia (Morgan and Blair, 1998b).

Sorensen *et al.* (1998) found population variability among ITS sequences from *E. trivolvis* (two North American isolates) and *E. revolutum* (North American isolates compared to a German isolate). Their result is surprising as ribosomal DNA is generally highly conserved within species (Hillis and Dixon, 1991). The three North American *E. revolutum* isolates were identical to each other but consistently different to the German strain suggesting that the two geographic strains have been isolated for some time. In contrast, the two *E. trivolvis* isolates were collected from the same site in northern Indiana. Further experiments are required to determine if the two strains interbreed before a decision can be made about their taxonomic status (Sorensen *et al.*, 1998).

2.3. USEFULNESS IN MATCHING LIFE CYCLE STAGES

The genome is not influenced by environmental conditions. Thus, phenotypic variation, which can confound morphological studies, is not a problem when using molecular approaches. Species with complex life histories can also be difficult to identify using traditional methods. Morphologically different life-cycle stages can be identified unambiguously using molecular methods.

Varying morphology and complex life cycles make echinostomes difficult to identify in the field. Due to the problems associated with obtaining all of the life stages of individual species their names have generally only been applied to adult specimens. The use of molecular markers to characterise species overcomes problems of phenotypic and environmental variation as well as life cycle stage variability. A molecular approach can be used to match intermediate and adult stages when it is impossible to complete life cycles experimentally.

It may be necessary to pool individual cercariae or metacercariae when extracting DNA from intermediate life stages. If the host only has a single infection the individuals should be genetically identical. DNA sequences from cercariae, collected over a four year period in Australia, were aligned to published sequences from known adults (Morgan and Blair, 1998b). The presence of *E. paraensei* was established, plus a further three unidentified Australian species were found. Fujino *et al.* (1997) matched metacercariae of *E. caproni* and *E. trivolvis* to adult specimens using specific primers designed from a variable region of genomic DNA. The ability to match stages without having to complete complex life cycles makes molecular characterisation a powerful tool for echinostome diagnostics.

2.4. ECHINOSTOME PHYLOGENETICS

Every heritable component of an organism is coded within its DNA. Thus, the largest possible set of characters for systematic analysis is stored in a single cell. For a cell to remain operational the DNA that codes for functional genes and proteins must be highly conserved. DNA sequences from these genes are being compared among distantly related taxa to investigate evolutionary relationships (Hillis and Dixon, 1991). In contrast non-coding DNA, which constitutes a large proportion of the genome, contains regions that evolve at such a rapid rate that they can be used to distinguish individuals from one-another (Hoelzel, 1992).

Establishing how species are related to one another is important. Phylogenetic trees enable predictions to be made about the evolutionary origins of species. Echinostome phylogenetics is in its infancy. Higher taxonomic studies group the Echinostomatidae with the Fasciolidae in the order Echinostomiformes (Brooks and McLennan, 1993).

Adult morphological characters vary so little among members of the genus *Echinostoma* that in the past it has not been possible to investigate systematic relationships within this group. Molecular techniques provide a greater number of informative characters for systematic studies. Using six fixed enzyme differences, Sloss *et al.* (1995) suggested that *E. trivolvis* and *E. caproni* are more closely related to each other than either is to *E. paraensei*.

DNA sequencing provides the largest number of variable characters to distinguish among 37 collar-spine species. Three phylogenetic trees have been generated from DNA sequence alignments originating from the ribosomal ITS (Fig. 1), mitochondrial CO1 (Fig. 2) and mitochondrial ND1 genes (Fig. 3). Branch lengths are indicated on all trees and bootstrap values, estimated from 1000 random trees, are also shown. The CO1 tree (Fig. 2) is the least informative of the three as the strict consensus results in a number of collapsed branches.

The ribosomal ITS and mitochondrial ND1 trees are largely congruent with the exception of *E.* sp.I (a parthenogenetic isolate from Africa) which shifts from a basal position in the 37 collar-spine cluster to an internal position depending on the data set. Published ITS trees, excluding *E.* sp (Aus), an Australian isolate, and *Fasciola hepatica*, place E. sp.I. in the basal position (Morgan and Blair, 1995; Sorenson *et al.*, 1998). This change of topology highlights the importance of outgroup selection and underlines the warning of Sorensen *et al.* (1998) not to place too much faith in a single tree.

Both the ND1 and ITS trees strongly support the monophyly of the 37 collar-spine species. Unlike the enzyme data, DNA sequences suggest a closer relationship between *E. trivolvis* and *E. paraensei*, with *E. caproni* more distant.

In their comparison of the three gene regions, Morgan and Blair (1998a) found the mitochondrial ND1 gene to be the most informative marker for echinostome population studies. They warn against using this gene for estimating relationships among distantly related species of trematodes due to high levels of saturation (multiple substitutions at the same site).

Phylogenetic studies within the family Echinostomatidae have a long way to go. Selecting an appropriate gene region, which displays sufficient variation to distinguish species at the correct taxonomic level, is possibly the most important factor to consider before embarking on a systematic study.

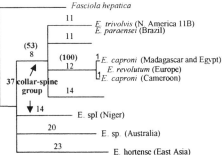

Figure 1. Strict consensus parsimony tree inferred from ribosomal ITS sequence data for a number of isolates of described echinostome species (2 equally parsimonious trees, 323 steps, *CI* = 0.879). *Fasciola hepatica* is included as an outgroup. Bootstrap values appear in brackets above branch lengths. The base of the 37 collar-spine group is indicated. Tree generated from sequences published by Morgan and Blair (1995), Sorensen *et al.* (1998) and Morgan (1997).

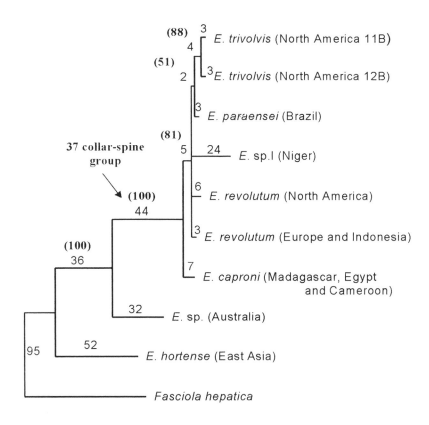

Figure 2. Strict consensus parsimony tree inferred from partial mitochondrial CO1 sequence data for described echinostome species and an Australian isolate (3 equally parsimonious trees, 124 steps, CI = 0.734). *Fasciola hepatica* is included as an outgroup. The tree is congruent with that produced by distance matrix analyses. Bootstrap values (PAUP) appear in brackets above branch lengths. The base of the 37 collar-spine group is indicated. Tree based on sequences published by Morgan and Blair (1998b).

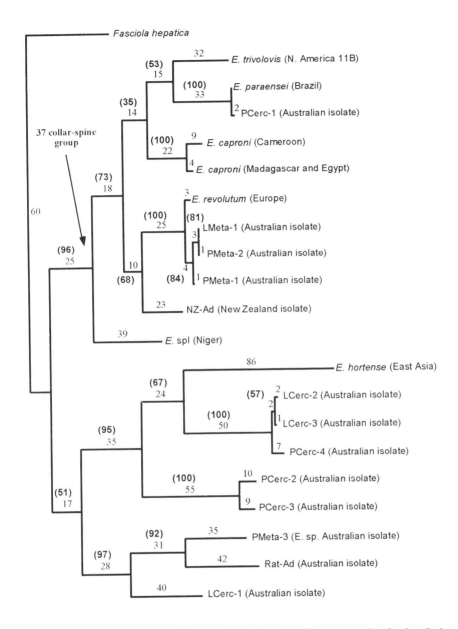

Figure 3. Parsimony tree inferred from partial mitochondrial ND1 sequence data for described echinostome species and Australian and New Zealand isolates (1 tree, 817 steps, CI = 0.561). *Fasciola hepatica* is included as an outgroup. Parsimony and distance matrix analyses produce identical trees. Bootstrap values (PAUP) appear in brackets above branch lengths. The base of the 37 collar-spine group is indicated. Tree slightly modified from that published by Morgan and Blair (1998b). Morgan and Blair (1998b) contains a key to the Australian and New Zealand isolates.

3. Importance of Molecular Biology to Modern Biology

3.1. THE FUTURE OF SPECIES IDENTIFICATION AND CLASSIFICATION

Molecular markers are proving to be sensitive indicators for the discrimination of closely related species in both plant and animal kingdoms (Coleman and Mai, 1997, Wesson et al., 1993). Our classification system is relying more heavily upon DNA sequences to provide variable characters to distinguish among cryptic species. Genbank is a molecular database, freely available via the internet, where DNA and protein sequences are catalogued and stored (the NCBI Entrez nucleotide sequence search page address is http://www.ncbi.nlm.nih.gov/entrez/nucleotide.html). Mitochondrial and ribosomal DNA sequences are currently available for a range of trematode species including representatives from the genera *Schistosoma*, *Fasciola*, *Paragonimus*, *Dolichosaccus* and *Echinostoma*. A search of the previous five genera identifies nearly 11,000 entries; of these over 95% are for *Schistosoma* species. As this database grows the potential uses for the system expand. Cryptic species identification in the future may be as simple as pressing a "match" button on your computer.

3.2. PHYLOGENETIC COMPARISONS

Scientists are starting to make use of the database of sequence information building up on different taxa. Molecular markers, in particular ribosomal coding regions, have proved helpful for studying deeper phylogenetic estimates. The ribosomal 18S gene has been sequenced extensively to study relationships among Platyhelminthes, (Blair et al., 1996; Littlewood et al., 1999). More recently the 28S gene has been utilised (Mollaret, et al., 1997; Litvaitis and Rohde, 1999). Both 18S and 28S ribosomal genes were found to be informative for inferring relationships at or below the level of family, however their resolution was poor for deeper level relationships. Although the various trees support many of the traditional morphological groupings none of them are strictly congruous. Increasing effort is being directed towards resolving these deeper phylogenetic levels.

   If the same phylogeny is predicted from independent gene regions then increasing confidence can be placed in the predicted tree. Directing attention to other gene regions will provide new sequence information for analysis. Blair et al. (1996) recommend focusing on single copy, protein coding nuclear genes for ease of alignment and to ensure that the gene region is strictly orthologous.

3.3. GENOME STRUCTURE AND FUNCTION

Although extensive regions of DNA have been sequenced we still have little understanding of genome organisation and function in trematodes. Homeobox genes have received a great deal of attention. The Hox/HOM class of homeobox genes are important in the anterior-posterior formation in early development of all major metazoan phyla (Bartels *et al.*, 1993). A study of the Hox/HOM homeobox gene family in *Echinostoma trivolvis*, along with other Platyhelminthes, suggest that the flatworm genes may be useful as an outgroup for investigating the evolution of the gene cluster in higher metazoans (Bartels *et al.*, 1993).

The ribosomal RNA internal transcribed spacer 2 is transcribed then removed from the maturing rRNA. Its presence and structural integrity appears to be essential in the process of ribosomal biogenesis (van der Sande *et al.*, 1992; Schlötterer *et al.*, 1994; van Nues *et al.*, 1995). Despite this importance ITS2 primary sequences are poorly conserved varying in both base composition and size. Morgan and Blair (1998c) used secondary structures to aid primary sequence alignments of ITS2 sequences spanning four trematode orders, including the Echinostomiformes, and a monogenean. Their alignment revealed two conserved strings. The first string aligned to a processing site identified in yeast and rat. The second string matched a site identified in the same structural location in plants. Although the precise function of these sites remains unknown they have now been identified for future research.

Another step in RNA processing involves joining (splicing) independently transcribed RNAs to generate mature mRNAs (Lewin, 1990). The spliced leader (SL) is a small exon that is spliced to the 5' end of a pre-mRNA to form a mature mRNA in a process called *trans*-splicing. Davis (1997) carried out a comparative study of flatworm SL RNAs, including those from *Echinostoma caproni*. He found that the sequence and length of the SL exon differed significantly among taxa suggesting that neither of these factors is important in the process of *trans*-splicing in flatworms. Davis (1997) also identified phylogenetically conserved regions suggesting that the SL exon is derived from a common ancestor.

4. Selected Molecular Biology Protocols

4.1. DNA EXTRACTION, PCR AMPLIFICATION AND SEQUENCING PROTOCOLS

Below are a number of molecular biology protocols that have been successfully used on echinostomes. They are by no means inclusive but provide enough information to extract, amplify and sequence DNA from a single specimen.

4.1.1. Genomic DNA Extraction
Samples may be fresh, frozen, stored in DMSO or stored in ethanol. Before extraction DMSO and ethanol stored specimens should be left overnight in extraction buffer (40mM Tris, pH 8.0, 10mM EDTA, 200mM NaCl). Specimens are then homogenised in a 1.5 ml microcentrifuge tube containing 500 μl of fresh extraction buffer, SDS (sodium dodecyl sulphate) to a final concentration of 1% and 200 μg of proteinase K.

Complete digestion of the lysis mixture varies from 30 min to 2 h at 37°C depending on the size of the specimen.

Lysis is followed by a phenol/chloroform extraction to remove protein. An equal volume of phenol:chloroform:isoamyl alcohol is added and the mixture is gently inverted then centrifuged at 9,000 rpm for 5 min. The upper aqueous layer is transferred to a new microcentrifuge tube. This step is repeated until the aqueous layer becomes clear. Then an equal volume of chloroform is added to the aqueous solution, inverted to mix then centrifuged at 9,000 rpm for 5 min. Again the aqueous solution is transferred to a new microcentrifuge tube.

To precipitate the DNA one tenth volume of 3 M sodium acetate, pH5, and 2.5 volumes of ethanol are added to the microcentrifuge tube, then the solution is gently mixed and left at –20°C overnight. The DNA is pelleted by centrifuging for 15 min at 14,000 rpm. After removing the supernatant the pellet is washed with 500 μl of 70% ethanol, respun for 5 min at 14,000 rpm, decanted and the pellet is left to air dry before resuspending in water (30-50 μl depending on the size of the original sample).

4.1.2. PCR Amplification

Until PCR reactions have been optimised an initial reaction volume of 20 μl is recommended to minimise the potential loss of genomic DNA. To a thin walled PCR tube 10-100 ng of DNA template is added to 10 pmol of each primer (see next section for successful primer pairs), 10x *Taq* buffer, 0.8 mM dNTP, 4mM $MgCl_2$ and 0.25 units of *Taq* polymerase. This mix is then thermocycled through 30 temperature cycles. The cycling conditions displayed in Table 1 are recommended to produce a product up to 1000 bases long between primers that may not match the template perfectly (the annealing temperature can be raised if mis-priming occurs and times can be reduced for smaller products) (Table 1). Amplified products can be visualised electrophoretically by loading 2 μl on a 0.6% agarose gel stained with ethidium bromide. PCR products should be stored at –20°C.

Table 1. Suggested thermocycling conditions for PCR amplification.

Cycle	Step	Temp °C	Time	Times through cycle
1	1	96	3 min	1
	2	50	2 min	
	3	72	90 sec	
2	1	95	15 sec	4
	2	50	15 sec	
	3	72	90 sec	
3	1	95	5 sec	25
	2	57	5 sec	
	3	72	90 sec	
4	1	72	5 min	1

4.1.3. Direct DNA Sequencing

Before sequencing, PCR products need to be purified to remove primers, nucleotides, polymerases and salts. A number of comercial purification kits are available for this purpose. Fluorescent cycle sequencing kits come with a ready reaction mix that contains all but template and primer. In a 20 μl reaction 30-90 ng of PCR product plus 3.2 pmol of primer are added to the ready reaction mix and thermocycled following the kits recommended protocol. This is followed by an ethanol/sodium acetate precipitation before the products are run on an automated sequencer. The chromatograms produced are then ready to align. Aligning two sequences produces a series of matched pairs. These pairs consist of either identical bases, different bases or a base and an alignment gap. The simplest method of aligning sequences is by eye.

4.2. PCR AND SEQUENCING PRIMERS

A number of successful PCR and sequencing primers have been used to amplify and sequence echinostome DNA. The sequences of these primers are given in Table 2 and their position in either the ribosomal or mitochondrial genome is indicated in Figs 4., 5.

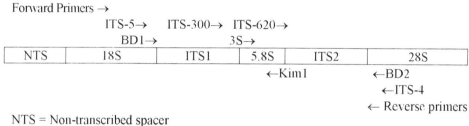

NTS = Non-transcribed spacer
18S. 5.8S. 28S = Coding regions
ITS1. ITS2 = Internal transcribed spacers 1 and 2

Figure 4. Diagram of Ribosomal DNA repeat unit showing location and direction of primers used in echinostome studies.

ND1. ND3 = Nicotinamide adenine dinucleotide dehydrogenase subunit 1 or 3
CO1 = Cytochrome *c* oxidase subunit 1

Figure 5. Schematic of mtDNA ND1 and CO1 genes, based on *Fasciola hepatica* sequence, showing location and direction of primers used for echinostome study.

Table 2. List of published PCR and sequencing primers successfully used on Echinostome DNA.

Site	Dir.	Name	Primer sequence (5'-3' direction)	Reference
18S	Forward	ITS-5	GGAAGTAAAAGTCGTAACAAG	Sorensen *et al.*, 1998
18S	Reverse	ITS-4	TCCTCCGCTTAGTGATATGC	Sorensen *et al.*, 1998
ITS1	Forward	ITS-300	TCACTGTTCAAGTGGTTG	Sorensen *et al.*, 1998
ITS2	Forward	ITS-620	GTCGGCTTAAACTATCAC	Sorensen *et al.*, 1998
18S	Forward	BD1	GTCGTAACAAGGTTTCCGTA	Morgan and Blair, 1995
28S	Reverse	BD2	TATGCTTAAATTCAGCGGGT	Morgan and Blair, 1995
5.8S	Forward	3S	GGTACCGGTGGATCACGTGGCTAGTG	Morgan and Blair, 1995
5.8S	Reverse	Kim1	GTGTTACCGCGGCG/TGCTG	Morgan and Blair, 1995
ND1	Forward	JB11	AGATTCGTAAGGGGCCTAATA	Morgan and Blair, 1998
ND1	Reverse	JB12	ACCACTAACTAATTCACTTTC	Morgan and Blair, 1998
CO1	Forward	JB3	TTTTTTGGGCATCCTGAGGTTTAT	Morgan and Blair, 1998
CO1	Reverse	JB13	TCATGAAAACACCTTAATACC	Morgan and Blair, 1998

4.3. ANALYSING RESULTS USING DIRECT COMPARISON TO FIND DIFFERENCES AND PHYLOGENETIC METHODS TO INVESTIGATE RELATIONSHIPS

Aligning two or more homologous sequences involves adding gaps, where appropriate, to obtain the correct alignment. The assumption of homology at each position is critical as minor changes in alignment can cause pronounced differences in inferred trees (Morrison and Ellis, 1997). Aligning sequences by eye is sufficient for aligning few taxa with similar sequences (Blair *et al.*, 1996). A number of computer programs have been designed to assist with more difficult alignments; two of these are CLUSTAL (Higgins and Sharp, 1988) and MALIGN (Wheeler and Gladstein, 1994). Similarity and distance methods minimise the number of gaps then either maximise the number of identical pairs (similarity), or minimise the number of different pairs (distance) (Li, 1997). An algorithm based comparison method will always find an alignment, even between completely random sequences. The human eye can usually detect if an alignment is sensible and at times adjustments may be required (Blair *et al.*, 1996).

The most common way of quantifying genetic distances between species is to produce a distance matrix indicating the degree of difference among all pairs of sequences. The simplest distance estimator is the p-distance. It measures the proportion of sites at which two sequences differ. Provided pairwise distances are small, this measure is accurate (Kumar *et al.*, 1993). As pairwise distances increase, back mutations become more frequent and the p-distance underestimates the true distance.

Different distance estimators have been devised to compensate for multiple substitutions at the same site. Some of the more commonly used distance estimators include the Jukes-Cantor distance, the Kimura 2-parameter distance, the Tamura distance and the Tamura-Nei distance (Kumar *et al.*, 1993). These estimators become more elaborate with additional parameters included to allow for unequal nucleotide ratios and different substitution rates (Hillis *et al.*, 1996). The majority of studies investigating trematode systematics have used the Kimura 2-parameter distance (Bowles *et al.*, 1995a, 1995b; Barker and Blair, 1996; Blair *et al.*, 1997b, Morgan and Blair, 1995, 1998a, 1998b, 1998c).

Although genetic distances can be used to gauge the degree to which species have separated, they are most readily interpretable when used to infer a phylogenetic tree. Two classes of tree building methods exist. The first class compares relationships among distances generated in a distance matrix. This method uses all variable sites. The second class is based on discrete characters. The two methods work well under different circumstances (Li, 1997).

A number of tree-making methods can be used to generate trees from a distance matrix. These include the UPGMA method (unweighted pair group method with arithmetic means), the Minimum Evolution method (ME) and the Neighbour Joining method (NJ). All of the methods aim to minimise branch lengths and distances between taxa. The NJ method is used most frequently for constructing trees in trematode studies (Bowles *et al.*, 1995a, 1995b; Barker and Blair, 1996; Blair *et al.*, 1997b).

Two commonly used discrete character methods exist to construct trees; they are maximum parsimony and maximum likelihood (Kumar *et al.*, 1993). Maximum parsimony uses discrete character states (at each nucleotide position) to construct a tree. It seeks to find the tree on which the fewest character changes are required to account for the observed data (Felsenstein, 1989). Maximum Likelihood calculates the probability, for every possible tree, of observing the original sequence configuration. The tree with the largest (maximum) likelihood value is chosen as the preferred tree (Li, 1997). This technique is the most general method available to infer phylogenies and it is also one of the slowest (Felsenstein, 1988). Hillis *et al.*, (1996) provide a detailed table of software packages available for conducting phylogenetic and population genetic analyses (pp510-514).

All of the tree building methods described above will produce an optimal tree (or trees) from a given data set. A collection of completely random sequences will still produce a tree (Nadler, 1990). Bootstrapping is the most common method used to measure the robustness of estimated trees. It can be applied to trees inferred using distance matrix or parsimony methods. The bootstrap test involves constructing a large number of trees by repeatedly resampling the original data set, with replacement. The reproducibility of the original estimate can be inferred from these replicates (Felsenstein, 1985).

A number of problems have been discovered with bootstrap values, reviewed in Swofford *et al.* (1996). Bootstrap repeatability estimates tend to be relatively unbiased but highly imprecise. Bootstrap accuracy estimates are biased tending to underestimate high values and overestimate low values. Finally, if more than one internal branch is being tested, or if the branches of interest can not be pre-specified, then a multiple tests problem arises which causes an inflation in the type 1 error, that is, the chance increases of rejecting the null hypothesis when it is true. For these reasons bootstrap values should be used as an index of support rather than a statistical test.

5. Future Research

There are still many avenues available for molecular research on echinostomes. Relatively few molecular based studies have been carried out to date. A number of morphologically cryptic groups require research. The genus *Echinoparyphium* contains a number of morphologically similar species. McCarthy (1990) identified two sympatric sibling species in the 45 collar-spined *Echinoparyphium recurvatum* complex. Like members of the *Echinostoma revolutum* complex these species were found to be morphologically indistinguishable in all major respects, yet each displayed distinct biological charactersitics. Revision of members within the 43 collar-spined *Echinoparyphium elegans* complex have also been carried out with a number of species being synonymised based on the morphology of their paratypes (Mouahid and Moné, 1988). Our understanding of species relationships within both of these complexes will be greatly aided when molecular characters are included in their descriptions.

Within the genus *Echinostoma* information relating to the cercarial stage is often the most useful for distinguishing among cryptic species. However, many collections only contain adults. The ability to match different life cycle stages using molecular characters will make greater use of these and future collections. As the number of genes sequenced and species sampled increases it will be possible to make more accurate hypotheses about phylogenetic relationships, both within the family and among trematodes in general.

Molecular techniques are continually being developed and improved. New areas of research that could be applied to echinostomes include microsatellite markers for studying population genetics; protein structure and function for developing vaccines and investigating drug resistence; and understanding gene expression and regulation during different developmental stages.

6. References

Adlard, R. D., Barker, S. C., Blair, D. and Cribb, T. H. (1993) Comparison of the second internal transcribed spacer (ribosomal DNA) from populations and species of Fasciolidae (Digenea), *International Journal for Parasitology* **23**, 423-425.

Alberts, B., Bray, D., Lewis, J., Raff, M., Roberts, K. and Watson, J. D. (1989) *Molecular biology of the cell.* Second edition. Garland Publishing, New York.

Arnheim, N. (1983) Concerted evolution of multigene families. In: *Evolution of genes and proteins.* Nei, M. and Koehn, R. K. (eds.) Sinauer Associates, Massachusetts, 38-61.

Barker, S. C. and Blair, D. (1996) Molecular phylogeny of *Schistosoma* species supports traditional groupings within the genus, *Journal of Parasitology* **82**, 292-298.

Baršiene, J. and Kiseliene, V. (1990) Karyological studies of *Echinoparyphium aconiatum* (Dietz, 1909), *Hypoderaeum conoideum* (Bloch, 1782) Dietz, 1909 and *Neoacanthoparyphium echinatoides* (Filippi, 1854) Odening, 1962 (Trematoda, Echinostomatidae), *Acta Parasitologia Polonica* **35**, 272-276.

Baršiene, J. and Kiseliene, V. (1991) Karyological studies of Trematodes within the genus *Echinostoma*, *Acta Parasitologica Polonica* **36**, 25-30.

Bartels, J. L., Murtha, M. T. and Ruddle, F. H. (1993) Multiple Hox/HOM-class homeoboxes in Platyhelminthes, *Molecular Phylogenetics and Evolution* **2**, 143-151.

Bergquist, N. R. (1995) Controlling schistosamiasis by vaccination: a realistic option? *Parasitology Today* **11**, 191.

Blair, D., Agatsuma, T., Watanobe, T., Okamoto, M. and Ito, A. (1997a) Geographical genetic structure within the human lung fluke, *Paragonimus westermani*, detected from DNA sequences, *Parasitology* **115**, 411-417.

Blair, D., van Herwerden, L., Hirai, H., Taguchi, T., Habe, S., Hirata, M., Lai, K., Upatham, S. and Agatsuma, T. (1997b) Relationships between *Schistosoma malayensis* and other Asian schistosomes deduced from DNA sequences, *Molecular and Biochemical Parasitology* **85**, 259-263.

Boore, J. L. (1999) Survey and Summary: Animal mitochondrial genomes, *Nucleic Acids Research* **27**, 1767-1780.

Bowles, J., Blair, D. and McManus, D. P. (1995a) A molecular phylogeny of the genus *Echinococcus*, *Parasitology* **110**, 317-328.

Bowles, J., Blair, D. and McManus, D. P. (1995b) A molecular phylogeny of the human schistosomes, *Molecular Phylogenetics and Evolution* **4**, 103-109.

Blair, D., Campos, A., Cummings, M. P., and Laclette, J. P. (1996) Evolutionary biology of parasitic platyhelminthes: the role of molecular phylogenetics, *Parasitology Today* **12**, 66-71.

Brooks, D. R. and McLennan, D. A. (1993) *Parascript - Parasites and the language of evolution.*

Smithsonian Institute Press, Washington.

Churchill, H. M. (1950) Germ cell cycle of *Echinostoma revolutum* (Froelich, 1802), *Journal of Parasitology*, **36** (6, Suppl. Sec. 2), 15.

Coleman, A. W. and Mai, J. C. (1997) Ribosomal DNA ITS-1 and ITS-2 sequence comparisons as a tool for predicting relatedness, *Journal of Molecular Evolution* **45**, 168-177.

Davis, R. E. (1997) Surprising diversity and distribution of spliced leader RNAs in flatworms, *Molecular and Biochemical Parasitology* **87**, 29-48.

Després, L., Imbert-Establet, D., Combes, C. and Bonhomme, F. (1992) Molecular evidence linking hominid evolution to recent radiations of schistosomes, *Molecular Phylogenetics and Evolution* **1**, 295-304.

Dover, G. (1982) Molecular drive: a cohesive mode of species evolution, *Nature* **299**, 111-117.

Felsenstein, J. (1985) Confidence limits on phylogenies: an approach using bootstrap, *Evolution* **39**, 783-791.

Felsenstein, J. (1988) Phylogenies from molecular sequences: Inference and reliability, *Annual Review of Genetics* **22**, 521-565.

Felsenstein, J. (1989) PHYLIP - Phylogeny inference package (Version 3.2), *Cladistics* **5**, 164-166.

Frölich, J. A. (1802) Beyträge zur Naturgeschichte der Eingeweidewürmer, *Naturforscher Halle* **29**, 5-96.

Fujino, T., Takahashi, Y. and Fried, B. (1995) A comparison of *Echinostoma trivolvis* and *E. caproni* using random amplified polymorphic DNA analysis, *Journal of Helminthology* **69**, 263-264.

Fujino, T., Zhiliang, W., Nagano, I., Takahashi, Y. and Fried, B. (1997) Specific primers for the detection of genomic DNA of *Echinostoma trivolvis* and *E. caproni* (Trematoda:Echinostomatidae), *Molecular and Cellular Probes* **11**, 77-80.

Higgins, D. G. and Sharp, P. M. (1988) CLUSTAL: A package for performing multiple sequence alignment on a microcomputer, *Gene* **73**, 237-244.

Hillis, D. M. and Dixon, M. T. (1991) Ribosomal DNA: Molecular evolution and phylogenetic inference, *Quarterly Review of Biology* **66**, 411-446.

Hillis, D. M., Moritz, C. and Mable, B. K. (1996) *Molecular Systematics*. Second edition. Sinauer Associates, Sunderland, MA, USA.

Hoelzel A. R. (ed.) (1992) *Molecular genetic analysis of populations a practical approach*. Oxford University Press, Oxford.

Kanev, I. (1985) On the morphology, biology, ecology and taxonomy of *E. revolutum* group (Trematoda: Echinostomatidae: *Echinostoma*) Ph.D. Dissertation, University of Sofia, Bulgaria. English abstract.

Kanev, I., Eisenhut, U., Ostrowski De Nunez, M., Manga-Gonzalez, M. Y. and Radev, V. (1993) Penetration and paraoesophageal gland cells in *Echinostoma revolutum* cercariae from its type locality, *Helminthologia*. **30**, 131-133.

Kumar, S., Tamura, K. and Nei, M. (1993) *MEGA: Molecular evolutionary genetics analysis*. Version 1.02. The Pennsylvanian State University, University Park, Pennsylvania.

Kristensen, A. R. and Fried, B. (1991) A comparison of *Echinostoma caproni* and *Echinostoma trivolvis* (Trematoda: Echinostomatidae) adults using isoelectric-focusing, *Journal of Parasitology* **77**, 496-498.

Lawrence, E. (1989) *Henderson's dictionary of biological terms*. 10th edition. Longman Scientific and Technical, England.

Lewin, B. (1990) *Genes IV*. Oxford University Press, Oxford.

Li, W. H. (1997) *Molecular evolution*. Sinauer Associates, Sunderland Massachusetts, USA.

Littlewood, D. T. J., Rohde, K. and Clough, K. A. (1999) The interrelationships of all major groups of Platyhelminthes: phylogenetic evidence from morphology and molecules, *Biological Journal of the Linnean Society* **66**, 75-114.

Litvaitis, M. K. and Rohde, K. (1999) A molecular test of platyhelminth phylogeny: inferences from partial 28S rDNA sequences, *American Microscopical Society* **118**, 42-56.

Luton, K., Walker, D. and Blair, D. (1992) Comparisons of ribosomal internal transcribed spacers from two congeneric species of flukes (Trematoda: Digenea), *Molecular and Biochemical Parasitology* **56**,

323-328.

McCarthy, A. M. (1990) Speciation of echinostomes: evidence for the existence of two sympatric sibling species in the complex *Echinoparyphium recurvatum* (von Linstow 1873) (Digenea: Echinostomatidae), *Parasitology* **101**, 35-42.

Mollaret, I., Jamieson, B. G. M., Adlard, R. D., Hugall, A., Lecointre, G., Chombard, C., and Justine, J. L. (1997) Phylogenetic analysis of the Monogenea and their relationships with Digenea and Eucestoda inferred from 28S rDNA sequences, *Molecular Biochemistry and Parasitology* **90**, 433-438.

Morgan, J. A. T. (1997) *Evaluation of DNA sequences for solving taxonomic problems in trematodes, specifically echinostomes.* Ph.D. Dissertation, James Cook University of North Queensland, Australia.

Morgan, J. A. T. and Blair, D. (1995) Nuclear rDNA ITS sequence variation in the trematode genus *Echinostoma*: an aid to establishing relationships within the 37 collar-spine group, *Parasitology* **111**, 609-615.

Morgan, J. A. T. and Blair, D. (1998a) Relative merits of nuclear ribosomal internal transcribed spacers and mitochondrial CO1 and ND1 genes for distinguishing among *Echinostoma* species (Trematoda), *Parasitology* **116**, 289-297.

Morgan, J. A. T. and Blair, D. (1998b) Mitochondrial ND1 gene sequences used to identify echinostome isolates from Australia and New Zealand, *International Journal for Parasitology* **28**, 493-502.

Morgan, J. A. T. and Blair, D. (1998c) Trematode and monogenean rRNA ITS2 secondary structures support a four-domain model, *Journal of Molecular Evolution* **47**, 406-419.

Moritz, C., Dowling, T. E. and Brown, W. M. (1987) Evolution of animal mitochondrial DNA: relevance for population biology and systematics, *Annual Review of Ecology and Systematics* **18**, 269-292.

Morrison, D. A. and Ellis, J. T. (1997) Effects of nucleotide sequence alignment on phylogeny estimation: A case study of 18S rDNAs of Apicomplexa, *Molecular Biology and Evolution* **14**, 428-441.

Mouahid, A. and Moné, H. (1988) *Echinoparyphium elegans* (Looss, 1899) (Digenea: Echinostomatidae): the life cycle and redescription of the adult with a revision of the 43-spined members of the genus *Echinoparyphium*, *Systematic Parasitology* **12**, 149-157.

Mutafova, T. (1994) Karyological studies on some species of the families Echinostomatidae and Plagiorchiidae and aspects of chromosome evolution in trematodes, *Systematic Parasitology* **28**, 229-238.

Nadler, S. A. (1990) Molecular approaches to studying helminth population genetics and phylogeny, *International Journal for Parasitology* **20**, 11-29.

Pace, N. R., Olsen, G. J. and Woese, C. R. (1986) Ribosomal RNA phylogeny and the primary lines of evolutionary descent, *Cell* **45**, 325-326.

Petrie, J. L., Burg III, E. F. and Cain, G. D. (1996) Molecular characterization of *Echinostoma caproni* and *E. paraensei* by random amplification of polymorphic DNA (RAPD) analysis, *Journal of Parasitology* **82**, 360-362.

Richard, J. and Voltz, A. (1987) Preliminary data of the chromosomes of *Echinostoma caproni* Richard 1964 (Trematoda: Echinostomatidae), *Systematic Parasitology* **9**, 169-172.

Ross, G. C., Fried, B. and Southgate, V. R. (1989) *Echinostoma revolutum* and *E. liei*: Observations on enzymes and pigments, *Journal of Natural History* **23**, 977-981.

Schechtmann, D., Ram, D., Tarrab-Hazdai, R., Arnon, R. and Schechter, I. (1995) Stage-specific expression of the mRNA encoding a 14-3-3 protein during the life cycle of *Schistosoma mansoni*, *Molecular Biochemistry and Parasitology* **73**, 275.

Schlötterer, C., Hauser, M.-T., von Haeseler, A. and Tautz, D. (1994) Comparative evolutionary analysis of rDNA ITS regions in Drosophila, *Molecular Biology and Evolution* **11**, 513-522.

Simpson, A. J. G., Scher, A. and McCutchan, T. F. (1982) The genome of *Schistosoma mansoni*: isolation of DNA, its size, bases and repetitive sequences, *Molecular Biochemistry and Parasitology* **6**, 125.

Skryabin, K. I. (1979) *Trematodes of animals and man: Fundamentals of trematology.* Vol. 1. Amerind Publishing Company, New Delhi.

Sloss, B., Meece, J., Romano, M. and Nollen, P. M. (1995) The genetic relationships between *Echinostoma caproni*, *E. paraensei* and *E. trivolvis* as determined by electrophoresis, *Journal of Helminthology* **64**, 243-246.

Sorensen, R. E., Curtis, J. and Minchella, D.J. (1998) Intraspecific variation in the rDNA ITS loci of 37-collar-spined echinostomes from North America: implications for sequence-based diagnoses and phylogenetics, *Journal of Parasitology* **84**, 992-997.

Swofford, D. L., Olsen, G. J., Waddell, P. J. and Hillis, D. M. (1996) Phylogenetic inference. In *Molecular systematics* 2nd edition. Hillis, D. M., Moritz, C. and Mable, B. (eds.), K. Sinauer Associates Inc., Sunderland, Massachusetts U.S.A., 407-514.

Terasaki, K., Moriyama, N., Tani, S. and Ishida, K. (1982) Comparative studies on the karyotypes of *Echinostoma cinetorchis* and *E. hortense* (Echinostomatidae: Trematoda), *Japanese Journal of Parasitology* **31**, 569-574.

van der Sande, C. A. F. M., Kwa, M., van Nues, R. W., van Heerikhuizen, H., Raué, H. A. and Planta, R. J. (1992) Functional analysis of internal transcribed spacer 2 of *Saccharomyces cerevisiae* ribosomal DNA, *Journal of Molecular Biology* **223**, 899-910.

van Nues, R. W., Rientjes, J. M. J., Morré, S. A., Mollee, E., Planta, R. J., Venema, J. and Raué, H. A. (1995) Evolutionarily conserved structural elements are critical for processing of internal transcribed spacer 2 from Saccharomyces cerevisiae precursor ribosomal RNA, *Journal of Molecular Biology* **250**, 24-36.

Vasilev, I., Komandarev, S., Mikhov, L. and Kanev, I. (1978) Comparative electrophoretic studies of certain species of genus *Echinostoma* with 37-collar spines, *Khelmintologia* **6**, 31-38.

Voltz, A., Richard, J. and Pesson, B. (1987) A genetic comparison between natural and laboratory strains of *Echinostoma* (Trematoda) by isoenzyme analysis, *Parasitology* **95**, 471-477.

Voltz, A., Richard, J., Pesson, B., and Jourdane, J. (1988) Isoenzyme analysis of *Echinostoma liei*: Comparison and hybridization with other African species, *Experimental Parasitology* **66**, 13-17.

Wesson, D. M., McLain, D. K., Oliver, J. H., Piesman, J. and Collins, F. H. (1993) Investigation of the validity of species status of *Ixodes dammini* (Acari: Ixodidae) using rDNA, *Proceedings of the National Acadamy of Sciences of the USA* **90**, 10221-10225.

Wheeler, W. C. and Gladstein, D. (1994) MALIGN: A multiple sequence alignment program, *Journal of Heredity* **85**, 417.

INDEX

DATE DUE